Clinical Cardiac Electrophysiology in the Young

Developments in Cardiovascular Medicine

232. A. Bayés de Luna, F. Furlanello, B.J. Maron and D.P. Zipes (eds.): *Arrhythmias and Sudden Death in Athletes.* 2000 ISBN: 0-7923-6337-X
233. J-C. Tardif and M.G. Bourassa (eds.): *Antioxidants and Cardiovascular Disease.* 2000 ISBN: 0-7923-7829-6
234. J. Candell-Riera, J. Castell-Conesa, S. Aguadé Bruix (eds.): *Myocardium at Risk and Viable Myocardium Evaluation by SPET.* 2000 ISBN: 0-7923-6724-3
235. M.H. Ellestad and E. Amsterdam (eds.): Exercise Testing: New Concepts for the New Century. 2001 ISBN: 0-7923-7378-2
236. Douglas L. Mann (ed.): The Role of Inflammatory Mediators in the Failing Heart. 2001 ISBN: 0-7923-7381-2
237. Donald M. Bers (ed.): Excitation-Contraction Coupling and Cardiac Contractile Force, Second Edition. 2001 ISBN: 0-7923-7157-7
238. Brian D. Hoit, Richard A. Walsh (eds.): Cardiovascular Physiology in the Genetically Engineered Mouse, Second Edition. 2001 ISBN: 0-7923-7536-X
239. Pieter A. Doevendans, A.A.M. Wilde (eds.): Cardiovascular Genetics for Clinicians 2001 ISBN 1-4020-0097-9
240. Stephen M. Factor, Maria A. Lamberti-Abadi, Jacobo Abadi (eds.): Handbook of Pathology and Pathophysiology of Cardiovascular Disease. 2001 ISBN: 0-7923-7542-4
241. Liong Bing Liem, Eugene Downar (eds.): Progress in Catheter Ablation. 2001 ISBN: 1-4020-0147-9
242. Pieter A. Doevendans, Stefan Kääb (eds.): Cardiovascular Genomics: New Pathophysiological Concepts. 2002 ISBN: 1-4020-7022-5
243. Daan Kromhout, Alessandro Menotti, Henry Blackburn (eds.): Prevention of Coronary Heart Disease: Diet, Lifestyle and Risk Factors in the Seven Countries Study. 2002 ISBN: 1-4020-7123-X
244. Antonio Pacifico (ed.), Philip D. Henry, Gust H. Bardy, Martin Borggrefe, Francis E. Marchlinski, Andrea Natale, Bruce L. Wilkoff (assoc. eds.): Implantable Defibrillator Therapy: A Clinical Guide. 2002 ISBN: 1-4020-7143-4
245. Hein J.J. Wellens, Anton P.M. Gorgels, Pieter A. Doevendans (eds.): The ECG in Acute Myocardial Infarction and Unstable Angina: Diagnosis and Risk Stratification. 2002 ISBN: 1-4020-7214-7
246. Jack Rychik, Gil Wernovsky (eds.): Hypoplastic Left Heart Syndrome. 2003 ISBN: 1-4020-7319-4
247. Thomas H. Marwick: Stress Echocardiography. Its Role in the Diagnosis and Evaluation of Coronary Artery Disease 2nd Edition. ISBN: 1-4020-7369-0
248. Akira Matsumori: Cardiomyopathies and Heart Failure: Biomolecular, Infectious and Immune Mechanisms. 2003 ISBN: 1-4020-7438-7
249. Ralph Shabetai: The Pericardium. 2003 ISBN: 1-4020-7639-8
250. Irene D. Turpie, George A. Heckman (eds.): Aging Issues in Cardiology. 2004 ISBN: 1-4020-7674-6
251. C.H. Peels, L.H.B. Baur (eds.): Valve Surgery at the Turn of the Millennium. 2004 ISBN: 1-4020-7834-X
252. Jason X.-J. Yuan (ed.): Hypoxic Pulmonary Vasoconstriction: Cellular and Molecular Mechanisms. 2004 ISBN: 1-4020-7857-9
253. Francisco J. Villarreal (ed.): Interstitial Fibrosis In Heart Failure 2004 ISBN: 0-387-22824-1
254. Xander H.T. Wehrens, Andrew R. Marks (eds.): Ryanodine Receptors: Structure, function and dysfunction in clinical disease. 2005 ISBN: 0-387-23187-0
255. Guillem Pons-Lladó, Francesc Carreras (eds.): Atlas of Practical Applications of Cardiovascular Magnetic Resonance. 2005 ISBN: 0-387-23632-5
256. José Marín-García : Mitochondria and the Heart. 2005 ISBN: 0-387-25574-5
257. Macdonald Dick II: Clinical Cardiac Electrophysiology in the Young 2006 ISBN: 0-387-29164-4

Previous volumes are still available

Clinical Cardiac Electrophysiology in the Young

Edited by
Macdonald Dick II M.D.

*With past and present
Fellows and Faculty
of the
Division of Pediatric Cardiology
University of Michigan*

Macdonald Dick II, MD
Professor of Pediatrics
University of Michigan
C.S. Mott Children's Hospital
Womens L1242, Box 0204
1500 East Medical Center Drive
Ann Arbor, MI 48109-0204
USA

Library of Congress Cataloging-in-Publication Data

Clinical cardiac electrophysiology in the young / edited by Macdonald Dick II; with past
 and present fellows and faculty of the Division of Pediatric Cardiology, University of Michigan.
     p. ; cm. – (Developments in cardiovascular medicine; v. 257)
   Includes bibliographical references and index.
 ISBN 978-1-4419-3973-9   e-ISBN 978-0-387-29170-3
                                 e-ISBN 0-387-29170-9
  1. Pediatric cardiology. 2. Electrophysiology. 3. Heart conduction system 4.
 Children—Diseases--Diagnosis. 5. Heart--Diseases--Diagnosis. I. Dick, MacDonald. II.
 University of Michigan. Mott Children's Hospital. Division of Pediatric Cardiology. III.
 Series.
   [DNLM: 1. Heart--physiology--Child. 2. Heart--physiology--Infant. 3.
 Electrophysiology--methods--Child. 4. Electrophysiology--methods--Infant. 5. Heart
 Conduction System--physiology--Child. 6. Heart Conduction System--physiology--Infant.
 7. Heart Diseases—physiopathology--Child. 8. Heart Diseases--physiopathology--Infant.
 WG 202 C641 2006]
 RJ421.C555 2006
 618.92'12--dc22   2005054106
 Printed on acid-free paper.

© 2006 Springer Science+Business Media, LLC.
 Softcover reprint of the hardcover 1st edition 2006
 All rights reserved. This work may not be translated or copied in whole or in part without
the written permission of the publisher (Springer Science+Business Media, LLC, 233 Spring
Street, New York, NY 10013, USA), except for brief excerpts in connection with reviews or
scholarly analysis. Use in connection with any form of information storage and retrieval,
electronic adaptation, computer software, or by similar or dissimilar methodology now
known or hereafter developed is forbidden.
The use in this publication of trade names, trademarks, service marks and similar terms,
even if they are not identified as such, is not to be taken as an expression of opinion as to
whether or not they are subject to proprietary rights.
While the advice and information in this book are believed to be true and accurate at the
date of going to press, neither the authors nor the editors nor the publisher can accept any
legal responsibility for any errors or omissions that may be made. The publisher makes no
warranty, express or implied, with respect to the material contained herein.

9 8 7 6 5 4 3 2

springer.com

*This book was written by all of us
because of our parents and teachers,
with our spouses and partners,
for our and all children.*

Contributors

Mohamad Al-Ahdab, M.D., Lecturer, University of Michigan Medical School, Ann Arbor, Michigan

David Bradley, M.D., Assistant Professor of Pediatrics, Pediatric Cardiology, University of Utah Medical School, Salt Lake City, Utah

Burt Bromberg, M.D., Pediatric Cardiologist and Electrophysiologist, St. Louis, MO

Craig Byrum, M.D., Associate Professor of Pediatrics, Upstate Medical School, New York University, State University of New York, Syracuse, New York.

Robert M. Campbell, M.D., Associate Professor of Pediatrics, Children's Heart Center, Emory University, Atlanta, Georgia

Macdonald Dick II, M.D., Professor of Pediatrics, University of Michigan Medical School, Ann Arbor, Michigan

Parvin Dorostkar, M.D., Associate Professor of Pediatrics, Rainbow Babies and Children's Hospital, University Hospitals Health Systems, Cleveland, Ohio

Peter S. Fischbach, M.D., Assistant Professor of Pediatrics and Pharmacology, University of Michigan Medical School, Ann Arbor, Michigan

Carlen Gomez, M.D., Associate Professor of Pediatrics, University of Michigan Medical School, Ann Arbor, Michigan

Ian H. Law, M.D., Associate Professor of Pediatrics, University of Iowa Medical School, Iowa City, Iowa

Sarah Leroy, Clinical Nurse Specialist and Nurse Practitioner, Pediatric Electrophysiology and Anti-Arrhythmia Device Clinics, University of Michigan Medical School, Ann Arbor, Michigan

Mark Russell, M.D., Associate Professor of Pediatrics, University of Michigan Medical School, Ann Arbor, Michigan

Elizabeth V. Saarel, M.D., Assistant Professor of Pediatrics, Cleveland Clinic Foundation, Cleveland, Ohio

William A. Scott, M.D., Professor of Pediatrics, Southwestern Texas Medical School, Dallas, Texas

Gerald Serwer, M.D., Professor of Pediatrics, University of Michigan Medical School, Ann Arbor, Michigan

Christopher B. Stefanelli, M.D., Pediatric Cardiologist, Tacoma, Washington

Margaret Strieper, D.O., Associate Professor of Pediatrics, Children's Heart Center, Emory University, Atlanta, Georgia

Stephanie Wechsler, M.D., Associate Professor of Pediatrics, University of Michigan Medical School, Ann Arbor, Michigan

Preface

It takes a certain hubris to come forth with a book entitled *Clinical Cardiac Electrophysiology in the Young*. There are a number of excellent texts, monographs, and reviews on cardiac arrhythmias in both adults and children—Josephson's and also Zipes and Jalife's comprehensive texts come to mind, as well as a number of others, including Deal, Wolff, and Gelband's, the several volumes from Gillette, and the recent text from Walsh, Saul, and Triedman, the latter three texts focusing on children.

Nonetheless the past three decades have witnessed enormous advances in the understanding and management of human cardiac arrhythmias. This development represents the fruits of both basic and clinical investigations in cardiac impulse formation and propagation at the organ, tissue, and more recently, cellular and molecular levels. This information explosion may result in information overload and frustrate the student, the young physician in training, as well as the seasoned practitioner. This book focuses on the practical (and theoretical when applicable) aspects of clinical electrophysiology of cardiac arrhythmias in the young. Our intention is that the young house officer or mature physician who is faced with a child with a cardiac arrhythmia will find this book useful in increasing their understanding, sparking their interest, and perhaps leading them to a therapeutic solution.

This book emerges from the clinical practice and research of the pediatric cardiac electrophysiology group in the Division of Pediatric Cardiology at the C.S. Mott Children's Hospital, the University of Michigan in Ann Arbor, and the former pediatric electrophysiology fellows from Michigan, now established electrophysiologists in their own right. It represents a compilation of the clinical course, electrocardiograms, electrophysiologic studies, pharmacological management, and transcatheter ablation therapy in patients from infancy through young adulthood seen in Ann Arbor and at the current clinical sites of the former Michigan fellows. Thus, while the product may be idiosyncratic, it is not provincial. We are interested in "how it is done" but not to the exclusion of other approaches. This is only one (or several) way to address the clinical problem of arrhythmias in children, and surely not the only way, especially as one views the future of emerging energy sources for ablation, non-ionizing radiation imaging techniques, and molecular diagnostic possibilities.

The book is divided into two parts. The first part, Background (Chapters 1–3), discusses the cardiac conduction system—development, anatomy, and physiology. Particular attention is directed to the clinical electrophysiology of the cardiac conduction system and the techniques of electrophysiologic study that are specific to children and that have been developed and practiced at the University of Michigan and at other centers. The second part, Cardiac Electrophysiology in Infants and Children (Chapters 4–23), focuses on the clinical science of cardiac arrhythmias in infants and children.

Chapters 4–12 discuss the mechanism, the ECG characteristics, the electrophysiologic findings, the treatment, and the prognosis of tachyarrhythmias. Chapters 13–16 focus on bradyarrhythmias. Chapters 17–20 address certain specialized subjects, including, syncope, cardiac pacemakers, implantable cardiac defibrillators, genetic disorders of the cardiac impulse, fetal arrhythmias, and sudden cardiac death as it occurs in the young. Chapters 21–22 center on the pharmacology of antiarrhythmic agents, indications for use, doses, side effects and toxicity, as well as on transcatheter arrhythmia ablation. Finally, what the practitioner can expect to see from the impact of cardiac arrhythmias on the life of the patient and family is discussed from the nursing point of view in Chapter 23.

The intent of the book is practical and thus the suggested readings are selected and not encyclopedic. They are meant as a starting place for the interested reader. Examples and tables are included in the anticipation that the reader will rapidly be able to match the clinical problem to the examples and the accompanying text.

A text or technical book is rarely the product of a single individual. With that in mind, any value or sense that can be made of this work is solely due to the terrific efforts of the authors; any error or fault can be correctly attributed to me. I am deeply grateful to all of the authors for their contributions, as well as their patience in bringing the project together. I want to recognize the generosity of my colleagues at Michigan in providing coverage when I would hide out (including a sabbatical) to work on the text. Thanks also to the medical electrophysiology group at Michigan for encouragement and support for the pediatric program. I also want to thank my local editor, Kathryn Clark, for all her efforts in keeping me on task, endlessly and repeatedly formatting the multiple revisions of the text, and finding and eliminating too many examples of "nonsense" to count. Finally, I want to thank Melissa Ramondetta at Springer for her great patience, great good humor, and sound advice throughout the course of the project. Carolin, my wife, graciously permitted me to weed the book of its unwanted wordage (probably missed a bit) rather than our yard of unwanted plant life on numerous weekends.

Macdonald Dick II, M.D.
Ann Arbor, MI
August, 2005

Foreword

The text of *Clinical Cardiac Electrophysiology in the Young* provides a systematic approach to the anatomy, pathophysiology, basic electrophysiology, diagnosis and therapy of atrial and ventricular arrhythmias as well as conduction abnormalities in the young. It elucidates the broad spectrum of rhythm disturbances that may occur from the fetus to young adult, as an isolated abnormality, in the presence of underlying congenital heart disease, both prior to and subsequent to surgical repair. The clinical manifestations, diagnosis and appropriate pharmacologic and interventional therapy by a trained healthcare team are fully discussed. Science is consistently used to explain the electrophysiologic diagnoses, pharmacologic, interventional and surgical treatment. Some prior knowledge and understanding of electrophysiology and rhythm disturbances is helpful and the information provided here may be utilized as a guidebook, resource and reference for residents, cardiology fellows, trained cardiologists and electrophysiologists as well as other allied health professionals. The rapid advances in the field in such areas as interventional and surgical cryoablation techniques, complexity of rhythm disturbances, new monitoring devices and pharmaceuticals make it an invaluable text.

Dr. Macdonald Dick as an author and editor of the book is an internationally recognized scholar and clinical pediatric electrophysiologist. A superb teacher and role model for trainees and faculty his affability and diligent effort have brought about the compilation and publication of the book. The majority of the knowledgeable and experienced contributors have received their training in pediatric cardiology at the University of Michigan. The authors are indebted to their medical and surgical colleagues, fellows, family members and respective institutions for the support and encouragement in the endeavor.

Amnon Rosenthal, MD
Professor of Pediatrics
University of Michigan Medical School
Ann Arbor, MI

Contents

I. BACKGROUND

1. Development and Structure of the Cardiac Conduction System — 1
   Parvin Dorostkar

2. Physiology of the Cardiac Conduction System — 17
   Peter S. Fischbach

3. Clinical Electrophysiology of the Cardiac Conduction System — 33
   Macdonald Dick II, Peter S. Fischbach, Ian H. Law, and William A. Scott

II. CLINICAL ELECTROPHYSIOLOGY IN INFANTS AND CHILDREN

4. Atrioventricular Reentry Tachycardia — 51
   Ian H. Law

5. Atrioventricular Nodal Reentrant Tachycardia — 69
   David J. Bradley

6. Persistent Junctional Reciprocating Tachycardia — 83
   Parvin C. Dorostkar

7. Sinoatrial Reentrant Tachycardia — 91
   Macdonald Dick II

8. Intra-atrial Reentrant Tachycardia—Atrial Flutter — 95
   Ian H. Law and Macdonald Dick II

9. Atrial Fibrillation — 115
   Peter S. Fischbach

10. Atrial Ectopic Tachycardias/Atrial Automatic Tachycardia — 119
    Burt Bromberg

11	Multifocal Atrial Tachycardia *David J. Bradley*	135
12	Ventricular Tachycardia *Craig Byrum*	139
13	Sick Sinus Syndrome *William A. Scott*	153
14	First- and Second-Degree Atrioventricular Block *William A. Scott*	163
15	Complete Heart Block—Third-Degree Heart Block *Mohamad Al-Ahdab*	173
16	Syncope *Margaret Strieper, Robert M. Campbell, William A. Scott*	183
17	Cardiac Pacemakers and Implantable Cardioverter-Defibrillators *Gerald S. Serwer and Ian H. Law*	195
18	Genetic Disorders of the Cardiac Impulse *Mark W.W. Russell and Stephanie Wechsler*	217
19	Fetal Arrhythmia *Elizabeth V. Saarel and Carlen Gomez*	241
20	Sudden Cardiac Death in the Young *Christopher B. Stefanelli*	257
21	Pharmacology of Antiarrhythmic Agents *Peter S. Fischbach*	267
22	Transcatheter Ablation of Cardiac Arrhythmias in the Young *Macdonald Dick II, Peter S. Fischbach and Ian H. Law*	289
23	Nursing Management of Arrhythmias in the Young *Sarah Leroy*	315

Index   327

I

Background

1

Development and Structure of the Cardiac Conduction System

Parvin Dorostkar

In the adult mammalian heart, the primary cardiac impulse is ultimately driven by the sinoatrial (SA) node, which contains the leading pacemaker cells of the mature human heart. The generated impulse then travels through the atrial myocardium to the atrioventricular (AV) node. Here a delay in conduction occurs, after which the impulse is rapidly transmitted from the AV node through the His-Purkinje system to the ventricular myocardium. It is the peripheral His-Purkinje system that transmits the impulse to the ventricular myocardium, which, in conjunction with electromechanical coupling, results in myocardial contraction from the apex of the heart toward the base of the heart, generating cardiac output with each beat.

Even though the cardiac conduction system in its function and anatomy is considered quite separate from the working myocardium of the heart, it is virtually indistinguishable from adjacent myocardial tissue by gross visualization. Even on a cellular and microscopic level the cells are indistinguishable. Cells of the conduction system as well as all other myocardial cells are capable of contraction, automaticity, intercellular conduction, and electromechanical coupling. These similarities in characteristics make a focused study of the specialized conduction system very difficult. However, the conduction system cells, once matured, exhibit subcellular elements that differentiate these cells from other working myocytes, such as connexins (in the myocardial cell membrane), and contractile and cytoskeletal proteins (intramyocardial).

A number of investigators have studied the embryologic formation and development of the heart. Animal models have been used extensively to study the development of the heart and the cardiac conduction system. However, interspecies variability and multiple challenges associated with the study of the early embryo have made the quest to unravel the many interrelated factors of cardiac development difficult.

In the chick embryo, which has been studied in detail, the earliest development of the heart occurs in the cardiac progenitor cells originating from the embryonic mesoderm. There are specific "heart-forming fields" where the cells will develop and produce beating tissue. In chicks, cells designated to form the heart arise lateral to the primitive streak, which can be identified during Hamburger Hamilton (HH) stage 3 of chick embryo

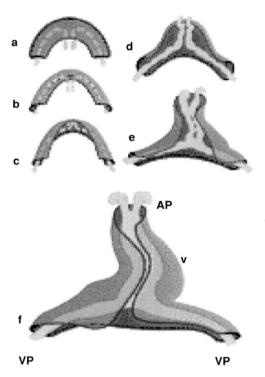

FIGURE 1. Formation of the cardiac tube. Transformation of the flat cardiogenic crescent into a cardiac tube is displayed. During this process the red outer contour of the myocardial crescent (grey) folds around the fusing endocardial vesicles (yellow) and passes the blue inner contour of the crescent, thereby forming the cardiac tube. AP = anterior pole, VP = venous pole, V = future ventricle. Reprinted with permission from AFN Moorman et al., Development of the cardiac conduction system, Circulation Research 1998; 82:629–644.

development (Figure 1). These cells migrate rostrolaterally to form the lateral plate. Color patterns in Figure 1 show the region of the embryo that gives rise to precardiac cells, which will eventually contribute to the development of myocardium. The relative anterior–posterior positions of precardiac cells in the primitive streak is maintained in the heart field in the mesoderm and continues during HH stage 5 through 7 of development. At HH stage 8, the embryo folds ventrally, generating the foregut and somatic and splanchnic layers of the mesoderm. The splanchnic layer of the mesoderm contains myocardial precursors. By HH stage 10, the chick heart primordia fuse to form a tubular heart with the anterior most region of the cardiac tube giving rise to an outflow region or conotruncus or bulbus cordis. In the chick embryo, the most caudal portion contributes to the most posterior end of the heart and will result in the inflow region or the sinoatrial region. This primary heart tube undergoes peristalsis-like contractions that support unidirectional flow. In the mammalian heart, the formation of the cardiac tube occurs in six, similar, successive stages, which support transformation of a flat cardiogenic crest into a cardiac tube. During this process the outer contour of the myocardial crescent folds around the fusing endocardial vesicles. This primitive heart tube consists of the endocardium with adjacent myocardium and has slowly conducting contractions that support peristaltic movements. Within this primitive heart tube fast-conducting atrial and ventricular myocytes develop. The fast-conducting atrial and ventricular myocytes, however, remain next to the slowly conducting tissue, which eventually gives rise to the inflow tract, AV canal area, and outflow tract. As development proceeds, the slow conducting areas give rise to the SA node (around the inflow area), the AV node, and the slowly conducting outflow tract, whereas the fast conducting atrial and ventricular cells give rise to the His-Purkinje system and its ramifications. During development, alternating slow and fast conducting segments support unidirectional flow and are responsible for the embryonic ECG. Alternating fast and slow conduction also prevent relaxation of the atrial or ventricular segments before contraction of a downstream segment, therefore, minimizing regurgitating blood. As polarity develops in the vertebrate heart, there is an increase in phenotypic atrial cells posteriorly and an increase of phenotypic ventricular cells anteriorly. Highest pacemaker activity (highest beat frequency) can then be observed in cells associated with the intake portion of the cardiac tube (atrial cells); this phenomenon occurs at the human embryonic age of about 20 days.

DEVELOPMENT OF IMPULSE GENERATION

The early cardiac mesoderm arises from ectodermal tissue and subsequently forms the cardiac tube, which has polarity along its anterior posterior axis. As mentioned previously, a straight heart tube is present at about day 20 of human embryonic life, while cardiac looping occurs at about day 21. The polarity of the mammalian heart is characterized by the predominance of an atrial tissue phenotype posteriorly at the inflow region of the heart or upstream side of the heart and of the ventricular tissue phenotype anteriorly at the outflow or downstream region of the heart. Dominant pacemaker activity and highest beat frequency are found at the intake of the heart tube. Here an efficient contraction wave is generated. Cells of the primary cardiac tube and the future sinus node region show action potentials resembling those of the adult pacemaker cells. These action potentials display slow depolarization and are similar to those of pacemaker cells that are associated with slow voltage-gated calcium ion channels. Cells of the future ventricles, however, show action potential behavior similar to that of adult ventricles exhibiting action potentials that have high amplitudes similar to action potentials associated with fast voltage-gated sodium channels.

As the heart develops, the frequency of the intrinsic beat rate increases along the inflow tract of the heart. In both birds and mammals, the leading pacemaker area is initially found on the left side but as soon as the sinus venosus has formed (at approximately 25 days in the human), the right side becomes more dominant. Both right and left inflow tracts will become incorporated into the future, mature right atrium. "Node-like cells" develop in the right atrium. Such cells have also been found in the myocardium surrounding the distal portion of the pulmonary veins in adult rats and appear to play a role in preventing regurgitation of blood into the pulmonary veins from the left atrium. How leading pacemaker cells develop into an anatomically distinct SA node is still unclear.

DEVELOPMENT OF IMPULSE PROPAGATION

The primary myocardium is characterized by action potentials that are primarily supported by slow voltage-gated calcium ion channels. As embryonic atrial and ventricular chambers develop, synchronous contractions of these chambers are characterized by higher conduction velocities, which are more likely associated with fast voltage-gated sodium channels. These variations in conduction are accompanied by the development of an adult type ECG that reflects the sequential activation of the atrial and ventricular chambers rather than the presence of a morphologically recognizable conduction system. For example, a noted AV delay is present before the development of a morphologically identifiable AV node. This AV delay occurs in the region of the AV canal, which is recognized as an area of slow conduction. Segments of slowly conducting primary myocardium also persist at both the inflow and outflow area of the heart.

Several theories exist regarding the development of the specialized conduction system (i.e., the AV node and its penetrating bundles and bundle branches). Using a monoclonal antibody against a specific neural marker (G1N2), several investigators have suggested that there is a process of differentiation that supports the development of a myocardial ring that encircles the presumptive foramen between the developing right and left ventricles. These investigators suggest that the dorsal portion of this ring will develop the AV bundle whereas parts covering the septum will give rise to the right and left bundles (Figures 2 and 3). However, several investigators suggest that ventricular depolarization undergoes a transition, where the myocardium undergoes a switch from base-to-apex depolarization of the ventricular myocardium to an apex-to-base depolarization

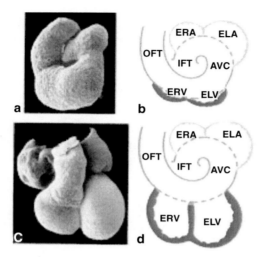

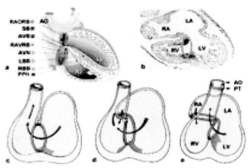

FIGURE 2. Formation of the cardiac chambers. Scanning electron photomicrographs (a and c) and schematic representations (b and d) of a 3-day embryonic chicken heart, where the first signs of the ventricles emerge (a and b), and of a 37-day embryonic human heart with clearly developed ventricles (c and d). ERA = indicates embryonic right atrium; ELA = embryonic left atrium; ELV = embryonic left ventricle; and ERV = embryonic right ventricle. The atrial segment is indicated in blue; the ventricular segment, in red; and the primary heart tube, encompassing the flanking segments, IFT, AVC, and OFT, as well as the atrial and ventricular parts, in purple. Reprinted with permission from AFN Moorman et al., Development of the cardiac conduction system, Circulation Research 1998; 82:629–644.

FIGURE 3. Development of the ventricular conduction system. a. Drawing of a prototypical heart, in which the position of the ventricular conduction system is indicated, including those parts that are only present in the fetal mammalian heart. The entire system persists in the adult chicken heart. b. Section of a 5-week human heart immunostained for the presence of GlN2 in which the position of the developing conduction system is represented in brick red. c–e. Drawings representing the development of the ventricular conduction system based on reconstructions of GlN2 expression in developing hearts at ≈5 (c), ≈6 (d), and ≈7 (e) weeks of development. See text for explanation. RAORB = retro-aortic root branch; SB = septal branch; RAVRB = right atrioventricular ring bundle; LBB and RBB = left and right bundle branches, respectively; LA = left atrium; LV = left ventricle; RA = right atrium; RV = right ventricle; AO = aorta; and PT = pulmonary trunk. Reprinted with permission from AFN Moorman et al., Development of the cardiac conduction system, Circulation Research 1998; 82:629–644.

in the mature, intact His-Purkinje system. A transition in the ventricular myocardial depolarization pattern was demonstrated using monoclonal antibodies to the polysialylated neural cell adhesion molecule and the HNK-1 sulfated carbohydrate epitope. In the chick embryo, HH stage 30 appears to represent a critical period in the morphogenesis of the heart. The primitive myocardium has slow conduction; however, faster conduction along ventricular myocardium has been observed as early as HH stage 23. This fast conduction is functionally distinct from slow conduction around HH stage 28, just before the transition period. Because activation of the ventricles occurs base-to-apex, the developing His-Purkinje system is still thought to be relatively immature at this stage. At HH stage 30, this sequence reverses and becomes apex-to-base.

The authors suggest that the switch may occur because the His-Purkinje system has matured, and the muscular AV junction is then able to support and limit the avenue of AV conduction to the His-Purkinje system, allowing rapid impulse propagation to the apical myocardium. The mature AV node as a nodal structure becomes only gradually identifiable after about Carnegie stage 15 (5 weeks of human development).

In summary, during the process of chamber formation, fast conducting atrial and ventricular segments are being formed within slowly conducting primary myocardium of the embryonic heart tube so that the cardiac tube becomes a composite of alternating slow and fast conducting segments that persist in a more specialized fashion in the mature heart. The molecular basis underlining such compartmentalization is still poorly understood and continues to be studied.

CELLULAR DEVELOPMENT OF "NODAL" PHENOTYPE

In the mature heart, the nodal myocytes display a number of embryonic characteristics. Nodal cells are poorly distinguishable from surrounding myocardium in the early embryonic heart as they exhibit many of the same characteristics as the surrounding myocardium. However, with development, it appears that nodal cells retain some of the same characteristics as early embryonic myocytes such as organized actin and myosin filaments and poorly developed sarcoplasmic reticulum. In addition, nodal cells express different structural and cellular markers which are species-specific. Several classes of markers have been identified including connexins, specific contractile proteins, desmin, and neurofilament that provide specific markers for the study of conduction system development. In addition, antibodies to carbohydrate markers such as the polysialylated neural cell adhesion molecule and HNK1 have been used to study the development of specific regions of the specialized conduction tissue.

Connexins

The transmission of the electrical action potential is thought to be primarily associated with gap junctions. Gap junctions are aggregates of membrane channels, composed of protein subunits named connexins that are encoded by a multi-gene family. Five different connexins are expressed in the mammalian heart including connexin 37, 40, 43, 45, and 46. In the early myocardium, both number and size of gap junctions are small but they increase during development. The number of gap junctions remains scarce in the developing SA node and the AV node. The low abundance of connexin expression in the nodes corresponds with both the slow conduction velocities observed in the nodes and the absence of fast sodium currents. The poor coupling of nodal cells appears to be a requirement for the expression of an action potential by these nodal myocytes, which differentiates itself from the much more abundant atrial and/or ventricular working myocardium. This difference in connexin concentration has been an important marker for nodal-specific tissue. An abrupt rather than gradual increase in the number of gap junctions is found at the transition of nodal tissue to working myocardium. This boundary is thought to be due to a decrease in the number of nodal cells towards the atrial working myocardium rather than a gradient due to a change in molecular phenotype.

Cytoskeletal Proteins

Nodal-specific developmental expression of contractile proteins such as myosin heavy chain and its isoforms, desmin and neurofilament, has been used to delineate the sinoatrial and AV nodes. However, interspecies variability in the staining of these markers does not allow enough consistent data to draw a definitive global conclusion in relation to development or morphologic changes that are specific to the conduction system or its development and differentiation.

Cell Markers

Nodal tissue seems to acquire unique characteristics during development, including the expression of higher amounts of calcium-release channel/type-1 inositol triphosphate receptor, gamma enolase, alpha 1 and alpha 2 of the sodium pump, G-protein alpha subunit, and angiotensin II receptor. The role of these differences remains to be studied.

ANATOMIC DEVELOPMENT OF THE SPECIALIZED CONDUCTION TISSUE

The AV node structure appears to arise from "primordia cells" that originate from the myocardium of the posterior wall of the AV canal or primitive endocardial cushions. The primordia later forms both the AV node and the His bundle. The Tendon of Todaro and the central fibrous body are later formed from the inferior endocardial cushions.

Although several authors have suggested that the specialized conduction tissue may originate from neural crest tissue, the origin of the myocytes of the specialized conduction system has been established by recent studies to be from the cardiomyocyte, rather than the neural crest. By a process that may involve ET-1 (endothelin-1) signaling, neural crest-derived cells have been reported to migrate to regions of the central conduction system and may play an as yet undefined role in the development of the definitive mature structure of the AV conduction system.

Genetic, molecular, functional, and morphologic evidence suggests that the ventricular conduction system develops separately and may originate from the trabecular component of the developing ventricles. While the trabecular portion of the heart contracts without a specialized conduction system via slow, homogeneous cell-to-cell propagation during early embryogenesis, a faster more mature form of conduction occurs when the His-Purkinje system is engaged and involved in contraction. An important transition time has been described (stage 30) where a specialized His-Purkinje system emerges (in both form and function), defining a critical time period in development at which time the electrical activation of the myocardium switches from base-to-apex to apex-to-base with preferential electrical activation over the His-Purkinje system.

In summary, recent evidence suggests that the specialized conduction system develops from further differentiation of local myocytes. The molecular signals for this differentiation are unknown. The exact stimulants for differentiation, selective cellular potency, and variable cell protein and channel expression and their roles in differentiation warrant further study. The variable role and interactions of these factors in embryogenesis and differentiation continue to be poorly understood and provide the groundwork for further investigation using advanced techniques to increase understanding of the developing conduction system.

ANATOMY OF THE MATURE CARDIAC CONDUCTION SYSTEM

The specialized conduction system of the human heart consists of a single SA node, atrial and intra-nodal pathways, the AV node, the His-Purkinje system, the right and left bundle branches of the His-Purkinje system, and the peripheral His-Purkinje system.

The SA node is the dominant pacemaker of the heart and lies in the right atrium at the superior vena cava/right atrial junction, one mm below the epicardium of the sulcus terminalus. It was first described in the early 1900s. The SA node appears to have the shape of an inverted comma, descriptively containing a head, body, and tail. It tapers both medially and laterally and bends backwards toward the left and then downward. The SA node is supplied by a relatively large artery, which courses through the node giving off branches to the sinus node and adjacent atrial myocardium. It originates from the right coronary artery about 55% of the time and from the left circumflex artery in about 45% of cases.

It is still somewhat controversial whether preferential intranodal pathways exist. Preferential conduction or impulse propagation may be associated with the underlying anatomic differences in muscle density or muscle fiber orientation and/or the thickness of the right atrial wall and its pectinate muscles. Some authors argue that "specialized pathways" are thought to consist of aggregations or concentrations of myocardial muscle fibers, bridging the SA and AV nodes or the right atrium to the left atrium. These authors propose three internodal tracts: the anterior internodal fibers, thought to have two components: Bachman's Bundle, which bridges right and left atrium and "descending branches," which descend in the intra-atrial septum. The middle internodal tracts, also known as Wenchebach's bundle, are thought to arise from the posterior portion of the sinus node and then descend within the inta-atrial septum, anterior to the fossa ovale. The posterior internodal tracts, also known as Thorel's pathway, are thought to exit the sinus node posteriorly and then descend within the crista terminalis, traversing through the Eustachian ridge, entering the AV node posteriorly in the mouth of the coronary sinus.

The AV node is located in the posteroseptal area, primarily on the right atrial side, in the region known as the triangle of Koch. This triangle is defined by the Tendon of Todaro, the edge of the tricuspid valve and the edge of the mouth of the coronary sinus, which marks the base of the triangle. In the adult, the triangle measures 14–20 mm in its longest apex-to-base dimension. The AV node is located mostly at the base of this triangle and on the right side of the central fibrous body. In children, the triangle of Koch varies with age and size of the child. It is considered to be a complex structure. Descriptively, the AV node abuts the mitral valve annulus and tricuspid valve annulus with its posterior margin abutting the coronary sinus. Unlike the bundle of His, the AV node cannot be seen visually, nor does it generate a distinct, recordable signal during electrophysiologic testing. Therefore, the knowledge of its location is discerned by implied mechanisms during electrophysiologic mapping techniques. The anterior portion or distal ends of the AV node blend with the bundle of His, which penetrates the central fibrous body. The AV node is thought to be a flattened, oblong structure with multiple extensions, some extending to the left atrium. The AV node is also thought to have extensions with a compact portion of the node existing more closely associated with the perimembranous portion of the ventricular septum. The AV node is usually supplied by an AV nodal artery, which arises from the right coronary artery in 90% of cases, and from the left circumflex artery in 10% of the cases.

The bundle of His consists of extensions of the AV node. These extensions occur distal to the compact AV node. The bundle of His is characterized by fibers, which are organized in parallel channels or strands. These fibers are surrounded by a fibrous sheath more proximally and are, therefore, well insulated. The bundle of His penetrates the fibrous body and proceeds anteriorly descending towards the AV septum where it divides into the right and left bundle branches.

The right bundle is a relatively well defined and easily dissectible structure situated beneath the epicardium on the right side. The right bundle branch proceeds along the free edge of the moderator band to the base of the anterior papillary muscles in the right ventricle and along the septal band to the apex of the right ventricle.

The left bundle passes down the left side of the intraventricular septum and emerges below the posterior cusp of the aortic valve. In contrast to the right bundle, the left bundle breaks up almost immediately into a number of small fan-shaped branches, which proceed down the smooth aspect of the left side of the intraventricular septum. The bundle contains two major branches including an antero-superior division and a postero-inferior division. The antero-superior division is relatively long and thin whereas the postero-inferior division is relatively short and thick. The antero-superior division is closer to the aortic valve whereas the postero-inferior division supplies the posterior and inferior aspect of the left ventricle.

Parasympathetic supply to the myocardium arises from branches from the right and left vagus nerves. The right vagus nerve supplies primarily the SA node; the left vagus nerve supplies primarily the AV node. The SA node is thought to originate from the right horn of the sinus venosus and is, therefore, connected with the right vagus nerve. The AV node is thought to originate from the left horn of the sinus venosus. The sympathetic system innervates the atrial and ventricular musculature as well.

With the advent of newer technology and the possibility of curative radiofrequency ablation of anomalous conduction, it is very important to have a sound understanding of the AV junction, since the description and treatment of arrhythmias is crucially dependent on an accurate understanding of the underlying anatomy. A consensus statement from the cardiac Nomenclature Study Group has been advocated and published in *Circulation*. This nomenclature divides the AV junction into anatomically distinct and separate regions for description of accessory pathway location and better health care professional communication. In addition, it is important to appreciate developmental changes, as these have important implications for the study of the electrophysiologic structures in the pediatric age group and associated approach to ablation of an underlying abnormal substrate, such as that seen in dual AV nodal physiology.

ANATOMY OF THE CONDUCTION SYSTEM IN CONGENITAL HEART DISEASE

With congenital heart disease, development of the AV node and His-Purkinje system depend on appropriate atrial and ventricular orientation and proper alignment of the atrial and ventricular septum with appropriate closure of septal defects.

Atrioventricular Septal Defects

A most obvious abnormality occurs in association with AV septal defects (otherwise known as canal defects or endocardial cushion defects) where abnormalities in AV conduction system occur in association with abnormal development of the endocardial cushions. In these defects, the AV node is inferiorly and posteriorly displaced. The AV node is situated anterior to the mouth of the coronary sinus at a site just below where the base of the triangle of Koch would have occurred if the crux of the heart were properly formed. A common His bundle extents along the lower rim of the inlet portion of the ventricular septal defect resulting in a posterior course of the intraventricular conduction network. The classic ECG pattern inscribes a superior axis (vector) associated with this course of the His-Purkinje system.

In patients with ventricular septal defects, the AV node is usually in its anatomically correct position. The exceptions include ventricular septal defects that are inlet in type and, therefore, support a more inferior and posterior propagation of initial ventricular activation. The course of the common bundle or its branches relative to the ventricular septal defect may exhibit a longer common bundle. In patients with either inlet, perimembraneous, or outlet ventricular septal defects, the His bundle and its branches will typically be found on the lower crest of the effect, and will tend to deviate slightly towards the left side of the defect. Therefore, in postoperative patients, a ventricular septal patch may overlay the region of interest where a His bundle could be recorded. In these patients, the amplitude and frequency of a His signal may be variable and perhaps diminished.

Atrioventricular Discordance

Other abnormalities of the conduction system are associated with AV discordance either in biventricular hearts or in hearts with single ventricle physiology. In these patients, the AV node is situated outside the triangle of Koch and is elongated in morphology. Usually, the conduction system extends medially and runs along the right-sided mitral valve and pulmonary valve. If there is a ventricular septal defect, conduction usually occurs along the upper border of the septal defect. It is well known that the AV conduction system is tenacious in these patients and is sometimes associated with the development of spontaneous AV block.

Another type of AV discordance occurs in atrial situs inversus with D-loop ventricles. Because there is evidence to suggest embryologic development of more than one

AV node with one main node predominantly remaining in this malformation, the posterior AV node seems to persist. These patients will, therefore, have a left-sided triangle of Koch, but will usually have an associated AV node located posteriorly and inferiorly and may be inferiorly displaced; if there is the presence of a ventricular septal defect, the conduction system will run along the inferior border of the septal defect. These findings suggest that the AV node follows or is associated primarily with the morphologic right atrium.

Ventriculoatrial Discordance or Transposition of the Great Arteries

Abnormalities of outflow, and other conotruncal abnormalities and septal defects remote from the crux of the heart, usually do not effect the position and the location of the conduction system. In isolated transposition of the great arteries without a ventricular septal defect, there is minimal influence on the location of the conduction system. The AV anatomy is normal and there is normal AV concordance allowing for normal AV conduction system development. These patients have abnormalities of the outflow tracts. Many patients with D-transposition of the great arteries have undergone surgical repairs with either a Mustard or a Senning procedure, during which the atrial blood is rechanneled via a baffle to the appropriate ventricle. There is a high incidence of SA node dysfunction late after Mustard or Senning procedures. These patients are also at higher risk for the development of atrial flutter in association with these surgeries. It is important to understand the atrial surgery performed in such patients at the time of electrophysiologic studies to maximize outcomes of ablation therapies.

For both Mustard and Senning procedures, superior caval blood and inferior caval blood are directed via an intra-atrial baffle to the left ventricle. This baffle usually excludes the AV conduction system. This precludes direct catheter access to the His bundle area by a standard transvenous approach. In these patients, a proximal left bundle may be recorded from the left ventricular septum from the medial aspect of the mitral valve annulus from the venous side. A His signal can also be recorded or obtained from the non coronary cusp of the aortic valve or from the right ventricle after the catheter has been advanced through the aortic valve into the right ventricle and back towards the right-sided tricuspid valve (via a transarterial approach) to a position near the central fibrous body. Formal landmarks of triangle of Koch remain present, but can be distorted by previous surgery, and by the fact that the eustachian valve is often cut or intersected as part of the Mustard or the Senning procedures. In these patients, the coronary sinus can be left to drain to the pulmonary venous atrium or the systemic atrium. Regardless, recognition of these postoperative changes is crucial for patients with supraventricular tachycardia in association with transposition of the great arteries, which may include atrial tachycardias or AV node reentry tachycardias.

Twin Atrioventricular Nodes

These anatomic variants can be associated with dual compact AV nodes where both an anterior and posterior nodal structures are present. In these cases, the posterior node seems to be a more developed structure and ultimately forms the connection to the His bundle. These patients may experience AV reentry tachycardia using both AV nodes.

Tricuspid Atresia

In tricuspid atresia, the AV node is typically associated with the atretic tricuspid valve in the right atrium. Studies confirm that the compact AV node in tricuspid atresia is situated in the right atrium inside the underdeveloped and diminutive triangle of Koch. The orifice of the coronary sinus can still be identified as the base of the triangle, but the tricuspid valve may be small and difficult to identify. A very short common bundle is described running towards the central fiber body, which then descends along the septum. Should a ventricular septal defect be present, the conduction system tends to travel along the lower margin of the ventricular septal defect on the side of the septum between the rudimentary right ventricle and left ventricle.

Ebstein's Anomaly

Ebstein's Anomaly is associated with a normal atrioventricular node and triangle of Koch. However, because of anatomic distortions associated with displacement of the septal and posterior leaflets of the tricuspid valve in association with right atrial and right ventricular enlargement, identification of the normal anatomy may be difficult. In these cases, the coronary sinus may serve as an especially useful marker for the delineation of the triangle of Koch. Because the anatomic and electrophysiologic atrioventricular groove may be discrepant in patients with Epstein's anomaly, it might be helpful to perform a right coronary artery angiogram to define the anatomic atrioventricular groove. This cardiac abnormality is often associated with one or more accessory pathways and carries with it a higher incidence of atrial arrhythmias as well. An understanding of the anatomy and an effort to delineate present distortions can be critical for successful ablation at the time of the electrophysiologic study.

Heterotaxy Syndromes

These syndromes encompass a complex set of defects associated with "sidedness" confusion of organs in the thorax and/or abdomen. Two general subgroups exist: those with right atrial isomerism or "bilateral right sidedness" and those with left atrial isomerism or "bilateral left sidedness." Typical cardiac features of bilateral right sidedness or asplenia include an intact inferior vena cava, unroofed coronary sinus, total anomalous pulmonary venous return, complete AV septal defect, ventricular inversion, transposition of the great arteries or double outlet right ventricle with pulmonary stenosis/atresia. Features of double left-sidedness include interrupted inferior vena cava, total or partial anomalous pulmonary venous return, common or partial AV septal defects, normally related great vessels or double outlet right ventricle with or without pulmonary stenosis. The mode of inheritance of heterotaxy syndromes remains uncertain, although there is some suggestion that there may be autosomal dominant and recessive forms, with the majority of cases being due to mutations in genes that encode sidedness in association with environmental insults. Wren and colleagues reviewed the electrocardiograms of 126 patients with atrial isomerism, 67 with left atrial isomerism and 59 with right atrial isomerism. The cardiac rhythm in patients with left atrial isomerism, with supposed "absence" of normal sinus nodal tissue, tends to exhibit an atrial rhythm with a variety of atrial pacemaker locations as manifested in the wide range of P-wave axes recorded. In contract, patients with right atrial isomerism, with supposed "bilateral" sinus nodes, tended to exhibit P-wave axes predictive of either a high right-sided (between 0 and 89°) or high left-sided (between 90° and 179°) atrial pacemaker location. In addition, patients with double left-sidedness exhibit sinus node dysfunction (at 10-year follow-up, 80%). In addition, there are instances of AV nodal abnormalities (15%); in contrast there were none noted in patients with double right-sidedness. In asplenia (double right-sidedness), ventricular inversion is common. The incidence of complete heart block in patients with L-transposition has been reported to be between 17% and 22%. Approximately 3% to 5% of patients with L-transposition are born with complete heart block; heart block thereafter occurs approximately 2% per year. All cases of AV block were reportedly spontaneous, with no cases as a consequence of heart surgery or other mechanical insult to the AV node.

Finally, an entity of twin AV nodes has been described where there is co-existence of two distinct AV nodes (Mönckeberg's sling).

Reciprocating tachycardias involving both nodes can be the source of supraventricular tachycardia in these patients. Catheter ablation can successfully treat these patients. This entity seems to be more common in patients with right atrial isomerism.

In summary, patients with complex congenital heart disease, with or without heterotaxy, represent both a challenge to the surgeon for repair and a window for the developmental biologist into the development of the cardiac conduction system.

SUGGESTED READING

Development

1. Anderson RH, Ho SY. The architecture of the sinus node, the atrioventricular conduction axis, and the internodal atrial myocardium. *J Cardiovasc Electrophysiol* 1998;9:1233–1248.
2. Moorman AF, de Jong F, Denyn MM, Lamers WH. Development of the cardiac conduction system. *Circ Res* 1998;82:629–644.
3. Fishman MC, Chien KR. Fashioning the vertebrate heart: earliest embryonic decisions. *Development* 1997;124:2099–2117.
4. Garcia-Martinez V, Alvarez IS, Schoenwolf GC. Locations of the ectodermal and nonectodermal subdivisions of the epiblast at stages 3 and 4 of avian gastrulation and neurulation. *J Exp Zool* 1993;267:431–446.
5. Sedmera D, Pexieder T, Vuillemin M, et al. Developmental patterning of the myocardium. *Anat Rec* 2000;258:319–337.
6. Ar A, Tazawa H. Analysis of heart rate in developing bird embryos: effects of developmental mode and mass. *Comp Biochem Physiol A Mol Integr Physiol* 1999;124:491–500.
7. Akiyama R, Matsuhisa A, Pearson JT, Tazawa H. Long-term measurement of heart rate in chicken eggs. *Comp Biochem Physiol A Mol Integr Physiol* 1999;124:483–490.
8. Moriya K, Hochel J, Pearson JT, Tazawa H. Cardiac rhythms in developing chicks. *Comp Biochem Physiol A Mol Integr Physiol* 1999;124:461–468.
9. Broekhuizen ML, Gittenberger-de Groot AC, Baasten MJ, et al. Disturbed vagal nerve distribution in embryonic chick hearts after treatment with all-trans retinoic acid. *Anat Embryol (Berl)* 1998;197:391–397.
10. Aranega AE, Velez MC, Gonzalez FJ, et al. Modulation of cardiac intermediate filament proteins in the chick heart by fibric acid derivatives. *Eur J Morphol* 1995;33:421–431.
11. Velez C, Aranega AE, Fernandez JE, et al. Modulation of contractile protein troponin-T in chick myocardial cells by catecholamines during development. *Cell Mol Biol (Noisy-le-grand)* 1994;40:1189–1199.
12. Anderson RH. The developing heart in chick embryos. *Circulation* 1990;82:1542–1543.
13. Arguello C, Alanis J, Valenzuela B. The early development of the atrioventricular node and bundle of His in the embryonic chick heart. An electrophysiological and morphological study. *Development* 1988;102:623–637.
14. Chuck ET, Freeman DM, Watanabe M, Rosenbaum DS. Changing activation sequence in the embryonic chick heart. Implications for the development of the His-Purkinje system. *Circ Res* 1997;81:470–476.
15. Van Mierop L. Location of Pacemaker in chick embryo heart at the time of initiation of heart beat. *Am J Physiology* 1967;212:407–415.
16. Anderson RH AB, AC Weninck, et al. The Development of the Cardiac Specialized Tissue. In: *The Conduction System of the Heart: Structure, Function and Clinical Implications, ed 1*: The Hague, Martinus Nijhoff Publishing; 1976:3–28.
17. Moorman AF, de Jong F, Lamers WH. Development of the conduction system of the heart. *Pacing Clin Electrophysiol* 1997;20:2087–2092.
18. Moorman A, Webb S, Brown NA, et al. Development of the heart: (1) formation of the cardiac chambers and arterial trunks. *Heart* 2003;89:806–814.
19. Manasek FJ, Burnside MB, Waterman RE. Myocardial cell shape change as a mechanism of embryonic heart looping. *Dev Biol* 1972;29:349–371.
20. Manasek FJ, Monroe RG. Early cardiac morphogenesis is independent of function. *Dev Biol* 1972;27:584–588.
21. Manasek FJ, Burnside B, Stroman J. The sensitivity of developing cardiac myofibrils to cytochalasin-B (electron microscopy-polarized light-Z-bands-heartbeat). *Proc Natl Acad Sci USA*. 1972;69:308–312.
22. Moorman AF, Lamers WH. A molecular approach towards the understanding of early heart development: an emerging synthesis. *Symp Soc Exp Biol* 1992;46:285–300.
23. Watanabe MDR, Libbus I, Chuck ET. The Form and Function of the Developing Cardiac Conduction System. In: Clark EB NM, Takao A., ed. *Etiology and Morphogenesis of Congenital Heart Disease: Twenty Years of Progress in Genetics and Developmental Biology*: Armonk, NY: Futura Publishing Co. Inc.; 2000.
24. Fromaget C, el Aoumari A, Gros D. Distribution pattern of connexin 43, a gap junctional protein,

during the differentiation of mouse heart myocytes. *Differentiation* 1992;51:9–20.
25. Davis LM, Rodefeld ME, Green K, et al. Gap junction protein phenotypes of the human heart and conduction system. *J Cardiovasc Electrophysiol* 1995;6:813–822.
26. Verheijck EE, Wessels A, van Ginneken AC, et al. Distribution of atrial and nodal cells within the rabbit sinoatrial node: models of sinoatrial transition. *Circulation* 1998;97:1623–1631.
27. Domenech-Mateu JM, Arno Palau A, Martinez Pozo A. The development of the atrioventricular node and bundle of His in the human embryonic period. *Rev Esp Cardiol* 1993;46:421–430.
28. Nakagawa M, Thompson RP, Terracio L, Borg TK. Developmental anatomy of HNK-1 immunoreactivity in the embryonic rat heart: co-distribution with early conduction tissue. *Anat Embryol (Berl)* 1993;187:445–460.
29. Development of the Cardiac Conduction System. Symposium, held at the Novartis Foundation, London, May 21–21, 2003. Editors: Chadwick DJ, Goode J. John Wiley & Sons, Ltd, Chichester, UK.

Structure

30. Schamroth L. The Disorders of Cardiac Rhythm in Disorders of Impulse Formation. Basic Principles. In: *The Disorders of Cardiac Rhythm*: Blackwell Scientific Publications; 1971.
31. Keith AFM. The form and nature of the muscular connections between the primary divisions of the vertebrate heart. *J Anat Physiol* 1907;41:172–189.
32. Anderson KR, Ho SY, Anderson RH. Location and vascular supply of sinus node in human heart. *Br Heart J* 1979;41:28–32.
33. Anderson RH, Ho SY, Smith A, Becker AE. The internodal atrial myocardium. *Anat Rec* 1981;201:75–82.
34. McGuire MA, Johnson DC, Robotin M, et al. Dimensions of the triangle of Koch in humans. *Am J Cardiol* 1992;70:829–830.
35. Ho SY. Embryology and Anatomy of the normal and abnormal conduction system. In: Gillette PCaGA, ed. *Pediatric Arrhythmias: Electrophysiology and Pacing*. Philadelphia: WB Saunders; 1990.
36. Goldberg CS, Caplan MJ, Heidelberger KP, Dick M, 2nd. The dimensions of the triangle of Koch in children. *Am J Cardiol* 1999;83:117–20, A9.
37. Anderson RH, Becker AE, Brechenmacher C, et al. The human atrioventricular junctional area. A morphological study of the A-V node and bundle. *Eur J Cardiol* 1975;3:11–25.
38. Anderson RH, Brown NA. The anatomy of the heart revisited. *Anat Rec* 1996;246:1–7.
39. Anderson RH, Ho SY, Becker AE. Anatomy of the human atrioventricular junctions revisited. *Anat Rec*. 2000;260:81–91.
40. Bharati S. Anatomy of the atrioventricular conduction system. *Circulation*. 2001;103:E63–4.
41. Dean JW, Ho SY, Rowland E, et al. Clinical anatomy of the atrioventricular junctions. *J Am Coll Cardiol* 1994;24:1725–1731.
42. Quan KJ, Lee JH, Van Hare GF, et al. Identification and characterization of atrioventricular parasympathetic innervation in humans. *J Cardiovasc Electrophysiol* 2002;13:735–739.
43. Quan KJ, Van Hare GF, Biblo LA, et al. Endocardial stimulation of efferent parasympathetic nerves to the atrioventricular node in humans: optimal stimulation sites and the effects of digoxin. *J Interv Card Electrophysiol* 2001;5:145–152.
44. Quan KJ, Lee JH, Geha AS, et al. Characterization of sinoatrial parasympathetic innervation in humans. *J Cardiovasc Electrophysiol* 1999;10:1060–1065.
45. Cosio FG, Anderson RH, Becker A, et al. Living anatomy of the atrioventricular junctions. A guide to electrophysiological mapping. A Consensus Statement from the Cardiac Nomenclature Study Group, Working Group of Arrhythmias, European Society of Cardiology, and the Task Force on Cardiac Nomenclature from NASPE. North American Society of Pacing and Electrophysiology. *Eur Heart J* 1999;20:1068–1075.
46. Ho SY, McComb JM, Scott CD, et al. Morphology of the cardiac conduction system in patients with electrophysiologically proven dual atrioventricular nodal pathways. *J Cardiovasc Electrophysiol* 1993;4:504–512.
47. Dick M, 2nd, Norwood WI, Chipman C, Castaneda AR. Intraoperative recording of specialized atrioventricular conduction tissue electrograms in 47 patients. *Circulation* 1979;59:150–160.
48. Feldt RE, Puga WD, Seward FJ, et al. Atrial Septal Defects and Atrioventricular Canal. In: Adams FE, GC, ed. *Heart Disease in Infants, Children and Adolescents*: Williams and Wilkins; 1983.
49. Allwork SP, Bentall HH, Becker AE, et al. Congenitally corrected transposition of the great arteries: morphologic study of 32 cases. *Am J Cardiol* 1976;38:910–923.
50. Anderson RH, Becker AE, Arnold R, Wilkinson JL. The conducting tissues in congenitally corrected transposition. *Circulation* 1974;50:911–923.
51. Anderson RH, Danielson GK, Maloney JD, Becker AE. Atrioventricular bundle in corrected transposition. *Ann Thorac Surg* 1978;26:95–97.
52. Suzuki K, Ho SY, Anderson RH, et al. Interventricular communication in complete atrioventricular

septal defect. *Ann Thorac Surg* 1998;66:1389–1393.
53. Moorman AF, de Jong F, Denyn MM, Lamers WH. Development of the cardiac conduction system. *Circ Res* 1998;82:629–644.
54. Mayer ACRJJ. Cardiac Surgery of the neonate and infant D-Transposition of the Great Arteries. Philadelphia: WB Saunders; 1994.
55. Rhodes LA, Wernovsky G, Keane JF, et al. Arrhythmias and intracardiac conduction after the arterial switch operation. *J Thorac Cardiovasc Surg* 1995;109:303–310.
56. Collins KK, Love BA, Walsh EP, et al. Location of acutely successful radiofrequency catheter ablation of intraatrial reentrant tachycardia in patients with congenital heart disease. *Am J Cardiol* 2000;86:969–974.
57. Triedman JK, Bergau DM, Saul JP, et al. Efficacy of radiofrequency ablation for control of intraatrial reentrant tachycardia in patients with congenital heart disease. *J Am Coll Cardiol* 1997;30:1032–1038.
58. Wenink AC. Congenitally complete heart block with an interrupted Monckeberg sling. *Eur J Cardiol* 1979;9:89–99.
59. Becker MDRAA. The conduction system of the heart. London: Butterworth-Heineman; 1983.
60. Kajitani M, Ezeldin AK, Neuhauser JH. Wolff-Parkinson-White syndrome in previously undiagnosed Ebstein's anomaly. *J Ark Med Soc* 2004;100:362–364.
61. Nikolic G. Tachycardia in Ebstein's anomaly. *Heart Lung* 2003;32:347–349.
62. Ai T, Ikeguchi S, Watanuki M, et al. Successful radiofrequency current catheter ablation of accessory atrioventricular pathway in Ebstein's anomaly using electroanatomic mapping. *Pacing Clin Electrophysiol* 2002;25:374–375.
63. Mabo P. Ebstein's anomaly. Rhythm disorders and their treatment. *Arch Mal Coeur Vaiss* 2002;95:522–524.
64. Chauvaud SM, Brancaccio G, Carpentier AF. Cardiac arrhythmia in patients undergoing surgical repair of Ebstein's anomaly. *Ann Thorac Surg* 2001;71:1547–1552.
65. Ho SY, Fagg N, Anderson RH, et al. Disposition of the atrioventricular conduction tissues in the heart with isomerism of the atrial appendages: its relation to congenital complete heart block. J Am Coll Cardiol 1992;20(4):904–910.

2

Physiology of the Cardiac Conduction System

Peter S. Fischbach

The diagnosis and management of cardiac arrhythmias has progressed rapidly as a science. Advances in the ability to diagnose and either suppress or eliminate arrhythmic substrates has taken an exponential trajectory. Whether utilizing three-dimensional electroanatomic mapping systems for examining complex arrhythmias in patients with palliated congenital heart disease or genetic analysis in a search for evidence of heritable arrhythmia syndromes, technological advances have improved our ability to observe, diagnosis, and manage rhythm disturbances in patients from fetal life through adulthood. To fully harness the possibilities offered by these new technologies, a detailed understanding of cardiac anatomy and cellular electrophysiology is imperative.

The orderly spread of electrical activity through the myocardium is a well-choreographed process involving the coordinated actions of multiple intracellular and membrane proteins. Abnormalities in the physical structure of the heart or the function of these cellular proteins may serve as the substrate for arrhythmias. Cardiac myocytes like other excitable cells maintain an electrical gradient across the cell membrane. Various proteins including ion channels, ion pumps, and ion exchangers span the membrane contributing to the voltage difference between the inside and outside of the cell. Because these integral membrane proteins, along with membrane receptors and regulatory proteins, form the basis of the electrophysiologic properties of the heart, a knowledge of their structure and function is necessary to understand fully cardiac arrhythmias, as well as for the appropriate selection of antiarrhythmic pharmacological agents.

RESTING MEMBRANE POTENTIAL

At rest, cardiac myocytes maintain a voltage gradient across the sarcolemmal membrane with the inside being negatively charged relative to the outside of the cell. The sarcolemmal membrane is a lipid bilayer that prevents the free exchange of intracellular contents with the extracellular space. The transmembrane potential is generated by an unequal distribution of charged ions between the intracellular and extracellular compartments. The maintenance of the resting membrane potential is an active, energy dependent process relying in part on ion channels, ion pumps, and ion exchangers, as well as by large intracellular non-mobile anionic proteins. The ions are not free to move across the membrane and can only do so through the selective ion channels or via the pumps and exchangers. The net result is a resting membrane potential generally ranging from −80 to −90 mV. The most important membrane proteins for establishing the resting membrane potential are the Na^+/K^+-ATPase (ion exchanger) and the inwardly rectifying potassium channel (I_K). The Na^+/K^+-ATPase is an electrogenic pump that exchanges three sodium ions from the inside of the cell for two potassium ions in the extracellular space, resulting in a net outward flow of positive charge.

The unequal distribution of charged ions across the sarcolemmal membrane leads to both an electrical and a chemical force causing the ions to move into or out of the cell. If the membrane is permeable to only a single ion at a time, than for each ion, there is a membrane potential, the "equilibrium potential," at which there is no net driving force acting on the ion. The equilibrium potential may be calculated if the ionic concentrations on both sides of the membrane are known using the Nernst equation:

$$E_x = RT/F \ln[X]_o/[X]_i.$$

In this equation, R = the gas constant, T = absolute temperature, F = the Faraday constant and X is the ion in question. As an example, the usual intracellular and extracellular concentration of potassium is 4.0 mM and 140 mM, respectively. Substituting these values into the Nernst equation gives the following values:

$$E_k = -61 \ln[4]/[140] = -94.$$

At rest, the sarcolemmal membrane is nearly impermeable to sodium and calcium ions while the conductance (conductance = 1/resistance) for potassium ions is high. It is not surprising therefore that the resting membrane potential of most cardiac myocytes approaches the equilibrium potential for potassium. The sarcolemmal membrane, however, is a dynamic structure with changing permeability to various ions with a resultant change in the membrane potential. If the cellular membrane were permeable only to potassium then the Nernst equation would suffice to describe the membrane potential for all circumstances. As the membrane becomes permeable to various ions at different moments in time during the action potential, the Nernst equation is insufficient to fully describe the changes in the alteration of the membrane potential. The membrane potential at any given moment may be calculated if the corresponding instantaneous intra- and extracellular concentrations of the ions and the permeability of the respective ion channels are known. The Goldman-Hodgkin-Katz equation describes the membrane potential for any given set of concentrations and permeability's (P). The equation is:

$$V_m = -RT/F \ln \{P_K[K^+]_i + P_{Na}[Na^+]_i + P_{Cl}[Cl^+]_i\}/\{P_K[K^+]_o + P_{Na}[Na^+]_o + P_{Cl}[Cl^+]_o\}.$$

The Goldman-Hodgkin-Katz equation more closely approximates the cellular potential than the Nernst equation because it accounts for the permeability of the membrane for all active ions. This equation can also be used to computer model single cells and cellular syncytia.

ION CHANNELS

The lipid bilayer that makes up the sarcolemmal membrane has a high resistance to the flow of electrical charge and therefore requires specialized channels to allow the selective movement of ions into and out of the cell. Ion channels are macromolecular proteins that span the sarcolemmal membrane and provide a low resistance pathway for ions to enter or exit the cell. The ion channels are selective for specific ions and upon opening provide a low resistance pathway that allows ions to pass down their electrochemical gradient. The ion channels have three general properties: (1) a central water filled pore through which the ions pass; (2) a selectivity filter; and (3) a gating mechanism to open and close the channel. The channels may be classified not only by their selectivity for specific ions, but also by the stimulus that causes the channel to open. Channels may open in response to changes in the transmembrane potential (voltage gated) in response to activation with various ligands, in response to mechanical forces (stretch activated), and in response to changes in the metabolic state of the cell (ATP gated potassium channels).

Sodium Channels

The sodium current is the principal current responsible for cellular depolarization in atrial, ventricular, and Purkinje fibers. The rapid flow of ions through the sodium channel permits rapid depolarization of the sarcolemmal membrane and rapid conduction of the electrical signal. Sodium channels are closed at normal hyperpolarized resting membrane potentials. When stimulated by membrane depolarization, they open allowing the rapid influx of sodium ions, which changes the membrane potential from −90 mV towards the equilibrium potential for sodium (+40 mV). The channel inactivates rapidly over a few milliseconds in a time dependent fashion. That is, even in the face of a sustained depolarized membrane potential, the channel will close after a short time.

The sodium channels are proteins composed of a large pore forming an alpha subunit and two smaller regulatory beta subunits. The alpha subunit consists of four homologous domains, each of which consists of six transmembrane segments (S1–S6, Figure 1), a motif that is consistent across the voltage-gated ion channels. The transmembrane segments are hydrophobic and have an alpha-helical

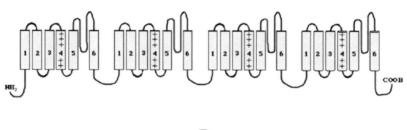

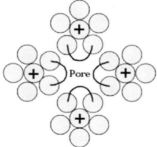

FIGURE 1. Top: Drawing of a voltage-gated sodium channel. The channel is composed of four domains, each of which has six membrane spanning hydrophobic helical segments. The fourth transmembrane segment is highly charged and acts as the voltage sensor for the channel. The linker segment between the 5th and 6th transmembrane segment in each domain bends back into the channel pore and is important in channel selectivity and gating. Bottom: This idealized drawing viewed from the extracellular surface demonstrates how the four domains organize to form a single pore with the S5–S6 linker segment of each domain contributing to the pore. (Downloaded from the internet).

conformation. The fourth transmembrane segment (S4) in each domain is highly charged with arginine and lysine residues located at every third position. The S4 segment acts as the voltage sensor for the channel with membrane depolarization causing an outward movement of all of the S4 domains leading to an opening of the transmembrane pore. The channel pore is formed by the S5 and S6 segments of each of the four domains in addition to the extracellular linker between S5 and S6. The transmembrane segments are linked by short loops, which alternate between intra- and extracellular. The extracellular linker loop between S5 and S6 is particularly long and curves back into the lipid bilayer to line the pore through which the ions pass. The four extracellular S5–S6 linker loops contribute to the selectivity of the channel.

The function of the beta subunit continues to be investigated. Much of the early work on beta subunits was in neuronal cells and recently attention has turned to cardiac cells. The beta subunits, in addition to modulating channel gating properties, are cell adhesion molecules that interact with the extracellular matrix serving an anchoring function. The beta subunits also regulate the level of channel expression in the plasma membrane.

The sodium channel opens rapidly in response to a depolarization in the membrane potential above a threshold value, reaching its maximal conductance within half a millisecond. After opening, the sodium current then rapidly dissipates, falling to almost zero within a few milliseconds. The inactivation of the sodium channel is the result of two separate processes, which may be differentiated based on their time constants. An initial rapid inactivation has a fast recovery constant and is, in part, caused by a conformational change in the intracellular linker between S3 and S4 that acts like a ball valve swinging into and occluding the pore-forming region. Rapid inactivation may occur without the channel opening, a process known as "closed state inactivation." A slower, more stable inactivated state also exists and may last from hundreds of milliseconds to several seconds. The mechanism(s) underlying slow inactivation are not well understood but likely results from the linker sequences between S5 and S6 in each domain bending back into the pore of the channel and occluding it.

The *SCN5A* gene located on chromosome 3 encodes the cardiac sodium channel. The cardiac sodium channel may be differentiated from the neuronal and skeletal muscle channel by its insensitivity to tetrodotoxin, which is isolated from puffer fish. Several diseases in humans resulting from sodium channel gene defects have been identified (Chapter 18).

Potassium Channels

Potassium channels are more numerous and diverse than any other type of ion channel in the heart (Figure 2). Over 200 genes have been identified that code for potassium channels. The channels may be categorized by their molecular structure, time, and voltage dependant properties, as well as their pharmacological sensitivities. Potassium channels are major components in the establishment of the resting membrane potential, automaticity, and the plateau phase of the action potential, as well as repolarization (phase 3, Figure 5). Within the heart a tremendous amount of heterogeneity exists in the density and expression of the potassium channels. The varied expression level of potassium channels contributes to the variability of the action potential morphology in different regions of the heart including transmural differences within the ventricular myocardium (Figure 3). In addition to the natural variability in the expression of potassium channels, many disease processes such as congestive heart failure and persistent tachyarrhythmias alter the density of these channels, as well as their functional properties, thereby leading to disruption of the normal electrical stability of the heart. This alteration in the density and function of these channels has been termed "electrical remodeling."

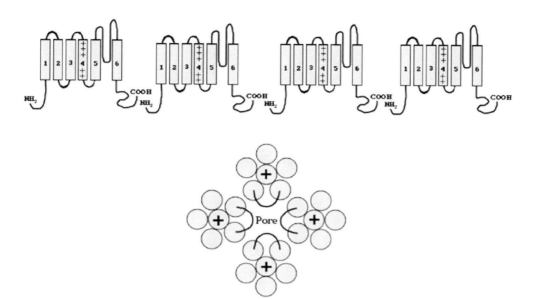

FIGURE 2. Similar to the voltage-gated sodium channels, the voltage-gated potassium channels are composed of four domains (α-subunits) composed of six membrane spanning segments. Unlike the sodium channels, the potassium channel domains are separate subunits that co-assemble to form a functional channel (compared with the sodium channel, which is a single large α-subunit composed of four domains. The voltage-gated potassium channels structurally are very similar to the sodium channels. The four α-subunits assemble to form a single pore with the S5–S6 linker from all α-subunits contributing to the pore. Similar to the voltage-gated sodium channel, the S4 subunit is also highly charged and serves as the voltage sensor leading to channel opening and closing. (Downloaded from the internet).

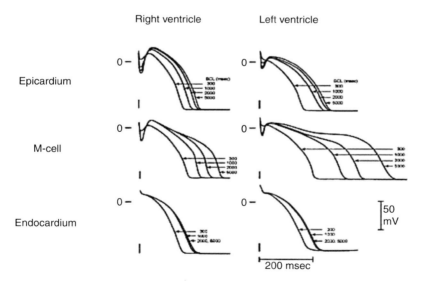

FIGURE 3. Action potential heterogeneity: The action potentials in this figure were recorded from strips of ventricular myocardium isolated from canine right and left ventricle. The difference in the morphology of the action potentials across the ventricular wall is obvious with a longer action potential found in the M-cells, which are located in the mid-myocardium. Additionally, the spike and dome configuration of the action potential generated by the activity of the transient outward current is prominent in the epicardial cells and nearly absent in the endocardium. Differences between the action potential morphology are also evident between the right and left ventricles. Finally, the action potentials were recorded at various paced cycle lengths. The rate adaptation of the cells from the different regions of the ventricle is strikingly different. (Reproduced from Antzelevitch C, Fish J. *Basic Res Cardiol* 2001;96(6):517–527, Springer Science + Business Media, Inc.)

Voltage-Gated Potassium Channel: I_{to} and I_K

Voltage-gated potassium channels are structurally very similar to the voltage-gated sodium channels. One major difference is that instead of the channels being composed of a single large α-subunit containing four domains, they are heteromultimeric complexes consisting of four α-subunits that form the channel pore and covalently attached regulatory β-subunits. The α-subunits are the equivalents of the separate domains of the sodium channel and are composed of six membrane spanning segments. The voltage sensor is contained in the S4 transmembrane segments that possess the same highly charged construct as in the sodium channel with alternating arginine and lysine in every third position. The mechanism supporting channel activation has not been as clearly delineated in potassium channels as in sodium channels. Identical to the sodium channel, the fifth and sixth transmembrane segments and the extracellular linker loop between S5 and S6 form the pore. While multiple sub-families of potassium channel α-subunits exist, only closely related sub-families of α-subunits are capable of co-assembling to form functioning channels.

I_{to}. The transient outward current is responsible for the early repolarization of the myocytes, creating the "spike and dome" configuration of the action potential noted in epicardial ventricular myocytes (Figure 5). The current may be divided into two components: a Ca independent and 4-aminopyridine sensitive current that is carried by K, and a Ca dependent and 4-aminopyridine insensitive current carried by the Cl ion. There is further evidence to suggest that the portion of I_{to} carried by potassium may be further subdivided into a fast and slow component. The level of expression of I_{to} is highly variable, with greater density in the atrium than the ventricles, greater density in the right ventricle than the left ventricle, and greater density in the epicardium than in the endocardium (Figure 3).

I_K (Delayed Rectifier). The delayed rectifier current is the ionic current primarily responsible for repolarization (phase 3) of the myocytes. It opens slowly relative to the sodium channel in response to membrane depolarization near the plateau potential of the myocytes (+10 to +20 mV). Following the initial description of the delayed rectifier current, two distinct components of the current were identified: a rapidly activating and a slowly activating portion. Recently a third component has been isolated. The different subsets may be differentiated based on their activation/inactivation kinetics, pharmacological sensitivity, and conductance. The rapidly activating component (I_{Kr}) has a large single channel conductance, demonstrates marked inward rectification, activates rapidly, and is selectively blocked by several pharmacological agents including sotalol and dofetilide. Non-cardiovascular drugs associated with heart rate corrected QT interval prolongation almost exclusively interact with I_{Kr} to produce their action potential prolonging effects. I_{Kr} is the product of the *KCNH2 (HERG)* gene located on chromosome 7 and abnormalities of this channel result in LQTS type 2 (Chapter 20). The slowly activating portion of the current (I_{Ks}) has a smaller single channel conductance and is selectively inhibited by chromanol 293b. I_{Ks} inactivates more slowly than I_{Kr} and becomes the dominant repolarizing current at more rapid heart rates. I_{Ks} is the product of the *KCNQ1 (KvLQT1)* gene on chromosome 11 and abnormalities of this channel result in LQTS type 1 (Chapter 20). The third subset of delayed rectifier current (I_{Kur}, the ultra-rapid delayed rectifier) has very rapid activation kinetics and slow inactivation kinetics with a single channel conductance that is close to that of I_{Kr}. I_{Kur} is significantly more sensitive to the potassium channel blockers 4-aminopyridine and TEA than either I_{Kr} or I_{Ks} and the channel may be selectively inhibited by the experimental

compound S9947. I_{Kur} is the product of the *KCNA5* gene on chromosome 12.

The expression of the subtypes of the delayed rectifier channel is heterogeneous. The expression of all three subtypes is greater in the atrium than in the ventricle, in part explaining the shorter action potential in atrial compared to ventricular myocytes. I_{Kur} is exclusively expressed in the atrium and has not been isolated from ventricular tissue. Within the ventricle, I_{Ks} density is low in the midmyocardium, the so-called M-cells, compared with cells in the epi- and endocardium. The transmural difference in the distribution of potassium channels across the ventricular wall accounts for the longer duration of the action potential in Purkinje fibers as compared to the action potential in the epicardium and underlies to the configuration of the T-wave in the surface electrocardiogram.

Inwardly Rectifying Currents: I_{K-Ach}, I_{K-ATP}, and I_{K1}

The inwardly rectifying channels are structurally distinct from the voltage-gated channels. As opposed to the four membrane spanning subunits in voltage-gated channels, the inwardly rectifying channels have two membrane spanning subunits (M1 and M2). The association of four subunits forms a pore. The ATP-gated channel (I_{K-ATP}) is more complex with the four pore forming subunits coassembling with four sulfonylurea receptors to form a functional channel. Inward rectification occurs via gating of the channels by magnesium and polyamines (spermine, spermidine, etc.) that block the inner opening of the pore.

I_{K1}. I_{K1} is the dominant resting conductance in the heart, setting the resting membrane potential in atrial, ventricular, and Purkinje cells. The heterogeneous density of channel distribution is greater in the ventricles relative to the atrium, but relatively sparse in nodal cells. I_{K1} has been demonstrated to inactivate at sustained depolarized potentials, such as during the plateau phase of the action potential.

I_{K-Ach}. Acetylcholine, which is released from the cardiac parasympathetic nerves, acts on type 2 muscarinic receptors to open the channels via a G-protein dependent mechanism. The channels are localized primarily in nodal cells and atrial myocytes. The presence of I_{K-Ach} channels in the ventricle has been identified, although interestingly, the sensitivity to ACh is less than in the nodal cells and atrial myocytes. Activation of the channels causes hyperpolarization of nodal cells and a slowing of the rate of spontaneous depolarization and shortening of the action potential in atrial and ventricular myocytes.

I_{K-ATP}. I_{K-ATP} is tonically inhibited by physiological concentrations of intracellular ATP. During periods of metabolic stress, when the ATP level decreases and the ATP/ADP ratio is altered, the inhibition on the channel is lost and the channel opens, providing a large conductance repolarizing current (outward movement of K^+). Two molecularly distinct populations of I_{K-ATP} have been described in the heart: one existing in the sarcolemmal membrane and the other in the inner mitochondrial membrane. I_{K-ATP}, and in particular the mitochondrial channel, has been demonstrated to be important in ischemic preconditioning.

Calcium Channels

Calcium channels, like the voltage-gated potassium channels and the voltage-gated sodium channels, share the same basic structural motif. The channels are composed of a single large pore forming α-subunit, with two regulatory subunits (β, α_2/δ). The α-subunit is similar in structure to that of the sodium channel, consisting of a single large protein with four domains composed of six membrane-spanning segments. The voltage sensor is also localized on S4 and the pore is composed of S5, S6, and the S5–S6 linker of each of the four domains. However, the pore-forming loop between S5 and S6 is significantly different between calcium and sodium channels. The calcium channel has several calcium-binding sites on the pore-forming loop and the presence of calcium at these sites blocks sodium from entering the channel pore. When these sites are devoid of calcium, the channel passes sodium ions freely.

There are two main types of sarcolemmal calcium channels, which are differentiated by their conductance, activation/inactivation, and pharmacological sensitivities. The L-type calcium channel ($I_{Ca,L}$–long acting) compared with the T-type channel ($I_{Ca,T}$–transient channel) activates at more positive membrane potentials, inactivates more slowly, has a larger single channel conductance, and is sensitive to dihydropyridines. $I_{Ca,L}$ is expressed abundantly in all myocytes while the expression of $I_{Ca,T}$ is more heterogeneous, being most prominent in nodal cells. The $I_{Ca,L}$ serves to bring calcium into working myocytes throughout the plateau phase of the action potential and leading to calcium-dependent calcium release from the sarcoplasmic reticulum. $I_{Ca,T}$ is more important for depolarization in nodal cells.

Chloride Channels

Far less is known about the structure and function of cardiac channels that carry anions. There are at least three distinct chloride channels in the heart. The first is the cardiac isoform of the cystic fibrosis transmembrane conductance regulator (CFTR), which is regulated by cAMP and protein kinase A. The second is a calcium-activated current that participates in early repolarization ($I_{to,2}$), and the third is a swelling activated channel. To date, no role for abnormal function of chloride channels has been found for arrhythmia formation, and the channels are not targeted by any of the currently available antiarrhythmic drugs. However, chloride channels do appear to play an important role in maintaining the normal action potential. While chloride con-

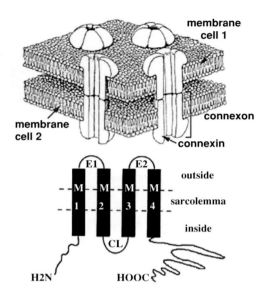

FIGURE 4. Schematic representation of part of a gap junction. The individual gap junction channels consist of two connexons that are non-covalently attached. Each connexon is composed of six connexins. The individual connexins has four membrane spanning regions (M1–M4), two extracellular loops (E1 and E2), and one cytoplasmic loop (CL). (Reproduced with permission from van der Velden et al. *Cardiovasc Res* 2002;54: 270–279, The European Society of Cardiology.)

ductance does not play a role in establishing the resting membrane potential, experiments in which the extracellular chloride was replaced with an impermeant anion resulted in markedly prolonged action potentials.

Gap Junctions

Gap junctions are tightly packed protein channels that provide a low resistance connection between adjacent cells, allowing the intercellular passage of ions and small molecules (Figure 4). These channels allow the rapid spread of the electrical signal from cell to cell. The gap junctions are produced by the non-covalent interaction of two hemi-channels (connexins) that are embedded in the plasma membranes of adjacent cells. The connexon is formed by the association of six connexin subunits, each of which have four transmembrane spanning domains and two extracellular

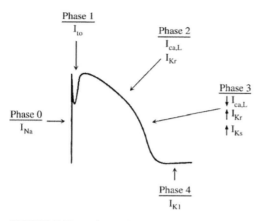

FIGURE 5. Shown is a typical action potential recorded from an epicardial ventricular myocyte. Phase 0 is the rapid depolarization of the membrane driven by opening of the voltage-gated sodium channel. Phase 1 is rapid initial repolarization resulting from closing of the sodium channel and opening of the transient outward current carried primarily by potassium. Phase 2, the plateau phase of the action potential, is notable for a balanced flow of inward (calcium) and outward (potassium) currents resulting in no significant change in the membrane potential. Phase 3 of the action potential, rapid repolarization, results from time dependent inactivation of the calcium channel in leaving the outward potassium current relatively unopposed. Phase 4 of the action potential is the resting membrane potential for myocytes without automaticity, and is principally defined by outward potassium current flow through I_{K1}.

loops. There are greater than 12 types of connexins expressed in myocardium. The transmembrane domains and extracellular loops in all of the connexins are highly preserved while the intracellular loop between the second and third domain and the carboxy-terminus are highly variable. The differences in the intracellular portions of the connexins account for the difference in molecular weight and physiological properties such as the junction conductance, pH dependence, voltage dependence, and selectivity.

THE ACTION POTENTIAL

Myocytes may be broadly divided into two distinct cell types; fast response cells (atrial, ventricular, and Purkinje cells) and slow response cells [sinoatrial (SA) and atrioventricular (AV) nodal cells]. The action potentials of fast response cells are notable for having a rapid upstroke (large V_{max}) generated by the large conductance voltage-gated sodium channel and hyperpolarized resting membrane potentials. The action potentials of slow response cells do not have a fixed resting membrane potential and have a slower upstroke driven by activation of L-type calcium channels. The rate at which the membrane is depolarized (V_{max}) determines how rapidly the electrical signal is conducted through tissue.

The action potential is divided into five phases, which may be delineated by the dominant membrane conductance. The action potential of fast response and slow response cells differ and will be addressed separately below:

Rapid Response Cells

Phase 0. This phase includes the rapid depolarization of the membrane. At rest, the membrane of a fast response cell is permeable almost exclusively to potassium. This drives the resting membrane potential towards the equilibrium potential of potassium (-94 mV). If the membrane potential is depolarized beyond a set threshold value, the voltage-gated sodium channels open and the membrane's dominant conductance changes to sodium. The membrane potential therefore moves towards the equilibrium potential of sodium ($\sim+40$ mV). The stimulus that generates the action potential elicits an all or nothing response. If the stimulus is sub threshold, the membrane is transiently depolarized and then quickly returns to the resting potential. If the stimulus is of sufficient intensity to raise the membrane potential above the threshold level, a maximal response is elicited and an action potential is generated. If the threshold potential is reached, phase 0 of the action potential is not altered by the intensity of the stimulus (i.e., a stimulus of greater intensity does not result in an increase in V_{max}). The number of available sodium channels is dependent on the resting membrane potential of the cell and determines the Vmax. If the resting membrane potential is depolarized relative to the normal value, fewer sodium channels are available to be recruited to participate in the action potential due to a greater number of sodium channels in inactivated states and the V_{max} is lower. Clinically this may occur during myocardial ischemia or other instances of stress. Phase 0 is also delayed during the use of sodium channel blocking antiarrhythmic drugs producing a decrease in V_{max}.

Phase 1. The maximal depolarized membrane potential reaches approximately +20 mV, which is below the equilibrium potential for sodium. The failure to reach the equilibrium potential of sodium is due to the rapid time dependent inactivation of sodium channels, changing to a non-conducting conformation, as well as the opening of hyperpolarizing currents, most significantly I_{to}. The heterogeneous expression of I_{to} throughout the myocardium explains the variability in the morphology of the early portion of the action potential. Those myocytes expressing relatively more I_{to} have a profound spike and dome conformation, such as epicardial ventricular myocytes. The inactivation of I_{to} is also rapid, and so it does not contribute significantly to the plateau phase, or later repolarization of the myocyte.

Phase 2. The plateau phase of the action potential represents a balance of inward depolarizing current carried primarily by calcium with a small contribution from a background sodium window current, and outward hyperpolarizing potassium current. The calcium and potassium currents are activated by depolarization, and both are inactivated in a time-dependent fashion.

Phase 3. At the completion of phase 2, the calcium channels close leaving the effects of the potassium conductance unopposed. The membrane potential moves once again towards the equilibrium potential of potassium. The delayed rectifier potassium currents (I_{Kr} and I_{Ks}) close during phase 3 and I_{K1} becomes the dominant conductance at the conclusion of phase 3.

Phase 4. Atrial and ventricular myocytes maintain a constant resting membrane potential awaiting the next depolarizing stimulus. The resting membrane potential is established by I_{K1}. The resting membrane potential remains slightly depolarized relative to the equilibrium potential of potassium due to an inward depolarizing leak current likely carried by sodium. During the terminal portions of phase 3 and all of phase 4, the voltage-gated sodium channels are recovering from the inactivated state into the resting state, and preparing to participate in the ensuing action potential.

Slow Response Cells

The action potential of slow response cells is morphologically distinct from fast response cells. Initial rapid depolarization (phase 0) in slow response cells is the result of current passing through voltage-gated calcium channels that activate at relatively depolarized potentials (−35 mV). The resultant V_{max} during phase 0 is considerably slower than that measured in fast response cells. Activation of phase 0 depolarization in slow response myocytes, similar to fast response myocytes, relies on the membrane potential surpassing a threshold potential with an all or none response in the depolarizing calcium current. Unlike the fast response myocytes, which require an external stimulus to raise the membrane potential past the threshold potential, the slow response myocytes generate their own depolarizing current. The slow response myocytes do not have a fixed resting membrane potential but rather hyperpolarize to a maximal diastolic potential (−50 to −65 mV for SA node cells) and then slowly depolarize towards the threshold potential of the calcium channels. Repolarization of slow response cells is the result of the time-dependent inactivation of the calcium channel in combination with the hyperpolarizing currents carried by the delayed rectifier potassium current.

Refractory Period

Following activation, all excitable cells enter a period in which repeat activation is impossible. This is referred to as the refractory period. The refractory period is longer in myocytes than in other excitable cells such as neurons or skeletal muscle. Physiologically, the refractory period allows for relaxation of the myocardium and filling of the cardiac chambers. Due to the long refractory period in cardiac muscle, tetanic contraction is not possible as it is in skeletal muscle. The refractory period may be divided into the absolute refractory period, during which time no action potential may be generated regardless of the stimulus, and the relative refractory period, in which an action potential may be induced following a supernormal stimulus. The refractory period is the result of the slow recovery from the inactivated state of the depolarizing current (I_{Na}) combined with the slow inactivation of the repolarizing currents.

Automaticity

Automaticity is the ability of cells to spontaneously depolarize and raise the resting membrane potential past the threshold potential for triggering an action potential (phase 4 depolarization). Automaticity normally is a property of cells localized within the SA and AV nodes, as well as Purkinje fibers. Spontaneous depolarization results from a net inward current resulting from the combination of several currents. The currents involved in phase 4 spontaneous depolarization continue to be debated. They appear to include activation of several inward cation channels including the pacemaker current (I_f), which is activated by hyperpolarization and closes shortly after the activation potential for the inward calcium channel is passed, I_{Ca-T}, I_{Ca-L} (mostly at the end of phase 4 depolarization as the activation potential is -40 mV for this current) and I_b, an inward time independent background current carried by sodium. The decay of the inward hyperpolarizing I_K also contributes to the net depolarizing current. Additionally, the Na^+/K^+ and Na^+/Ca^{2+} exchangers, that are electrogenic, provide additional inward current as K^+ and Na^+ are extruded from the cell. It is important to note that I_{K1} is not expressed in nodal cells and hence the hyperpolarizing effects do not influence phase 4 depolarization. The mechanism of phase 4 depolarization in Purkinje cells appears to be somewhat different compared to nodal cells, with I_f playing a far more prominent role. The maximal diastolic potential is approximately -85 mV in Purkinje cells rather than -60 mV in nodal cells, so the contribution of calcium current is thought to be less.

There is a hierarchical pattern to the rate of spontaneous depolarization, with the most rapid depolarization occurring in the SA node followed by the AV node with the Purkinje fibers being the slowest. The rate of automatic discharge is under tight autonomic control. Vagal input into the heart via release of acetylcholine activates the muscarinic receptor-gated potassium channel (I_{K-Ach}), which leads to hyperpolarization of the nodal cells. The hyperpolarization of the nodal cells creates a greater difference between the maximal diastolic potential and the threshold potential that is unchanged by the vagal stimulation. If the slope of diastolic depolarization remains unchanged, the time it takes to reach the threshold potential will increase and the rate of depolarization will decrease. Vagal stimulation, however, also inhibits I_f and I_{Ca-L} leading to a decrease in the slope of phase 4 depolarization, and further slowing the rate of spontaneous depolarization. Sympathetic stimulation conversely, via cAMP-dependent pathways, enhances I_f and I_{Ca-L} increasing the slope of phase 4 depolarization and increasing the rate of spontaneous depolarization.

Signal Propagation

Electrical conduction through the myocardium may be considered on the level of the signal transiting along a single myocyte, cell-to-cell conduction, and conduction through the whole organ. The electrical signal passes along a myocyte by incrementally depolarizing the membrane. At rest, an electrical gradient is maintained across the sarcolemmal membrane. When the local membrane potential is depolarized, sodium and calcium channels open, establishing a small region of positive charge on the inner surface of the membrane and leaving a small zone of negative charge extracellularly. This depolarizes the membrane in the immediately adjacent regions and leads to initiation of the depolarizing currents in the bordering regions. This forms a self-perpetuating reaction with the spread of depolarization. The more rapidly the local region of the membrane is able to change its potential, the more rapid is conduction down the length of the myocyte. Therefore, cells that depend on the rapid sodium current for the upstroke of phase 0 of the action potential conduct the signals rapidly, while cells that are dependent on the slower calcium current for phase 0 conduct signals more slowly. This concept is demonstrated by comparing the conduction velocity in the AV node (calcium dependent action potentials) with atrial and ventricular tissue (sodium channel dependent).

The myocardium is not a perfect syncytium, and conduction is discontinuous. Gap junctions that form a low resistance passageway to allow for the rapid spread of excitation from cell to cell attach the myocytes to one another. It has also been appreciated that the tissue passes electrical current more rapidly along the long axis of the myocytes than transversely. The difference that exists between the conduction velocities axially versus transversely is referred to as anisotropy. The degree of anisotropic conduction varies throughout the myocardium. In the ventricle, the ratio of conduction velocity parallel as opposed to transverse to the long axis of the cell is approximately 3:1, while along the crista terminalis in the atrium, the same ratio is 10:1.

On a macroscopic level, once the electrical signal escapes from the SA node, it rapidly spreads throughout the atrium. A debate remains as to whether or not there exist anatomically distinct internodal tracts that connect the SA and AV nodes. Histologic studies have failed to demonstrate the presence of these tracts. The SA node is located high in the right atrium adjacent to the orifice of the superior vena cava. In this location it is susceptible to injury at the time of cannulation for cardiopulmonary bypass. Additionally, the blood supply to the SA node may be damaged during surgery that involves extensive atrial manipulation such as the Mustard operation or the hemi-Fontan procedure. The electrical signal activates the right atrium initially with conduction spread into the left atrium preferentially via Bachmann's bundle and along the coronary sinus. The atria and ventricles are electrically isolated from one another by the fibrous AV ring, with electrical continuity provided by the AV node. The compact AV node is located in the right atrium at the apex of the triangle of Koch. The AV node is composed of three regions: the atrionodal (AN), nodal (N), and nodal-His (NH). The cells in these three regions differ in the shape of their respective action potentials and their conduction velocities. The cells in the AN region have action potentials that are intermediate between atrial and SA nodal cells with a more depolarized maximal diastolic potential, slower phase 0 depolarization, and the presence of phase 4 depolarization. The N cells are similar to the SA nodal cells. The NH cells transition between the N cells and the His bundle with action potentials that reflect this transition. Conduction through the entirety of the AV node is slower than that through the atria or ventricles and is not detectable on a routine surface ECG. The AV node is richly innervated by the autonomic nervous system and receives its blood supply from the posterior descending coronary artery. Atrioventricular block may result from interruption of the blood supply, which may result from atherosclerotic disease, as well from vasospasm following the delivery of radiofrequency energy in the posteroseptal region on the tricuspid valve annulus.

After passing through the AV node, the signal passes through the His bundle and enters the Purkinje fibers that divide into the bundle branches. The Purkinje cells have large diameters and preferential end-to-end connections rather than side-to-side connections, both of which lead to accelerated conduction velocities (200 cm/sec compared with 5 cm/sec in the SA node). The right bundle branch is cord-like and passes to the apex of the right ventricle prior to ramifying into the ventricular mass initially along the moderator band. The left bundle branch is fan-like and initially activates the ventricular septum from the left ventricular side progressing towards the right. This accounts for the Q-waves inscribed in the left lateral precordial leads in D-looped hearts, and the Q-waves in the right precordium in L-looped hearts.

MECHANISM OF ARRHYTHMIA FORMATION

Arrhythmias occur when the orderly initiation and conduction of the electrical signal is altered. This may result in abnormally fast or abnormally slow heart rates. Bradycardia may result from either a failure of initiation of impulse formation such as occurs in sick sinus syndrome, or failure of the signal to be conducted (SA node exit block, AV node block). Bradycardia is not amenable to chronic pharmacological therapy and when symptomatic, requires an implantable pacemaker (Chapter 17).

Tachycardias arise from one of three general mechanisms: abnormal automaticity, triggered activity, or reentry.

Abnormal Automaticity

Abnormal automaticity implies either abnormally fast activation in cells that normally possess automatic function (enhanced automaticity), or the development of spontaneous depolarization in cells that normally do not posses automaticity. Abnormally enhanced automaticity may result from hyperactivity of the autonomic nervous system, fever, thyrotoxicosis, or the exogenous administration of sympathomimetic agents.

Under normal physiological conditions, atrial and ventricular cells do not demonstrate phase 4 spontaneous depolarization. If the resting membrane potential is decreased (made less negative) to less than -60 mV, as may occur with ischemia, cells may develop spontaneous depolarization. Clinical examples of arrhythmias supported by abnormal automaticity include atrial ectopic tachycardia, junctional ectopic tachycardia (Chapter 10), accelerated idioventricular rhythm, and some ischemic ventricular tachycardias (Chapter 12).

Triggered Activity

Triggered activity arises from oscillations in the membrane potential, which if large enough, may reach threshold potentials and lead to additional action potentials. Triggered activity by definition is dependent on a preceding action potential or electrical stimulus to generate the oscillations in the membrane potential. An action potential generated via a triggered complex may serve as the stimulus for an ensuing action potential leading to a sustained arrhythmia. Triggered activity may result from oscillations in the membrane potential that occur either during phase 2 or 3 of the action potential (early afterdepolarization or EAD) or following full repolarization of the action potential (delayed afterdepolarization or DAD).

EADs. These oscillations occur during the plateau phase or late repolarization of the action potential. EADs may result from a decrease in outward current, an increase in inward current, or a combination of the two. During the plateau phase of the action potential, the absolute ionic flow across the membrane is small, hence a small change in either the inward or outward current may result in a large change in the membrane potential.

DADs. DADs occur after the cell has fully repolarized and returned to the resting membrane potential. The depolarization of the membrane resulting in DADs is caused by activation of the transient inward current (I_{ti}), which is a non-specific cation channel activated by intracellular calcium overload.

Reentry

Reentry is far and away the most frequent mechanism supporting arrhythmias. First described by George Mines in 1914, reentry describes the circulation of an electrical impulse around an electrical barrier leading to repetitive excitation of the heart. For reentry to occur, three conditions are required: a continuous circuit, unidirectional conduction block in part of the circuit, and conduction delay in the circuit allowing tissue previously activated to regain its excitability by the time the advancing wavefront returns. These conditions allow for a sustained circuit to be established.

Critical to the maintenance of reentrant arrhythmias are the impulse conduction velocity and the refractory period of the myocardium in the circuit. The interaction of these two properties of the myocardium determine whether the leading edge of electrical excitation will encounter myocardium capable of generating an action potential, or whether refractory tissue will block the circuit and cause the reentrant loop to extinguish. These two measurable quantities may be combined to calculate the "wavelength" of the arrhythmia.

Conduction Velocity (m/sec) × Refractory Period (sec)
 = Wavelength (m).

If the calculated wavelength of the arrhythmia exceeds the pathlength of the circuit, then reentry cannot occur. If the wavelength is less than the pathlength, then reentry may be sustained. This concept is the basis for the use of class III antiarrhythmic agents, which work by increasing the duration of the refractory period of the tissue by delaying repolarization. In reentrant arrhythmias with a fixed pathlength, an excitable gap exists, which is the time interval between the return of full excitability of the tissue following depolarization and the head of the returning electrical wavefront traversing the circuit. The presence of an excitable gap allows for external stimuli to enter the reentrant circuit and either advance (speed up), delay, or terminate the tachycardia.

Many forms of tachycardia result from reentry with fixed anatomic pathways. Examples include accessory pathway mediated tachycardia (Chapter 3), AV node reentrant tachycardia (Chapter 4), atrial flutter intra-atrial reentrant tachycardia (scar mediated atrial flutter)(Chapter 8), and ischemic ventricular tachycardia. Reentry has also been demonstrated to occur in tissue in which there are no fixed physical barriers. Reentry may be supported in this case by the development of functional barriers to conduction that serve as a focal point for the circuit to rotate around. These functional centers are not fixed and continually change and move with time. This principle referred to as "leading circle reentry" is thought to underlie atrial fibrillation. These reentrant circuits do not have an excitable gap with tissue becoming activated as soon as it is no longer refractory. The cycle length in arrhythmias supported by leading circle reentry is determined by the refractory period of the tissue that determines the circuit length.

SELECTED READING

1. Hille B, *Ion Channels of Excitable Membranes.* Sunderland, MA; Sinauer, 2001.
2. Zipes DP, Jalife J, (eds). Cardiac electrophysiology: from cell to bedside. Philadelphia, PA; Saunders, 2000.
3. Roden DM, Balser JR, Geroge AL, Anderson ME. Cardiac ion channels. *Annu Rev Physiol* 2002;64:431–475.
4. Schram G, Pourrier M, Melnyk P, Nattel S. Differential distribution of cardiac ion channel expression

as a basis for regional specialization in electrical function. *Circ Res* 2002;90:939–950.
5. Members of the Sicilian Gambit. New approaches to antiarrhythmic therapy: emerging therapeutic applications of the cell biology of cardiac arrhythmias. *Eur Heart J* 2001;22:2148–2163.
6. Singh BN, Sarma JS. Mechanisms of action of antiarrhythmic drugs relative to the origin and perpetuation of cardiac arrhythmias. *J Cardiovasc Pharm Therap* 2001;6:69–87.
7. Barry DM, Nerbonne JM. Myocardial Potassium channels: electrophysiological and molecular diversity. *Annu Rev Physiol* 1996;58:363–394.
8. Anderson ME, Al-Khatib SM, Roden DM, Califf RM. Duke Clinical Research Institute/American Heart Journal Expert Meeting on Repolarization Changes. *Am Heart J* 2002;144:769–781.
9. Roden DM, George AL. Structure and function of cardiac sodium and potassium channels. *Am J Physiol* 1997;273:H511–H525.
10. Snyders DJ. Structure and function of cardiac potassium channels. *Cardiovasc Res* 1999;42:377–390.
11. Liu DW, Gintant GA, Antzelevitch C. Ionic bases for electrophysiological distinctions among epicardial, midmyocardial, and endocardial myocytes from the free wall of the canine left ventricle. *Circ Res* 1993;72:671–687.
12. Allessie MA, Bonke FI, Schopman FJ. Circus movement in rabbit atrial muscle as a mechanism of tachycardia. III. The "leading circle" concept: a new model of circus movement in cardiac tissue without the involvement of an anatomical obstacle. *Circ Res* 1977;41:9–18.
13. Antzelevitch C, Shimizu W, Yan GX, et al. The M cell: its contribution to the ECG and to normal and abnormal electrical function of the heart, *J Cardiovasc Electrophysiol* 1999;10:1124–1152.
14. Allessie M, Ausma J, Schotten U. Electrical, contractile and structural remodeling during atrial fibrillation. *Cardiovasc Res* 2002;54:230–246.
15. Armoundas AA, Wu R, Juang G, et al. Electrical and structural remodeling of the failing ventricle. *Pharmacol Ther* 2001;92:213–230.
16. Antzelevitch C, Fish J. Electrical heterogeneity within the ventricular wall. *Basic Res Cardiol* 2001;96:517–527.

3

Clinical Electrophysiology of the Cardiac Conduction System

Macdonald Dick II, Peter S. Fischbach, Ian H. Law, and William A. Scott

Recording the human cardiac biological signal from the body surface, the electrocardiogram (ECG) has been used for at least seven decades in the diagnosis and management of cardiac disorders in infants and children, primarily by examining for criteria of hypertrophy or dilatation imposed by congenital cardiac malformations, rheumatic heart disease or other acquired heart conditions. Other than the ECG findings of congenital heart block and supraventricular tachycardia, originally called paroxysmal atrial tachycardia, little attention was directed to either the clinical or fundamental electrophysiology underlining arrhythmias in the young. Since the recording of the His bundle electrogram, first in the dog in 1957, then in the human in the operating room in 1959, and by means of the transcatheter technique in adults in 1969 and children in 1971, there has been a rapid advancement in the understanding of the clinical disorders of impulse formation and propagation. This advance has been both contemporary with and driven by major technological improvements including catheter design, computer-assisted recording, analysis and archival systems, and innovative transcatheter therapy. Because these major advances have evolved from adult populations, the application to young patients calls for special considerations.

CARDIAC ELECTROPHYSIOLOGY IN THE YOUNG

The intracardiac bioelectric signal, called an electrogram, is recorded through either bipolar, the usual configuration for electrophysiologic study, or unipolar electrodes strategically placed on intravascular catheters. The closely spaced bipolar electrodes (interelectrode distance is 1–5 mm; Figure 1) record activation of excitable tissue adjacent to the electrode pair, thus anatomically localizing the origin of the electrogram within the heart.

Thus, by virtue of this proximity effect, the electrical activity of localized portions of the atrial and ventricular myocardium, and the His bundle potential can be recorded by the bipolar pair of intracardiac electrodes (Figure 2). When the bipolar pair of electrodes is parallel to the direction of the wavefront, a large and fast-action electrogram is generated;

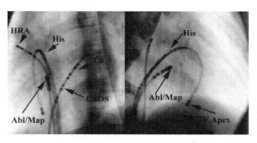

FIGURE 1. Chest radiographs in two projections (Left panel: left anterior oblique; Right panel: right anterior oblique) demonstrating placement of four bipolar electrode pair electrodes; one each in the high right atrium (HRA), coronary sinus (CS), His bundle region (His), and right ventricular apex (RV apex; the last two sites are recorded through a single catheter). The mapping catheter (Abl/Map) is in the mid septal region in the triangle of Koch. The CS catheter in the right anterior oblique projection is superimposed on and obscured by the His catheter.

however, the amplitude of the electrogram diminishes and its duration prolongs as the direction of the wavefront shifts to the perpendicular relative to the axis of the bipolar pair. In contrast, the unipolar electrode, using a single electrode tip catheter and an indifferent electrode placed within the bioelectric field (within the blood pool such as the IVC), is most useful for precise timing of the electrical wavefront as it abruptly reverses its polarity ("intrinsic deflection") when it passes the site of the intracardiac electrode.

The electrophysiologic properties of the heart consist of impulse formation, conduction velocity, and refractoriness. These properties are evaluated in several components of the conduction system, including the sinoatrial (SA) node, atria, inter-nodal tracts, atrioventricular (AV) node, His bundle, bundle branches, Purkinje network, and ventricle by clinical electrophysiologic study.

The SA node, AV node, and His–Purkinje system are excitable and capable of spontaneous impulse formation [depolarization, pacemaker activity; working atrial and ventricular tissue develop automaticity only during pathological conditions (i.e., ischemia, etc.)]. These spontaneous impulses are dependent upon phase 4 depolarizing outward potassium current (Chapter 2). This activity is rate variable, relative to the site of origin; the SA node has the fastest spontaneous discharge rate (140–160 bpm at birth, decreasing with age), and thus is the dominant pacemaker in the normal heart. A progressive decrease in spontaneous impulse formation distinguishes different cardiac excitable tissues as the pacemaker moves from atrium to AV node, His-Purkinje system, and lastly to ordinary working ventricular myocardium. When diastolic depolarization of the dominant pacemaker reaches threshold, it initiates cell to cell propagation (conduction), generating an excitation wavefront through ordinary atrial and ventricular myocardium and the specialized conduction tissue of the AV node and His Purkinje system. This complex process is modulated by circulating catecholamines, the autonomic nervous system, and both the metabolic milieu and state (i.e., disease) of the myocardium. Although the discharge of the SA node is not discernable on the ECG, other spontaneous impulse discharge rates are readily detected on the surface ECG and often the site of this discharge can be inferred.

Evaluation of the SA node may be performed in the electrophysiology laboratory. Impulse formation is indirectly evaluated by delivering a fixed train (usually 30 sec) of external stimuli to the SA node by pacing the atrium. When atrial pacing is terminated, the recovery interval of the SA node is measured (i.e., time from the last pacing stimulus to the first spontaneous signal emerging from the SA node recorded as an atrial electrogram in the catheter that delivered the pacing stimuli). This procedure is performed at a pacing rate of 90 bpm, 120 bpm, 150 bpm, 180 bpm, and 286 bpm. The maximal SA node recovery time (SNRT) is that maximal escape interval that follows termination of pacing at any pacing rate; it is corrected for the normal spontaneous discharge rate by subtracting the basic cycle length from it (Figure 3). Interestingly, because of potential SA conduction

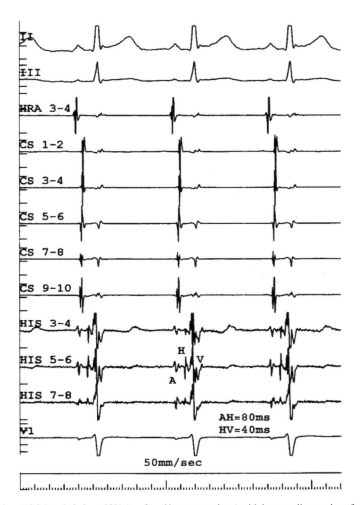

FIGURE 2. Surface ECG Leads 2, 3, and V1 (top 2 and bottom tracings) with intracardiac tracings from a 15-year-old boy whose catheter placement is shown in Figure 1. The intracardiac recordings are from the high right atrium (HRA), the coronary sinus (CS 1–2 most distal with CS 9–10 at the mouth of the CS). The His catheter with the HIS pair 3–4 most distal and the HIS pair 7–8 most proximal. The His catheter records an atrial signal from the low septal right atrium (A on the His bundle electrogram), the His bundle (H) potential and the ventricle (V) are shown. The AH interval is measured from the beginning of the first fast action deflection of the LSRA electrogram to the beginning of the His bundle potential (H)); the HV interval is measured from the beginning of the His bundle electrogram to the beginning of the V electrogram or QRS complex (best seen in V1), whichever comes first. Reprinted with permission from M Dick et al. Use of the His/RVA electrode catheter in children. J Electrocardiography 1996; 29:227–333.

delay (i.e., delay of the atrial pacing stimulus entering the SA node and then escaping it), the measured SA node recovery time may shorten at faster rate (i.e., not as many of the stimuli penetrate the SA node to suppress the spontaneous firing pacemaker cells). Pharmacologic blockade with atropine and propranolol is useful to isolate the intrinsic SA node recovery time from that modulated by extrinsic factors.

Conduction across the SA border can also be indirectly assessed by pacing techniques; this latter procedure is usually reserved for research purposes only. In practice, an abnormality in impulse discharge from the

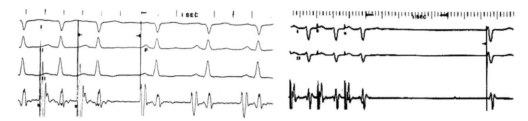

FIGURE 3. Left panel: Atrial pacing at 400 ms (150 bpm). Upon termination of pacing, the normal recovery time of the SA node (P-wave) is 620 ms. Subtracting that interval from the basic sinus interval of this patient of 500 ms, yields a normal Maximal Corrected Sinus Node Recovery Time (MCSNRT) of 120 ms. Right panel: In contrast, cessation of pacing in a 15-year-old patient following the Mustard operation for transposition of the great arteries, yields a junctional escape beat with an escape interval of 1450 ms. Both the escape mechanism and the escape interval are abnormal.

SA node is evaluated largely by the ECG and the bradycardia is correlated with the reported symptoms for optimal management (Chapter 13).

Electrophysiologic study also examines conduction through excitable tissue within the heart. Conduction velocity varies within the different segments of the cardiac conduction system but remains virtually constant with age (Table 1).

Because it is difficult to measure distance and time simultaneously within the heart, or it was until the recent advent of computer-assisted electro-anatomic mapping techniques, conduction intervals serve as surrogates for conduction velocity (Table 2).

The P-A interval denotes the shortest time, in milliseconds, from the exit of the sinus impulse from the SA node to its arrival at the low septal right atrium; its duration is a function of the size and state (fibrosis, etc.) of the right and left atria. The A-H interval assesses the conduction time from the low septal right atrium through the AV node. Because of the physiologic property of delay within the AV node, this interval lengthens as the preceding cycle length shortens to a critical length (relative refractory period; Tables 2 and 3). Also due to the physiologic function of the AV node, the A-H interval demonstrates variability dependent on the basic metabolic state, vagal tone, circulating catecholamines, and drug administration. The H-V interval marks the conduction time from the His–Purkinje system to the working myocardium. This interval is fairly constant but does increase slightly from approximately 25 ± 10 ms to 45 ± 10 ms from infancy to adulthood. An H-V interval greater then 60ms in the young is abnormal and suggests an abnormality in the His–Purkinje system. It is unusual in the child, even following surgery for congenital heart disease, to have a prolonged H-V interval. Top normal H-V intervals are not uncommon, however, in patients with AV septal defects (AV canal defects) and older postoperative tetralogy of Fallot and ventricular septal defect patients. In addition, parents with familial atrial septal defects and heart block have recently been linked to mutations in the cardiac homeobox *Nkx2.5* gene (Chapter 18).

In contrast to skeletal muscle, the action potential of cardiac muscle is long with a delay in recovery of excitability (long refractory period). As such, cardiac muscle cannot sustain prolonged isometric contraction (tetany) (Chapter 2). The refractory period of cardiac

TABLE 1. Conduction Velocity

Location	Conduction velocity
Atrium	0.8 msec
AV Node	0.5 msec
His–Purkinje System	1–2 msec
Ventricular Myocardium	0.8 msec

TABLE 2. Conduction Intervals

Interval	Measurement (normal values)
P-A	Atrial conduction time—beginning from the earliest P-wave on the ECG to the earliest fast atrial deflection on the low septal right atrial electrogram (approximately 25 msec; Table 3).
A-H	Atrioventricular nodal conduction time—from earliest fast atrial deflection on the low septal right atrial electrogram to the beginning of the His bundle potential recorded on the His bundle electrogram. Varies with heart rate and autonomic tone (Table 3).
H-V	His-Purkinje conduction time—from beginning of the His bundle potential recorded on the His bundle electrogram to the onset of ventricular activation recorded on the ECG (V1) or ventricular electrogram, whichever comes first ($35–40 \pm 9$ msec; Table 3).

tissue is determined on a beat-to-beat basis by the preceding cycle length; thus, as the heart rate increases and the cycle length shortens, the refractory period shortens more or less proportionately—up to a point. In clinical practice, refractory periods are determined by programmed extrastimulation at electrophysiologic study. This technique delivers a drive stimulus for 8–10 beats (S1); following the last drive beat, a premature stimulus (S2) is delivered at a cycle length less than the drive cycle length. This process is reiterated, sequentially shortening the premature beat (S2) until it fails to depolarize (or propagate to) the tissue in question (Figures 4–6; Tables 3 and 4). This process yields the relative, functional, and effective refractory periods of the tissue in question. The refractory periods increase as the preceding cycle lengthens (i.e., heart rate slows), a normal change with growth and development (Table 3).

The clinical implications of these different and varying electrophysiologic properties observed among developing children are critical for optimal analysis and management of their arrhythmias. Ninety percent of clinical arrhythmias in children are due to reentry through either anatomic or functional circuits (Chapters 2, 4–9); the remainder arise by way of abnormalities in impulse formation (abnormal automaticity, Chapters 2, 10–12). Table 5 and Chapter 2 outline the conditions required for reentry; Figure 7 depicts the mechanisms of arrhythmias.

Although conduction velocity varies with the type of cardiac tissue, it does not vary much with age (except for the changes imposed by acquired disease, drugs and advanced age, accounting for most of the arrhythmias seen in later adulthood). Thus, the chief determinants of the reentry circuit length in infants and children are the effective refractory period and the size of the heart (i.e., sufficient mass to support a reentry circuit). In short, despite short refractory periods in the infant, there is insufficient atrial or ventricular mass or volume, in most cases, to support reentry tachyarrhythmias. In other words, because the critical mass necessary to provide sufficient length (requirement #3 for reentry, Table 5) for a reentry circuit to be established is not present, both atrial fibrillation and ventricular fibrillation are rare in the infant and small child. On the other hand, when anatomic conditions exist, such as accessory pathways or dilated atria or ventricles providing a circuit of sufficient length, often aggravated by pathophysiologic factors such as heart or respiratory abnormalities, reentry-dependent AV reentry tachycardia and atrial flutter may be appear, even in infants. Even if the reentry circuit is present, the tachycardia must be initiated—usually by a premature ectopic beat. This spontaneous premature impulse must be appropriately timed, usually with a patient-specific interval range, so that it is blocked in one limb of the circuit (requirement #1 for reentry, Table 5) and conducted slowly down the other limb of the circuit (requirement #2 for reentry, Table 5) so that requirement #3 is fulfilled. The irregularity and spontaneity of these

TABLE 3. Electrophysiologic Data in Children

Group	<0.5 YR	N	0.05–1 YR	N	1–5 YR	N	5–10 YR	N	>10 YR	N
Age	0.28	19	0.66	18	3.1	37	6.9	17	16.2	21
Range	0.01–0.18		0.52–0.98		1.0–4.9		5.1–9.9		10–25	
SCL (x ± sd) (msec)	456 ± 50	19	471 ± 66	18	547 ± 37	37	610 ± 83	17	814 ± 132	21
Panel A: Conduction Intervals (x ± sd) (msec)										
PA	24 ± 7	10	22 ± 14	12	29 ± 15	30	29 ± 12	11	29 ± 14	13
AH	72 ± 14	11	75 ± 14	14	71 ± 20	30	75 ± 18	12	91 ± 21	14
HV	33 ± 9	11	35 ± 4	14	36 ± 8	30	34 ± 6	13	41 ± 8	17
RVA	24 ± 8	5	23 ± 11	11	26 ± 5	22	23 ± 7	10	29 ± 10	18
RVO	31 ± 5	4	41 ± 11	8	41 ± 10	14	52 ± 7	5	47 ± 14	7
RVI	33 ± 5	4	41 ± 11	9	47 ± 16	16	48 ± 9	3	47 ± 5	5
Panel B: Refractory Periods (x ± sd) (msec)										
AT. ERP	166 ± 25	17	163 ± 29	12	203 ± 26	22	239 ± 28	7	269 ± 39	4
AT. FRP	205 ± 35	17	212 ± 18	12	243 ± 29	22	264 ± 24	7	301 ± 45	4
AVCERP	221 ± 24	15	238 ± 40	4	244 ± 44	8	375	1	–	
AVCFRP	284 ± 30	15	305 ± 39	11	329 ± 35	18	381 ± 32	6	454 ± 85	4
VENT ER			226	1	228	2	250	1	225	1
Panel C: Sinus Node (x ± sd) (msec)										
TSACT	111 ± 28	17	107 ± 23	10	123 ± 36	17	153 ± 38	6	120 ± 62	2
MCSNRT	123 ± 44	18	104 ± 39	10	127 ± 50	23	163 ± 47	7	<250	

Notes: Adapted from Campbell RM, Dick M, Rosenthal A. Cardiac arrhythmias in children. Ann Rev Med 1984 35:397–410.
Abbreviations: AT. ERP = atrial effective refractory period; AT. FRP = atrial functional refractory period; AH = atrial-His bundle interval; AVCERP = atrioventricular conduction system effective refractory period; AVCFRP = atrioventricular conduction functional refractory period; HV = His bundle-ventricle interval; MCSNRT = maximal corrected sinus node recovery time; PA = high right atrial to low right atrial interval; RVA = right ventricular apical activation time; RVI = right ventricular inflow activation time; RVO = right ventricular outflow activation time; TSACT = total sinoatrial conduction time; VERP = ventricular effective refractory period.

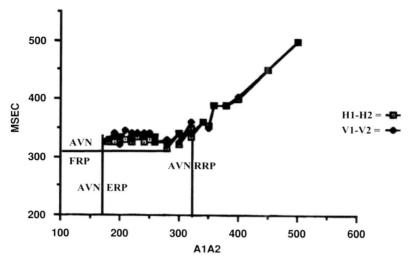

FIGURE 4. Plot of atrial premature interval (A1A2) versus H1H2 and V1V2.

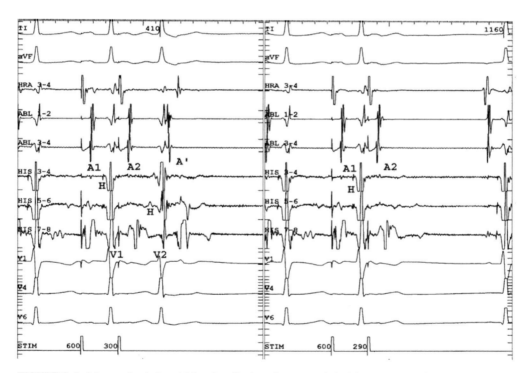

FIGURE 5. Atrial extrastimulation yielding the effective refractory period of the AV node. Left panel: A premature atria stimulus (A2) is delivered at a coupling interval of 300 ms at a basic drive of 600 ms. Conduction through the AV node to the His bundle and the ventricle is intact. In addition, there is an echo beat (A′) earliest in the Abl 1–2 located in the coronary sinus. This last observation indicates a left-sided accessory pathway. Right panel: The coupling interval of the premature impulse (A2) was shortened to 290 ms with capture of atrial tissue and conduction block in the AV node (i.e., no His electrogram following the atrial electrogram), indicating the effective refractory period of the AV node.

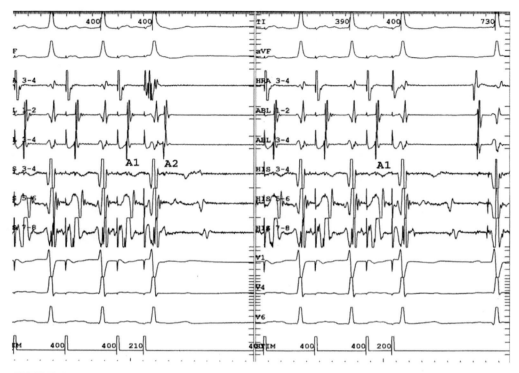

FIGURE 6. Atrial extrastimulation yielding the effective refractory period in the atrium. Left panel: A premature atrial stimulus (A2) is delivered at a coupling interval of 210 ms at a basic drive of 400 ms. Capture of the atrium is present. Right panel: The coupling interval of the premature impulse (A2) is shortened to 200ms and capture is blocked in the atrium (i.e., no A electrogram), indicating the effective refractory period of the atrium.

premature ectopic beats, in part, accounts for the unpredictability of reentry arrhythmias. At the same time, their frequent association with intercurrent illnesses such as respiratory infections and other stressful events can account for the concomitant appearance of the tachyarrhythmia.

When an infant reaches early childhood (approximately 5 years of age), increased heart size and longer refractory periods increase the potential for other reentry circuits. The shift from almost exclusively AV reentry tachycardia dependent on an accessory pathway to predominately AV nodal reentry tachycardia reflects these developmental changes. Both the loss of conductivity in the accessory pathway, presumably due to molding within the AV junction accompanying growth, and the larger size and longer refractory permit the conditions for reentry

TABLE 4. Refractory Periods

Refractory period	Definition
Effective Refractory Period (ERP)	The longest input interval (S1–S2) that fails to depolarize (propagate to) the tissue in question (Figure 6).
Functional Refractory Period (FRP)	The shortest output interval (e.g., H1–H2 or V1–V2) that propagates to the tissue in question (Figure 4).
Relative Refractory Period (RRP)	The shortest output interval (e.g., H1–H2 or V1–V2) equal to the input interval (e.g., A1–A2; Figure 4).

TABLE 5. Conditions for Reentry

1. Unidirectional block in one limb of the circuit.
2. Slow conduction in the other limb of the circuit.
3. Recovery of excitability of the site of origin.

(Table 5). Similarly, atrial flutter occurring after palliation of congenital heart disease involving extensive atrial surgery such as the Mustard, Senning, or Fontan operations create regions of slow conduction and conduction block resulting from scarring and suture lines, which can lead to complex reentrant circuits (Chapter 8).

Differential expression in the development of automatic arrhythmias is less clear. Multifocal atrial tachycardia and junctional ectopic tachycardia (Chapters 10 and 11) are, in general, arrhythmias of infants. Ventricular tachycardias unrelated to surgery span the developmental divide.

This schema of the mechanism and time course of arrhythmias and their relation to the developing heart and its conduction system is subject to multiple other factors that are in constant flux; among them are electrolyte balance, the external environment, and the autonomic nervous system that regulates the circulatory system. Nonetheless, the natural history of arrhythmias is, in large part, dependent on the underlying structural and functional changes in the developing child. Consideration of these changes is essential when managing children with cardiac impulse formation and conduction disorders.

Special Considerations

When planning an electrophysiologic study in an infant or child, a number of considerations surface. First, the operator should be experienced in advancing and navigating an electrode catheter through the cardiac chambers of young persons (i.e., ≤12 years old). The value of this experience is inversely related to the size of the patient to be studied. Using fluoroscopic guidance, the operator should always have in his/her mind's eye the exact location of the catheter tip and shaft relative to the cardiac structures, such as the

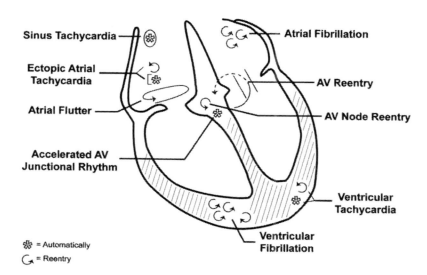

FIGURE 7. Drawing of cardiac arrhythmias: their origin and their mechanism. Reentry arrhythmias are by atrioventricular (AV) reentry, AV node reentry, atrial flutter, and some forms of ventricular tachycardia, as well as atrial and ventricular fibrillation. Automatic arrhythmias consist of ectopic atrial tachycardia, accelerated AV junctional rhythm and some forms of ventricular tachycardia (VT), particularly the VT of chronic ischemic heart disease. It is important to note that the reentry arrhythmias often require an ectopic beat for initiation.

His bundle, the coronary sinus, the atrial septum, the foramen ovale, the level and plane of the AV groove(s) and valves, and the right ventricular outflow tract.

Second, the operator should be fully informed regarding the clinical course and structural features of cardiac malformations. The possible presence of congenital heart disease underscores the need for operator experience in invasive cardiac studies in the young. Structural congenital heart lesions may place cardiac structures in unfamiliar locations, displace the conduction system, or be associated with unsuspected conducting pathways and spontaneously discharging ectopic foci. Operations for these congenital heart lesions can further alter the electrical properties of the atrial and ventricular myocardium, resulting in complex rhythm disturbances.

Third, the hemodynamic state of the heart affects the study of the cardiac electrophysiology in the young. A young, otherwise healthy heart with intact inotropic properties and vigorous contraction can often lead to catheter instability and dislodgement of catheters from recording sites. Faster heart rates in young patients also contribute to this technical confounder. In addition, the autonomic tone varies with age. The newborn and infant are highly susceptible to excessive parasympathetic input that can be aggravated by analgesia, anesthetics, or manipulation of the airway, resulting in a deleterious slowing of the heart rate. Beyond the first several months of life, until adolescence, autonomic tone is fairly stable and well balanced. However, the circulating volume as well as its reserve remain highly variable and are directly related to patient size. Thus, blood loss or volume depletion or contraction by prolonged tachycardiac episodes during an extended electrophysiologic study can adversely alter hemodynamic and metabolic states. Intravascular arterial monitoring of the blood pressure and immediate availability of volume replacement are prudent for infants and children, as well as all young patients studied for ventricular arrhythmias.

Techniques

Several principles are important in the application of electrophysiologic study to children. It is generally recognized that invasive electrophysiologic study and radiofrequency ablation carries a greater, although not prohibitive risk in children less than 15 kg. The enhanced risk includes difficulty of vascular and intracardiac access, alterations in hemodynamic state, potential perforation and pericardial effusion with tampanode, potential for excessive lesion size (and possible expansion), and injury to the AV node. Between 15 and 35 kg, the risks are diminished and beyond 35–50 kg, the risk of electrophysiologic study is essentially equivalent to that in adult patients. When considering electrophysiologic study in the small infant or child, the number of catheters, the size of the catheters, and the stiffness of the catheters should be minimized. Clearly the risk of trauma to and perforation of the myocardium is inversely related to the size of the patient and directly related to the size of the catheter, the number and size of electrodes on it, and other features such as cooling, thermisters, and stiffness. A number of technological advances in catheter design including increased flexibility and compliance (with the possible loss of torque control), multi-electrode array on small catheters, steerability, multiple functions catheters (e.g., His-RVA catheter), delivery sheaths, and proximity recording from the esophagus have reduced the risk and facilitated successful electrophysiologic studies in the smallest of patients. Four French (5 French in larger patients; Figure 1) quadripolar catheters are commonly used for high right atrial recording and stimulation. In addition, an especially designed soft 4 French bipolar (Figure 8) or quadripolar catheter can be advanced within the esophagus to the level of the intra-atrial groove. Recording through this transesophageal pair of electrodes reflects activation from the medial aspect of the left atrium and provides access to local activation of the heart without the risk to arteries or veins. Fluoroscopy may assist in the

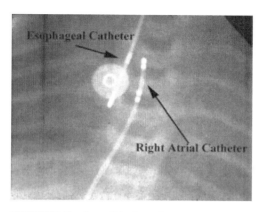

FIGURE 8. Esophageal bipolar catheter at the level of the left atrium and an intravascular quadripolar catheter in the right atrium. The esophageal catheter lies behind the left atrium near the intraatrial sulcus and records electrograms from the medial aspect of the left atrium.

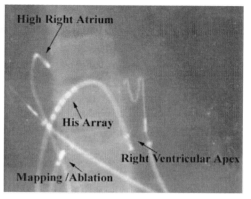

FIGURE 9. Radiograph displaying position of His-RVA catheter in right heart of a 5-year-old girl. Note the distal bipolar pair in the right ventricular apex and the His electrode array in the region of the tricuspid valve and AV septum. Reprinted with permission from M Dick et al. Use of the His/RVA electrode catheter in children. J Electrocardiography 1996; 29:227–333.

placement of the esophageal lead with final localization guided by the quality of the recorded signal.

Multiple approaches to record the His bundle potential at the AV junction have been utilized. Initially, a single 6 or 7 French quadripolar catheter was placed across the tricuspid valve in the low atrial septal area to record the His bundle potential. A third bipolar 6 or 7 French catheter was then passed to the right ventricular apex for ventricular stimulation.

To reduce the number of catheters placed in children, we have combined the functionality of the His bundle catheter and the right ventricular apex catheter into a single catheter. The dual function 5 or 6 French catheter (Figure 9) was advanced so that a distal pair of electrodes placed in the right ventricular apex and a proximal array of 6 electrodes (2–5 mm inter-electrode distance) abutted the atrial and ventricular septal surfaces across the tricuspid valve. This technique is facilitated by a family of catheters (Figure 10) ranging in interelectrode distance between the second and third electrode of 4–8 cm.

Utilizing a regression equation relating the 4–8 cm distance between the second and third electrode as the dependent variable and the patient's weight as the independent variable (Figures 11 and 12), an optimal catheter dimension can be estimated prior to placement.

Optimal His bundle electrograms are obtained when the His bundle array of electrodes on the His-RVA catheter is abutting the septum, best visualized in the LAO position (Figure 13). This technique is particularly useful in smaller children; in larger individuals it is more difficult to "fix" the His electrode array against the septum at the level of the tricuspid valve.

Often other catheters and sheaths will lie against the His-RVA catheter so that it

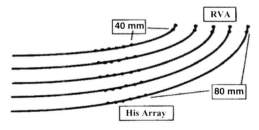

FIGURE 10. Family of His/RVA catheters with a 40–80 mm distance between the 2nd and 3rd electrode and an interelectrode distance between the His electrodes of 2–5 mm. Reprinted with permission from M Dick et al. Use of the His/RVA electrode catheter in children. J Electrocardiography 1996; 29:227–333.

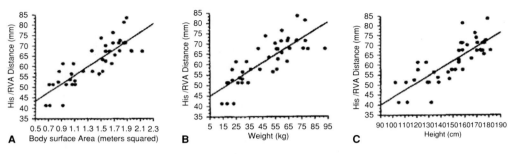

FIGURE 11. Relationship between body surface area (A), weight (B), height (C), and the catheter dimension best recording) the His bundle potential. Reprinted with permission from M Dick et al. Use of the His/RVA electrode catheter in children. J Electrocardiography 1996; 29:227–333.

is pressed against the AV septum, improving the proximity advantage and, *pari passu*, the amplitude of the His bundle signal. Other techniques using very fine, multiple electrode French catheters advanced into the coronary arteries or veins or even the cardiac chambers in small children are limited, in our experience, by the risk to small vessels and by the lack of stability, aggravated often by the faster heart rate and vigorous myocardial contractility common in young patients. Even with the firmer 6 French His-RVA catheter such myocardial action may transiently displace the His electrode array from the optimal recording site during an extrasystolic beat such as induced by extrastimulation.

A number of other technical advances have facilitated electrophysiologic study in the young. Long stabilizing sheaths allow for more secure positioning of the catheter to map target sites along the right AV groove and in the right atrium. To avoid injury to the femoral arteries of small (≤25 kg) patients, left-sided structures are best approached by the transseptal technique (Figure 14). This technique requires optimal selection of the transseptal needle, dilator, sheath, and electrode catheter. Deployment of the transseptal system in an antegrade direction across the atrial septum into the left atrium can be complicated by the small size of young patients and the presence of congenital cardiac malformations and calls for special training and extensive experience in heart catheterization. Although biplane fluoroscopy is ideal, monoplane fluoroscopy mounted on a computer-driven frame that rotates around the long axis of the patient's thorax is reasonable, fully functional, and considerably less expensive for imaging placement of the transseptal system. Several laboratories have also used transesophageal or intravascular ultrasound to assist in the procedure.

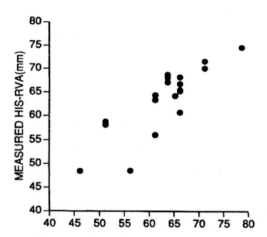

FIGURE 12. Plot between His-RVA dimension estimated from patient's weight on the abscissa and His-RVA dimension used for His recording on the ordinate. Because the majority of pediatric patients for electrophysiological study are in the mid range in weight (40–60 kg, mean 50 kg), the plot reflects the mid-range His-RVA dimension (i.e., 5–6 cm). Reprinted with permission from M Dick et al. Use of the His/RVA electrode catheter in children. J Electrocardiography 1996; 29:227–333.

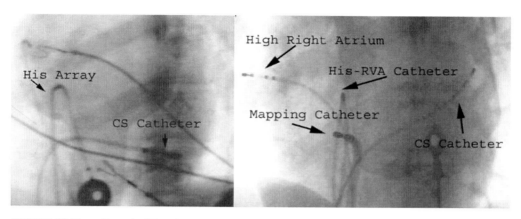

FIGURE 13. The radiograph (35°, left anterior oblique, LAO) on the right demonstrates that the His–RVA catheter is in profile and abuts the AV septum, the optimal position for recording the His bundle potential. The radiograph on the left (LAO) demonstrates that the His array is not in profile and thus not abutting the AV septum, a position less likely to record the His bundle potential.

Delivery of the transseptal catheter begins with passage of a J-tipped guide wire into the superior vena cava. A long dilator and sheath are then passed over the wire into the superior vena cava. The wire is then removed and the transseptal needle, which has been flushed with radio-opaque contrast, is advanced through the sheath with the tip remaining inside the dilator. With the needle retracted into the dilator and the needle's curve oriented posteriorly and leftward, the apparatus is withdrawn into the heart. The guide on the transseptal needle should be directed towards 5:00 as viewed by the operator on an imaginary clock face with 12:00 pointing directly upwards and 6:00 directly down. This approach brings the tip of the dilator and needle into the proper plane to approach the

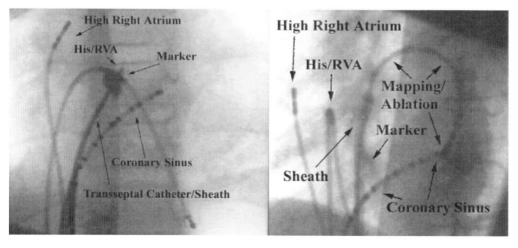

FIGURE 14. Right anterior oblique (right panel) and left anterior oblique (left panel) demonstrate marking of fossa ovale (transseptal marker) for optimal location to penetrate the atrial septum and enter the left atrium, yielding successful entry into the left atrium.

atrial septum while avoiding other intracardiac structures such as the aorta. When the apparatus enters the heart, the catheter is usually noted to jump posteriorly, and will frequently induce tachycardia in susceptible patients. The sheath is withdrawn farther until the tip passes into the fossa ovale, often sensed (visualized and felt) as a slight "bump." At that point the needle is advanced out of the dilator to engage the septum primum. Contrast may be infused through the needle to "stain" the atrial septum, which confirms the appropriate location of the needle and marks the septum to facilitate passing the dilator and sheath into the left atrium (Figure 14). After confirmation of engagement of the needle with the atrial septum, the needle is advanced across the septum into the left atrium, and its location confirmed by infusion of dye into the left atrial chamber. It is important to confirm correct needle tip location at this point to ensure that the needle has not entered the pericardium or aorta. The needle is then held in position as the dilator and sheath are advanced over it into the left atrium. The operator must remain cognizant that the dilator is stiff and the tip is relatively sharp and thus poses a risk for perforating the back wall of the left atrium as the apparatus passes through the atrial septum into the left atrium. Once in the left atrium, the needle and dilator are withdrawn, and a left atrial saturation and pressure are recorded. The mapping (and ablation) catheter is then passed up the long sheath into the left atrium for electrical mapping. In small children, the anterior–posterior dimension of the left atrium is particularly shallow, so avoidance of entering the posterior pericardial space is critical. In addition, passage of the transseptal system across the atrial septum must be sufficiently superior (cephalad) to avoid the proximal AV node (at the beginning) and the mitral valve as the system enters and is advanced into the left atrium. Following completion of the transseptal procedure, heparin should be administered to ensure appropriate anti-coagulation while catheters are present in the systemic chambers of the heart.

With the advent of the steerable multiple electrode catheter, entry into the coronary sinus has shifted from a superior vena cava approach through the internal jugular or the subclavian vein to an inferior vena cava approach (Figure 14). This technique is especially useful in small children as it delivers a 5 or 6 French catheter from the relatively large femoral vein and preserves the small subclavian and jugular veins. Only in the face of IVC obstruction is an alternative entry required. A transhepatic approach to the right atrium and then transseptally to the left atrium, if necessary, is an alternative approach to conserve the smaller jugular and subclavian veins (Figure 15).

Arterial cannulation for monitoring of systemic blood pressure can be used selectively. For most patients with supraventricular tachycardia, monitoring of the blood pressure by sphygmomanometer is sufficient, avoiding the need for an intra-arterial cannula. In patients with poor ventricular function, suspected ventricular arrhythmias, or in infants (≤ 15 kg), prudence suggests the use of an intra-arterial cannula to monitor blood pressure continuously.

Other practices and procedures significantly contribute to the stability and safety of electrophysiologic study (as well as ablation)

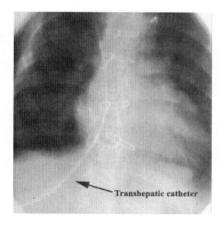

FIGURE 15. Passage of transhepatic sheath and catheter through the liver and into the right atrium and superior vena cava in an 8-kg infant.

of arrhythmias in children. Conscious sedation, and, in small children general anesthesia, when administered by a skilled pediatric anesthesiology team, is useful for thorough patient comfort and safety, and for rigorous monitoring of small children whose hemodynamic status is vulnerable to small shifts in fluid distribution and heart rate. Current techniques require fluoroscopy to safely place catheters within the heart. We have shown that children undergoing electrophysiologic study and radiofrequency ablation, on the average, are not exposed to more radiation energy than children undergoing diagnostic and interventional catheterization. Finally, the development of computer-assisted EP monitoring, mapping, and archival and analysis systems provides for on-line real time evaluation of complex arrhythmias. These systems greatly enhance the detection and precision of mapping complex arrhythmias, particularly in patients with complex intracardiac and postoperative anatomy.

CONCLUSION

Careful consideration and understanding of the underlying anatomy and electrophysiology of the heart in developing children will provide a sound foundation for the safe and effective management of arrhythmias in the young. Advances in new energy sources such as cryotherapy as well as development of new imaging techniques such as real time MRI will most likely improve these strategies.

SUGGESTED READING

1. Josephson ME. *Clinical Cardiac Electrophysiology: Techniques and Interpretations*. Second Edition, Lea and Febiger: Philadelphia, PA, 1993.
2. Deal BJ, Wolff GS, Gelband H. *Current Concepts in Diagnosis and Management of Arrhythmias in Infants and Children*. Futura Publishing Co, Inc: Armonk, NY, 1998
3. Cohen MI, Wieand TS, Rhodes LA, Vetter VL. Electrophysiologic properties of the atrioventricular node in pediatric patients. *J Am Coll Cardiol* 1997;29(2):403–407.
4. Blaufox AD, Saul JP. Influences on fast and slow pathway conduction in children: does the definition of dual atrioventricular node physiology need to be changed? *J Cardiovasc Electrophysiol* 2002;13(3):210–211.
5. Dick II M, Law IH, Dorostkar PC, Armstrong B. Use of the His/RVA catheter in children. *J Electrocardiology* 1996;29(Suppl):227–233.
6. Saul JP, Hulse JE, De W, et al. Catheter ablation of accessory atrioventricular pathways in young patients: use of long vascular sheaths, the transseptal approach and a retrograde left posterior parallel approach. *J Am Coll Cardiol* 1993;21(3):571–583.
7. Fischbach P, Campbell RM, Hulse E, et al. Transhepatic access to the atrioventricular ring for delivery of radiofrequency ablation. *J Cardiovasc Electrophysiol* 1997;8:512–516.

II

Clinical Electrophysiology in Infants and Children

4

Atrioventricular Reentry Tachycardia

Ian H. Law

Supraventricular tachycardia (SVT) is the most common tachyarrhythmia in children, occurring in perhaps as many as one in 1,000 to 2,500 children. Of these, 90% involve a reentrant circuit (Chapter 3) between the atria and ventricles [atrioventricular (AV) reentrant tachycardia, AV node reentrant tachycardia (Chapter 5), or persistent junctional reciprocal tachycardia (Chapter 6)]; the majority (50%–60%) present in the first year of life. While common, the percentage of children with SVT having AV reentrant tachycardia diminishes with age; 85% in children less than one year; 82% in children between one and five years; between six and 10 years of age the incidence decreases to 56%. Despite the frequency of the arrhythmia in children, between 15% and 41% of infants under six months of age having documented SVT will not have a recurrence beyond the first year of life.

WOLFF-PARKINSON-WHITE SYNDROME

Manifest versus Concealed Pathways

An anomalous strand of myocardium, most likely atrial, bridges the groove between the atria and ventricle, supporting atrioventricular reentrant tachycardia (AVRT). Antegrade activation of the myocardium through the strand of myocardium can be either present (i.e., manifest) or absent on an electrocardiogram (ECG). Manifest conduction, as in the Wolff-Parkinson-White syndrome, conducts antegrade (anterograde) reaching the ventricles in advance of activation through the AV node-His–Purkinje system (AVN-HPS), thus producing ventricular pre-excitation (Figures 1 and 2). The incidence of Wolff-Parkinson-White syndrome in children is low—between 1.6 per 1,000 to 4 per 100,000. In addition, ventricular pre-excitation can be quite subtle, apparent only as a lack of a Q-wave in the lateral precordial leads due to either slow conduction in the accessory pathway, accelerated conduction through the AV node, as is common in children, or a left lateral pathway location in the mitral valve area, the most frequent site of a pathway. Because a concealed accessory pathway has retrograde but no antegrade conduction (unidirectional conduction), a concealed pathway produces no findings on the ECG.

While both manifest and concealed pathways can mediate reentrant tachycardia, patients with Wolff-Parkinson-White syndrome have a greater rate of recurrence. In

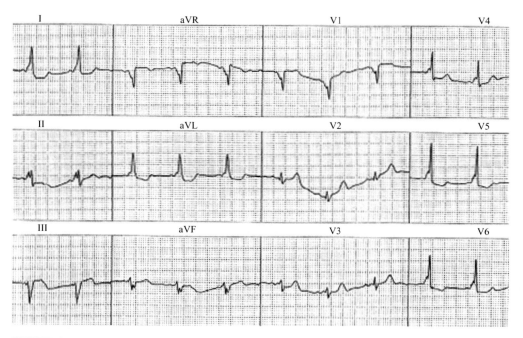

FIGURE 1. 12-Lead ECG of patient with Wolff-Parkinson-White syndrome with a right posteroseptal accessory pathway. The PR interval is short and positive delta waves are seen in leads I, II, aVL, and V2–V6; the transition is after V3.

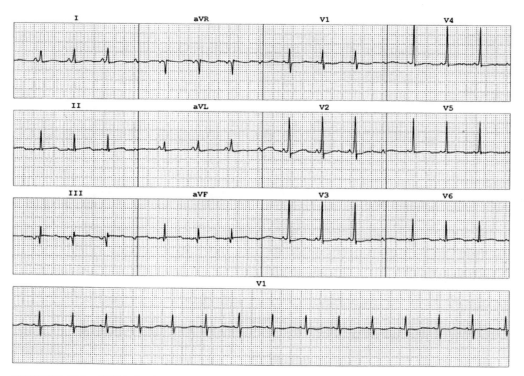

FIGURE 2. 12-Lead ECG of patient with Wolff-Parkinson-White syndrome with a left posterior accessory pathway. The PR interval is short and positive delta waves are seen in leads I, aVL, and V2; the transition is before V2.

addition, the course of AVRT is age related. The Wolff-Parkinson-White syndrome, even when manifest, may not consistently be seen on an individual's ECG, particularly when the pathway is on the left, the most common site. The pre-excitation can progress to complete disappearance as a child grows, usually by 18–24 months of age. Finally, it is well known that recurrence of AVRT is much more common if the initial presentation is after the first year of life. In one report the greatest risk factor was documentation of SVT after one year of age (94% recurrence rate), significantly greater than those diagnosed at less than one year of age (29% rate of recurrence).

The majority of accessory pathways are seen in patients having structurally normal hearts; however, patients who have Wolff-Parkinson-White syndrome have a somewhat increased risk of associated congenital heart disease. Approximately 10%–15% of patients with Wolff-Parkinson-White syndrome have congenital heart disease (L-transposition of the great arteries, cardiomyopathy, Ebstein's anomaly); the most common is Ebstein's anomaly (23%). Patients who have Wolff-Parkinson-White syndrome and congenital heart disease tend to have multiple pathways, reported in 6% of patients. In this review, those with normal anatomy had only one accessory pathway. A similar review of 317 Wolff-Parkinson-White patients found that 9% had multiple (2–4) accessory pathways; in patients with Ebstein's anomaly, 43% had multiple pathways and in patients with other organic heart disease, 25% had multiple pathways. Pathway location is also associated with congenital heart disease: right-sided pathways are more common in congenital heart disease (63%) whereas left-sided pathways are more common in the structurally normal heart (61%).

DIAGNOSIS

When ventricular pre-excitation is present on the surface ECG during sinus rhythm, the mechanism of the tachycardia is usually AVRT. However, ventricular pre-excitation is also seen in patients, which may or may not be clinically important. Sinus rhythm on the ECG in those patients with a concealed accessory pathway is of little diagnostic value. For those patients with manifest pathways, the ventricular pre-excitation pattern in sinus rhythm can be helpful in determining the location of the accessory pathway. A number of algorithms have been developed to aid in pathway location. Briefly, if the QRS transition (R greater than S-wave) is after V2 in the precordial leads, the pathway is most likely on the right side. If the transition is before V2, it is most likely on the left side. Further analysis of the delta wave polarity in the ECG leads can identify the location of the accessory pathway even more exactly.

An ECG obtained during SVT can be very helpful, especially when compared to sinus rhythm. Retrograde P-waves (inverted in leads II, III, and aVF; R-P retrograde interval >65–70 msec) in the T-wave can often be seen during AVRT, confirming the diagnosis. In addition, a negative retrograde P-wave observed in lead I during narrow QRS tachycardia suggests that AVRT is supported retrogradely through a left-sided accessory pathway.

MECHANISM OF ATRIOVENTRICULAR REENTRANT TACHYCARDIA

Atrioventricular reentrant tachycardia can be either orthodromic or antidromic. Orthodromic AVRT, comprising 90% of AVRT, is characterized by antegrade conduction through the AV node and His-Purkinje system and retrograde conduction through the accessory pathway. This results in narrow QRS tachycardia and is seen in patients who have either concealed or manifest pathways (Figure 3).

If the patient has a manifest accessory pathway, ventricular pre-excitation is

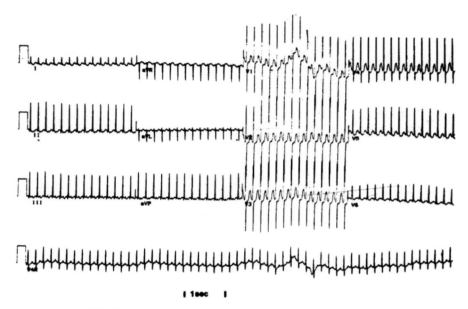

FIGURE 3. 12-Lead ECG with narrow QRS tachycardia in a 2-day-old boy.

no longer present during the tachycardia as the antegrade ventricular activation occurs rapidly through the His-Purkinje system. In contrast, antidromic AVRT characterized by anterograde conduction through the accessory pathway and retrograde conduction through the His-Purkinje-AV nodal system, results in a wide QRS complex tachycardia, mimicking ventricular tachycardia (Figure 4).

ATRIOVENTRICULAR AND MAHAIM FIBERS

Mahaim fibers, as first described in 1938, are direct histologic links between either the AV node and the right ventricle (nodoventricular fibers) or the His bundle and the right ventricle (fasciculoventricular fibers). These fibers were later advanced as the cause of pre-excitation characterized by left bundle branch block QRS morphology and wide QRS tachycardia. Intracardiac electrophysiologic studies subsequently demonstrated that this form of pre-excitation and the wide QRS tachycardia are, in fact, due to atriofascicular accessory pathways coursing from the lateral right atrial tricuspid valve area to insert deeper in the right bundle rather than at the AV groove. The nodo or fascicular ventricular fibers are not anatomically uncommon (though presumably often inactive, i.e., not resulting in pre-excitation) and generally, when manifest, behave as "by-standers" (i.e., they do not participate in a reentrant circuit). There are two forms. The first is when the nodo-fascicular fiber arises in the AV node and inserts into the right ventricular myocardium, producing a left bundle-branch block pattern when it is engaged. The PR interval varies relative to the fiber's take-off from the node. The second form is the fascicular-ventricular fiber arising from the His bundle and inserting into the ventricular myocardium. The PR interval is normal. Rarely, these nodo or fascicular ventricular fibers can cause wide QRS tachycardia by providing the atrioventricular limb as a "by-stander" if the patient has, upstream to the "by-stander" fiber, AVNRT (Figure 5). Because the Mahaim fiber does not participate in the reentry circuit of the tachycardia, it is deemed a bystander.

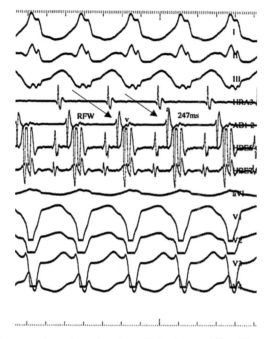

FIGURE 4. 8-Lead ECG demonstrating antidromic tachycardia in a 5-year-old boy. The top three tracings are recorded through the limb leads I, II, III; the bottom 4 tracings are aVL, V1, V2, V3, V4. The tracing 4th from the top is recorded through the high right atrial catheter. The tracing 5th from the top is recorded through the mapping catheter located in the right atrial free wall near the tricuspid valve. The 6th and 7th tracings are recorded through the His bundle catheter located at the low right atrial-septal area. Note the wide QRS tachycardia with maximal preexcitation of a left bundle branch configuration, compatible with a right-sided pathway. Note the tight AV relationship between the right atrial free wall electrogram (arrows) and the surface QRS complex (and intracardiac ventricular electrogram) indicating antidromic conduction down the right lateral accessory conduction to the ventricle. Retrograde conduction in this boy (not shown) supporting the tachycardia was through a left-sided concealed accessory pathway. Ablation of both pathways was successful.

The uncommon right-sided atriofascicular fiber (as noted, often and somewhat confusingly referred to as a Mahaim fiber), is located in the right posterior AV groove (Figure 6). This accessory pathway has several features: (1) it typically conducts in only an antegrade direction producing preexcitation in a left bundle branch block

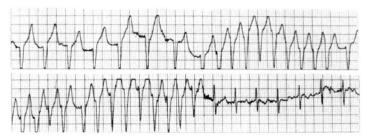

FIGURE 5. Wide QRS rhythm and tachycardia followed by narrow QRS sinus rhythm in a 9-year-old girl. Electrophysiology study demonstrated a nodoventricular fiber, which was ablated. Narrow QRS tachycardia returned due to AVNRT. The slow pathway and arrhythmia were ablated.

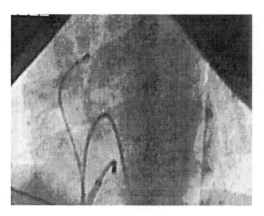

FIGURE 6. Left anterior oblique projection in a 15-year-old boy with atrio-fasicular fiber. Note catheter in high right atrium (top catheter), His-RVA catheter (middle catheter), and mapping/ablation catheter (bottom). Correlating with Figure 10, the mapping catheter is separated in space from the His-RVA catheter and in time from the atrio-fascicular electrogram (Mahaim, Figure 11, right-hand panel) during the wide QRS tachycardia.

configuration; (2) it inserts deep (not along the AV ring) in the right ventricle, often near the distal fascicle of the right bundle (moderator band); (3) the pathway demonstrates decremental conduction; and (4) the tachycardia is antidromic resulting in a wide QRS tachycardia (i.e., antegrade through the accessory pathway and retrograde through the His-Purjinke system) resulting in antidromic AVRT with a wide QRS left bundle branch pattern SVT.

The hallmark of identifying the site for ablation is the identification of a fast action specialized conduction tissue electrogram (Figure 7) distant in both location and timing from the His bundle electrogram. Ablation at that site terminates conduction within this pathway, as well as the wide QRS tachycardia.

Table 1 summarizes the findings for manifest and concealed pathways characteristically seen on the surface electrocardiogram during either sinus rhythm or SVT.

DIAGNOSTIC TESTING

Exercise Treadmill Testing

Exercise testing has been used as a noninvasive method to evaluate the conduction properties of a manifest accessory pathway. Abrupt disappearance of ventricular pre-excitation on the electrocardiogram during exercise suggests an accessory pathway effective refractory period perhaps as long as 360–390 msec, placing the patient in a lower risk category. However, in practice, disappearance of ventricular pre-excitation exercise testing can be difficult to interpret due to movement artifact on the ECG. In addition, it may be less diagnostic in patients with left-sided accessory pathways due to the enhanced AV node conduction from increased adrenergic tone during exercise, masking ventricular pre-excitation through the more remote left-sided accessory pathway.

Electrophysiology Study

The electrophysiology study is an essential tool in the assessment of the patient with documented SVT (Chapter 3). Although atrial extrastumli can be delivered through a transvenous catheter or transesophageal catheter (particularly useful for infants and small children), the transcatheter intracardiac technique is necessary for a definitive examination and transcatheter ablation.

At the start of the electrophysiology study baseline surface electrocardiogram and

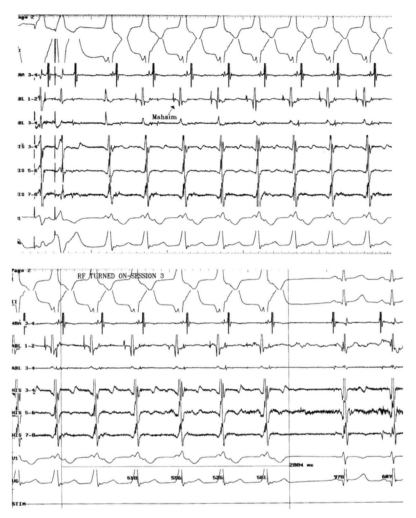

FIGURE 7. Same patient as in Figure 6. Top panel: Initiation of wide QRS tachycardia with antegrade conduction through the Mahaim fiber (note fast action deflection of atrio-fascicular fiber activation) and retrograde conduction through the normal His-Purkinje–AV node system. Bottom panel: Radiofrequency ablation of antegrade conduction in the atrio-fascicular fiber, terminating the tachycardia. Note the His bundle electrogram in the two sinus beats, as well as the absence of fast action deflection in the mapping/ablation catheter post ablation.

electrogram recordings are obtained. Baseline intracardiac intervals measured in patients with manifest conduction reveal an HV interval less than 40 seconds, at times negative (ventricular activation occurring before the His electrogram). When the patient has manifest accessory pathway conduction, it is possible to determine the location of ventricular insertion by looking for areas of early ventricular activation (prior to the onset of the surface QRS complex) during sinus rhythm (Figure 8). After the baseline recordings are taken, atrial and ventricular pacing thresholds are obtained. As a standard, stimuli are delivered at twice the diastolic pacing threshold.

Antegrade electrophysiologic properties of the AV node and manifest accessory pathway can then be determined. Atrial pacing is performed at a cycle length slightly shorter than the sinus rate and this is gradually

TABLE 1. Summary of the Findings For Manifest and Concealed Pathways

	Sinus			SVT				
	Concealed	Manifest WPW	Mahaim (Manifest)	Concealed	WPW Orthodromic	WPW Antidromic	Mahaim Atriofascicular	Nodo-or fasciculoventricular
Rate—beats/min	Within normal limits for age	Within normal limits for age	Within normal limits for age	Infants: 200–300; Children: 160–250	Infants: 200–300; Children: 160–250	Infants: 200–300; Children: 160–250	Infants: 200–300; Children: 160–250	Infants: 200–300; Children: 160–250
Rhythm P-wave	Regular Normal	Regular Normal	Regular Normal	Regular Absent or negative, within T-wave of lead II, aVF	Regular Absent or negative, within T-wave of lead II, aVF	Regular Absent	Regular Absent	Regular Absent or negative, buried within T-wave of lead II
PR interval & QRS complex	Normal	Pre-excited (delta-wave, short PR interval)	Pre-excited (delta-wave, short PR interval)	Normal or wide when aberrantly conducted	Normal or wide when aberrantly conducted	Wide with QRS pattern similar to or wider than sinus rhythm	Wide (ART-AVRT) with LBBB QRS pattern	Wide (ORT-AVRT) with QRS pattern

Comparison of concealed and manifest accessory pathways, and Mahaim fibers during sinus rhythm and SVT. WPW = Wolff-Parkinson-White syndrome; ART = antidromic reentrant tachycardia; ORT = orthodromic reentrant tachycardia; AVRT = atrioventricular reentrant tachycardia; LBBB = left bundle branch block.

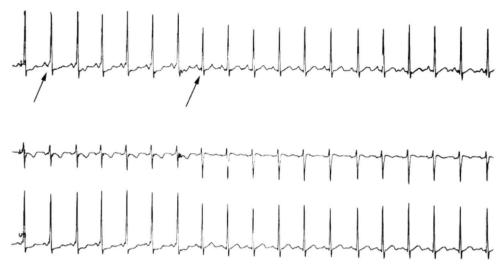

FIGURE 8. Leads II, V1, and V5 during an exercise treadmill test. Note the delta wave in the onset of the QRS (left arrow) and abrupt loss over 2-3 beats (right arrow and next 2 beats) as the heart rate slightly increases, suggesting a slowly conducting accessory pathway.

decreased in 10–20 msec increments until 1:1 AV conduction is no longer seen (Wenckebach cycle length). In patients with manifest conduction, it may be possible to achieve antegrade block in the accessory pathway before the AV node (Figure 9), or visaversa, aiding in pathway localization. In rare cases, it may be possible to differentiate

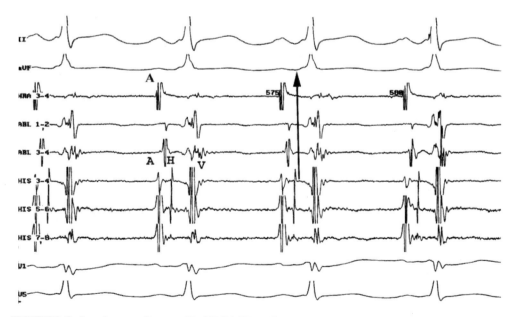

FIGURE 9. Surface electrocardiograms (II, aVF, V1, V5) and intracardiac electrograms obtained during sinus rhythm in a patient with a left lateral accessory pathway. A very short HV interval (H to arrow) is seen in the His electrogram tracings indicting that ventricular activation is reaching the ventricles earlier than expected if the wavefront from the sinus impulse were only traversing the atrioventricular nodal–His Purjinke axis (i.e., the ventricles are preexcited). A = atrial electrogram; HRA = high right atrium; HIS = H = His; Abl = ablation catheter in RA.

between two separate antegrade conducting pathways. The antegrade accessory refractory of the AV node and accessory pathway can then be determined by delivering extrastimuli after an 8–10 beat drive cycle length. Typically, two different drive cycle lengths are used (e.g., 600 msec, 400 msec). The S2 stimulus is decremented by 10–20 msec, observing for abrupt changes in the HV interval and ventricular activation consistent with block in the accessory pathway (accessory pathway effective refractory period) or loss of AV conduction, the accessory pathway and/or the AV node effective refractory period (Figures 10 and 11). The atrial effective refractory period is determined in a similar manner. At times the accessory pathway and/or AV node effective refractory period may be less than or equal to that of the atrial effective refractory period.

In patients with Wolff-Parkinson-White syndrome, the antegrade accessory pathway effective refractory period has been used for risk stratification. The properties of the accessory pathway in a group of young patients who suffered a sudden cardiac event were determined to explore for risk factors. Somewhat surprisingly, 20% of patients who experienced sudden cardiac death had no previous history of SVT but several had previous syncopal episodes prior to their life-threatening event. In addition, a short R-R interval during rapid atrial fibrillation (less than 220 msec) had a 100% sensitivity for sudden cardiac death; the positive predicted value was 68%. In some patients, atrial fibrillation may not be inducible, in which case rapid atrial pacing may be used as a surrogate.

The evaluation and treatment of asymptomatic patients having ventricular

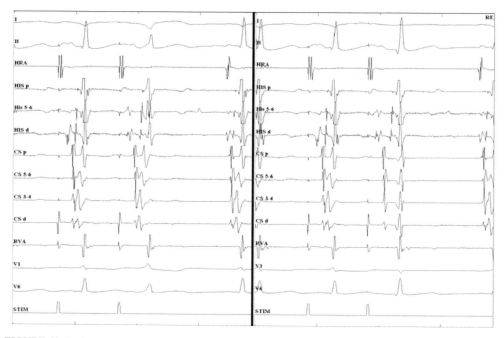

FIGURE 10. Surface electrocardiograms and intracardiac electrograms obtained during an atrial extrastimulus in a patient with a left lateral accessory pathway. In the left panel, the premature atrial stimulus (S2) conducts antegrade through the accessory pathway resulting in more pronounced ventricular pre-excitation and a short AV interval in the distal CS electrode pair. By shortening the atrial extrastimulus by 10 msec, the accessory pathway becomes refractory, resulting in normal conduction through the AV node, a narrow QRS complex, and lengthening of the AV interval in the coronary sinus electrodes. HRA = high right atrium; HIS = His; CS = coronary sinus; RVA = right ventricular apex.

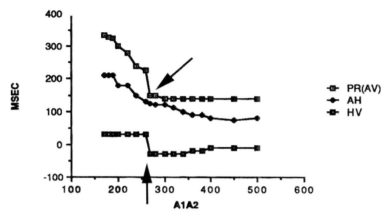

FIGURE 11. Plot of atrial extrastimulation (A1A2) versus the PR (or atrioventricular AV) interval, the AH interval, and the HV interval in a 10-year-old boy with Wolff-Parkinson-White syndrome. Note the increasingly negative HV interval and the stable PR interval with shortening of the A1A2 premature interval indicating increasing preexcitation. At a premature interval (A1A2) of 260 msec (lower arrow), the accessory pathway blocks (its effective refractory period) and conduction entirely shifts to the AV node (upper arrow).

pre-excitation, asymptomatic Wolff-Parkinson-White syndrome, is controversial. The relatively low risk of a sudden cardiac event must be weighed against the risk of performing an electrophysiologic study and an ablation procedure. A prospective, randomized, controlled study evaluated prophylactic ablation of high risk (inducible AVRT, atrial fibrillation, less than 35 years old) asymptomatic Wolff-Parkinson-White patients. In comparison to a matched control group, the patients who underwent prophylactic radiofrequency ablation had a significantly lower incidence of subsequent arrhythmias (5% vs. 60%). One patient in the control group had documented ventricular fibrillation following a syncopal event while jogging and underwent a radiofrequency ablation procedure during which multiple septal pathways were found on the right and left sides. It was concluded that prophylactic radiofrequency ablation in a high-risk population markedly reduced the incidence of arrhythmic events. While the median age in the study population was 23 years (range 15–30 years), and the definition of "high risk" was very broad, extrapolation of these results to an active adolescent and teenage population is not indicated at this time.

Following antegrade assessment by atrial pacing, retrograde AV node and accessory pathway properties are assessed by ventricular pacing. A relatively straightforward method to determine the presence of an accessory pathway is performed by ventricular pacing during a bolus infusion of IV adenosine (200 mcg/kg up to 12 mg) followed by a rapid flush. A ventricular pacing cycle length of 500 msec is often required to assure that the pacing rate is faster than the reflex sinus tachycardia often seen after adenosine infusion. As adenosine usually blocks retrograde conduction in the AV node but rarely blocks retrograde conduction in the accessory pathway (except in slowly conducting pathways), a change in the atrial activation sequence suggests a shift of retrograde conduction from the AV node to the accessory pathway, particularly evident with a left-sided accessory pathway (Figure 12). At times, a significant change in the retrograde atrial activation may not be seen due to either a para-Hisian accessory pathway or an adenosine-resistant AV node.

Similar to antegrade assessment, the retrograde Wenckebach cycle length and

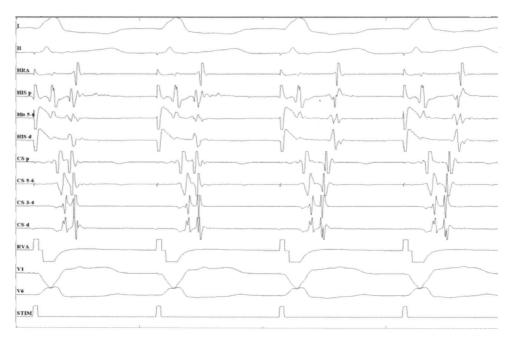

FIGURE 12. Change in retrograde atrial activation sequence following adenosine infusion during ventricular pacing. At the far left the earliest atrial activation is seen in the proximal CS (CSp) and the His bipolar electrode pairs (His d, His 5-6, His p). As the adenosine blocks retrograde AV node conduction the earliest activation shifts (compare second and third complexes in CSp) to the distal coronary sinus bipolar electrode pairs (CS 5-6, CS 3-4). HRA = high right atrium; HIS = His; CS = coronary sinus; RVA = right ventricular apex.

effective refractory periods of the AV node and accessory pathway are determined by rapid ventricular pacing and extrastimulus pacing. The presence of multiple concealed pathways can be determined during extrastimulus pacing, observing for multiple changes in the retrograde atrial activation sequence.

During the atrial and ventricular pacing maneuvers, SVT is often induced; however, on occasion an isoproterenol continuous infusion is required. The typical starting dose of isoproterenol is 0.01–0.02 mcg/kg/minute; rarely is more than 0.1 mcg/kg/minute needed. When SVT cannot be induced with the previously described atrial and ventricular pacing maneuvers, two and three extrastimuli in the atrial or ventricle may be tried. Vagolysis with intravenous atropine may be administered (0.01–0.03 mg/kg).

If a bundle branch block pattern is seen during the initiation of the SVT, changes in the cycle length as the bundle block branch pattern resolves can be helpful in determining the location of the accessory pathway. An initially wide bundle block branch pattern during the tachycardia that prolongs the tachycardia cycle length is indicative of an ipsilateral accessory pathway (e.g., right bundle branch block lengthens the cycle length of orthodromic AVRT using the right-sided accessory pathway). This finding is explained by the slower conduction from myocyte to myocyte as compared to that through the rapidly conducting bundle system, which prolongs the ventricular transit time to the ipsilateral accessory pathway, thus lengthening the cycle length.

The corollary to this finding is that if the bundle branch block pattern does not alter the SVT cycle length, the bundle branch block is contralateral to the accessory pathway (e.g., the right bundle branch block pattern with left-sided accessory pathway does not alter SVT cycle length). Occasionally this observation is confounded by a compensating faster conduction (slower return of the impulse via the accessory pathway allowing AV conduction to improve) through the AV node neutralizing the cycle length change. This confounder is corrected by examining the AV conduction (time)—it remains constant regardless of cycle length change in the presence of an ispilateral accessory pathway.

Following induction of the SVT, close attention must be paid to the hemodynamic status to assure that there is not significant compromise. If there is significant hemodynamic compromise, rapid pacing maneuvers, adenosine, or, rarely, direct current cardioversion may be required.

Various pacing maneuvers are important diagnostic procedures. Premature ventricular beats introduced into the SVT can be helpful in distinguishing between AVRT and AVNRT and can help determine the location of the accessory pathway. The pre-excitation index may help in deciphering between left- and right-sided accessory pathways. To ensure that retrograde conduction is not through the His bundle but rather the accessory pathway, the premature beats are given during His refractoriness (i.e., not earlier than 10–25 msec prior to the His deflection). If the ventricular stimulus is delivered during SVT while the His bundle is refractory in the presence of an accessory pathway, the retrograde conduction is through the accessory pathway, shortening the atrial cycle length. The pre-excitation index is determined by subtracting the longest premature ventricular pacing interval that shortened the atrial cycle length from the SVT cycle length (Figure 13). The shorter the pre-excitation index (longest PVC coupling interval), the more likely the pathway is to be on the right side. Posterior

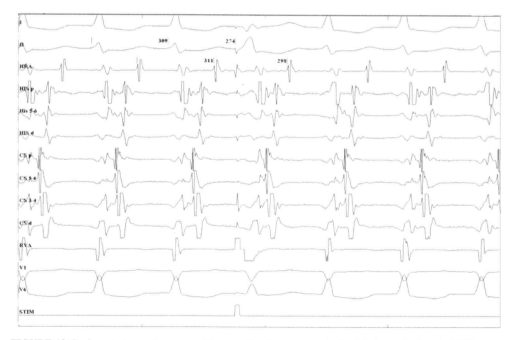

FIGURE 13. Surface electrocardiograms and intracardiac electrograms obtained during orthodromic AVRT. A premature ventricular extrastimulus introduced during His refractoriness shortens the atrial cycle length by greater than 10 msec. The pre-excitation index is calculated by subtracting the premature interval from the tachycardia cycle length (309–274 msec), resulting in a pre-excitation index of 35 msec, consistent with a right-sided accessory pathway. HRA = high right atrium; HIS = His; CS = coronary sinus; RVA = right ventricular apex; STIM = stimulation.

septal accessory pathways have a mean pre-excitation index of 38 msec, whereas anteroseptal septal accessory pathways have a mean pre-excitation of 17 msec. The longer the pre-excitation index (shortest PVC coupling interval), the more likely the pathway is left-sided, but the possibility of conduction through the His-Purkinje system/AV node should be considered, suggesting AVNRT.

While the pre-excitation index is helpful in achieving the general location of the accessory pathway, the more precise method is to determine the location of the pathway by examining for the shortest ventriculoatrial conduction time during orthodromic AVRT or the shortest AV conduction time during antidromic AVRT. If the pathway is on the left side, a coronary sinus catheter advanced further out into the coronary sinus may "bracket" the pathway (i.e., when there are bipolar recordings from either side of the pathway that have longer AV or ventriculoatrial conduction times than the bipolar electrograms located near the accessory pathway) (Figure 14). If the accessory pathway is right-sided, a multipole (e.g., duodecapolar) catheter can be positioned around the tricuspid valve during SVT, again observing for the shortest AV or ventriculoatrial conduction time.

On occasion sustained SVT cannot be induced. In these situations, mapping during tachycardia can be extremely difficult. A surrogate to mapping during SVT is to map during ventricular pacing (Chapter 22). It is important to remember that when pacing the ventricle, it is possible to have conduction

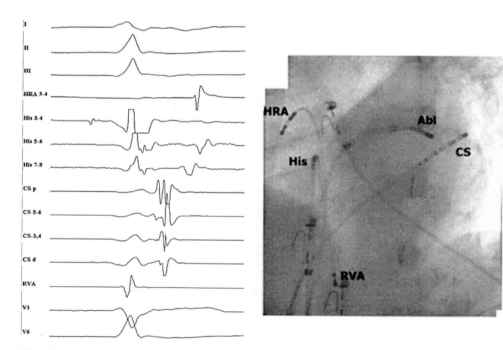

FIGURE 14. Surface electrocardiograms and intracardiac electrograms (left panel) obtained in a patient with a left posterior accessory pathway during orthodromic AVRT. The earliest atrial activation is seen in CS 5-6 (first low amplitude atrial signal, after the slow waveform from the ventricle and before second component of atrial signal (arrow)), "bracketed" by the slightly later atrial activation in CSp, CS 3-4, and CSd. The corresponding fluoroscopic images showing electrophysiologic catheter positions are shown in the right panel (left anterior oblique view). HRA = high right atrium; HIS = His; CS = coronary sinus; RVA = right ventricular apex; Abl = ablation.

through not only the accessory pathway but also through the AV node. Occasionally it is possible to pace sufficiently fast to block conduction through the AV node but not the accessory pathway. However, if this results in significant hemodynamic compromise, other pacing maneuvers may be tried. If the pathway is located on the left side, pacing from either the right ventricular outflow tract, LV apex or LV free wall (retrograde through the aorta) may preferentially activate the left-sided accessory pathway prior to retrograde conduction through the AV node. As mentioned previously, ventricular pacing at incrementally faster pacing rates may also be helpful in determining if there are multiple accessory pathways. At successively shorter pacing cycle lengths, a sudden change in the atrial activation sequence, eccentric from the AV node, suggests the presence of multiple accessory pathways.

Complex anatomy and/or para-Hisian pathways can make determination of the accessory pathway location more difficult. In these cases, three-dimensional mapping computer-based systems can prove to be extremely valuable in delineating the anatomy and precisely locating the area of His potentials.

TREATMENT

Observation

As AVRT is rarely life threatening, and only rarely the cause of syncope (when orthodromic), several different treatment options can be considered. For patients who have infrequent episodes that result in little hemodynamic compromise, observation is indicated, especially in smaller children. In these patients, vagal maneuvers can be effective at terminating tachycardia (Chapter 23). If the tachycardia persists after vagal maneuvers, the patient should seek the assistance of a healthcare provider. In trained hands, esophageal overdrive pacing has been proven to be very effective, but is seldom used outside the infant age group. Chemical cardioversion can also be very effective. Adenosine is 90% to 100% effective but has recurrence rates between 25% and 30%. The advantage of adenosine is a very short half-life; its effects seldom last more than 15 seconds. It is important to be prepared for atrial fibrillation that can occur after adenosine infusion. Patients with Wolff-Parkinson-White syndrome and a fast conducting accessory pathway can develop atrial fibrillation after adenosine intervention and degenerate into ventricular fibrillation. Although this combination is rare (probably 1% of patients who present with narrow QRS tachycardia), an external defibrillator should be readily available. Another option for chemical cardioversion is verapamil, which is 80% to 95% effective. Verapamil should not be used in infants with congestive heart failure as the calcium channel blocker effect can have a profound negative inotropic effect resulting in severe hypotension.

For those patients who can tolerate their SVT but do not respond to vagal maneuvers, periodic use of a beta-blocker or calcium channel blocker when the SVT begins can be effective in slowing and terminating the SVT.

Antiarrhythmic Medications

For those patients who have frequent events or do not tolerate their SVT, chronic antiarrhythmic medication is an option (Chapter 21). Digoxin for infants with SVT or a beta-blocker with or without digoxin have a long history and are safe when properly administered. As digoxin can shorten the accessory pathway effective refractory period, these medications should be avoided in patients who have Wolff-Parkinson-White syndrome. In older children, both have been largely replaced by others agents. Sodium channel blockers and potassium channel blockers have also been proven to be effective and there are liquid forms of all these antiarrhythmic medications. Amiodarone is an effective agent when used in infants and small children before they are candidates for ablation; however, duration of amiodarone should not exceed 18–24 months. As noted previously, calcium channel blockers should not be used in children under the age of one year. Some patients who have drug refractory SVT may require combination therapy. An effective combination is a sodium channel blocker (flecainide) and a potassium channel blocker (sotalol with beta-blocking properties) is reported up to a 100% efficacy in refractory SVT in children less than one year of age.

Transcatheter Ablation

When anti-arrhythmic therapy is ineffective or not desired, the ablation procedure offers the possibility of a definitive cure. Disruption of the accessory pathway was first performed surgically, followed by direct current ablation. Radiofrequency ablation was first introduced in 1987 and was reported in a series of pediatric patients in 1991.

Transcatheter ablation is discussed in Chapter 22. Unusual pathways warrant further consideration. For patients who have para-Hisian pathways or complex anatomy, three-dimensional mapping systems can be a useful adjunct to the electrophysiology study and the radiofrequency ablation procedure. Advances in mapping technology have resulted in 3-D spatial resolution of 1–2 mm, allowing for more precise localization of accessory pathways and reproducible targeting of an identical point within the heart. This technology cannot only increase the success rate and safety of the procedure but decreases the fluoroscopic time and radiation dose. In patients with para-Hisian pathways, accurate mapping of the AV node and accessory pathways using a 3-D system can allow for ablation in a more controlled environment. As an example, using the non-contact mapping system to map the AV node during ventricular pacing, then mapping the accessory pathway during adenosine infusion (blocking the retrograde conduction through the AV node) has enabled the ablation of patients with previously unsuccessful procedures. In patients who have congenital heart disease and AV tachycardia, 3-D mapping techniques can be used to better determine the complex anatomy and location of the His-Purkinje system, which can be displaced secondary to the congenital defect or previous surgery.

Care should be taken when applying radiofrequency energy during SVT as the intracardiac morphology can change abruptly upon termination, which affects catheter position. In general, temperatures greater than 50EC are desired. However, temperatures as low as 48EC have been shown to cause irreversible damage. Seldom is it necessary to use temperatures greater than 70EC. It is important to recognize that accessory pathways are in close proximity to other important structures (AV valve, coronary artery, coronary vein); therefore, excess temperature and time of energy radiofrequency application should be weighed against collateral injury. The duration of the radiofrequency application is usually dependent on the temperature obtained and catheter stability.

Cryotherapy has added another energy source for ablation (Chapter 22). While experience is limited, it is employed for AV nodal reentrant tachycardia (Chapter 5). It has been used for ablation in which the arrhythmia substrate is adjacent to and thus potentially jeopardizes the AV node—such as junctional tachycardia and septal accessory pathways.

Prognosis

Spontaneous resolution of SVT beyond 1–2 years of age in patients with preexcitation and who have AVRT is uncommon. Observation and medical management play a role in the care of very young patients. However, transcatheter ablation offers the advantage of a lifelong cure. Practice indications for ablation have expanded from life-threatening arrhythmias to patient or parental preference. The success rate of this procedure for accessory pathway ablation is now between 90% and 95%; the highest success rates are found in patients having left-sided accessory pathways. Patients with para-Hisian pathways have the lowest success rate, due to more cautious radiofrequency applications, fearing risk of AV block. Data from the pediatric radiofrequency ablation registry has recorded a risk of heart block at 1.2% with a risk as high as 10.4% for patients with mid-septal ablation sites. The overall risk of complications in patients undergoing an ablation is reported to be between 3% and 4%. The risk of early death has been reported at 0.05% with the risk of late death between 0.07% and 0.18%. These data were obtained in the era of conventional mapping and may be biased by early cases performed on the steeper portion of the learning curve.

Risk of sudden death in patients with the Wolff-Parkinson-White syndrome is a well recognized but infrequent occurrence. The sequence of events is most likely the development of atrial fibrillation with rapid antegrade conduction down a fast conducting accessory pathway leading to ventricular fibrillation (Figure 3, Chapter 20). The risk of sudden cardiac death in children with Wolff-Parkinson-White syndrome has been estimated to be as high as 3% to 6% over one's lifetime. Surprisingly, 20% of patients who experienced sudden cardiac death and recovered had no previous history of SVT but many had previous syncopal episodes. A short R-R interval during rapid atrial fibrillation (less than 220 msec) had a 100% sensitivity of predicting sudden cardiac death. These data are used to help risk stratify the "asymptomatic" Wolff-Parkinson-White syndrome patient, and if present, make a strong case for "prophylactic" ablation.

The pioneering development of the pediatric radiofrequency ablation registry has allowed for the compilation of more than 10,000 electrophysiology studies and ablations, enabling pediatric electrophysiologists to modify their approach based on the accrued knowledge. This knowledge coupled with advances in technology will allow for continued improvement in outcome for young patients with AVRT.

SUGGESTED READING

1. O'Connor BK, Dick Md. What every pediatrician should know about supraventricular tachycardia. *Pediatric Annals* 1991;20:368, 371–376.
2. Perry JC, Garson A, Jr. Supraventricular tachycardia due to Wolff-Parkinson-White syndrome in children: early disappearance and late recurrence. [Comment.] *J Am Coll Cardiol* 1990;16:1215–1220.
3. Lundberg A. Paroxysmal atrial tachycardia in infancy: long-term follow-up study of 49 subjects. *Pediatrics* 1982;70:638–642.
4. Ko JK, Deal BJ, Strasburger JF, Benson DW, Jr. Supraventricular tachycardia mechanisms and their age distribution in pediatric patients. *Am J Cardiol* 1992;69:1028–1032.
5. Giardina AC, Ehlers KH, Engle MA. Wolff-Parkinson-White syndrome in infants and children. A long-term follow-up study. *Br Heart J* 1972;34:839–846.
6. Wolff L, Parkinson J, White PD. Bundle-branch block with short P-R interval in healthy young people prone to paroxysmal tachycardia. *Am Heart J* 1930;10:685–704.
7. Munger TM, Packer DL, Hammill SC, et al. A population study of the natural history of Wolff-Parkinson-White syndrome in Olmsted County, Minnesota, 1953-1989. *Circulation* 1993;87:866–873.
8. Riggs TW, Byrd JA, Weinhouse E. Recurrence risk of supraventricular tachycardia in pediatric patients. *Cardiology* 1999;91:25–30.
9. Anderson RH, Becker AE, Arnold R, Wilkinson JL. The conducting tissues in congenitally corrected transposition. *Circulation* 1974;50:911–923.
10. Weng KP, Wolff GS, Young ML. Multiple accessory pathways in pediatric patients with Wolff-Parkinson-White syndrome. *Am J Cardiol* 2003;91:1178–1183.
11. Chiang CE, Chen SA, Teo WS, et al. An accurate stepwise electrocardiographic algorithm for localization of accessory pathways in patients with Wolff-Parkinson-White syndrome from a comprehensive analysis of delta waves and R/S ratio during sinus rhythm. *Am J Cardiol* 1995;76:40–46.
12. Fitzpatrick AP, Gonzales RP, Lesh MD, et al. New algorithm for the localization of accessory atrioventricular connections using a baseline electrocardiogram. [Erratum appears in *J Am Coll Cardiol* 1994

Apr;23(5):1272.] *J Am Coll Cardiol* 1994;23:107–116.

13. Mahaim I, Benatt A. Nouvelles recherches sur les connexions superieures de la branche gauche du faisceau de His-Tawara avec la cloison interventriculaire. *Cardiologia* 1938;1:61–73.

14. Haissaguerre M, Cauchemez B, Marcus F, et al. Characteristics of the ventricular insertion sites of accessory pathways with anterograde decremental conduction properties. *Circulation* 1995;91:1077–1085.

15. Sternick EB, Timmermans C, Rodriguez LM, Wellens HJJ. Mahaim fiber: an atriofascicualr or long ventricular atrioventricular pathway. *Heart Rhythm* 2004:1(6):724–727.

16. Bricker JT, Porter CJ, Garson A, et al. Exercise testing in children with Wolff-Parkinson-White syndrome. *Am J Cardiol* 1985;55:1001–1004.

17. Miles WM, Yee R, Klein GJ, et al. The pre-excitation index: an aid in determining the mechanism of supraventricular tachycardia and localizing accessory pathways. *Circulation* 1986; 74:493–500.

18. Bromberg BI, Lindsay BD, Cain ME, Cox JL. Impact of clinical history and electrophysiologic characterization of accessory pathways on management strategies to reduce sudden death among children with Wolff-Parkinson-White syndrome. *J Am Coll Cardiol* 1996;27:690–695.

19. Pappone C, Manguso F, Santinelli R, et al. Radiofrequency ablation in children with asymptomatic Wolff-Parkinson-White syndrome. *N Engl J Med* 2004;351(12):1197–1205.

20. Pappone C, Santinelli V, Manguso F, et al. A randomized study of prophylactic catheter ablation in asymptomatic patients with the Wolff-Parkinson-White syndrome. *N Engl J Med* 2003;349(19): 1803–1811.

21. Bink-Boelkens MT. Pharmacologic management of arrhythmias. *Ped Cardiol* 2000;21:508–515.

22. Kugler JD, Danford DA, Houston K, Felix G. Radiofrequency catheter ablation for paroxysmal supraventricular tachycardia in children and adolescents without structural heart disease. Pediatric EP Society, Radiofrequency Catheter Ablation Registry. *Am J Cardiol* 1997;80:1438–1443.

23. Erickson CC, Walsh EP, Triedman JK, Saul JP. Efficacy and safety of radiofrequency ablation in infants and young children < 18 months of age. *Am J Cardiol* 1994;74:944–947.

24. Bokenkamp R, Wibbelt G, Sturm M, et al. Effects of intracardiac radiofrequency current application on coronary artery vessels in young pigs. *J Cardiovascul Electrophysiol* 2000;11:565–571.

25. Law IH, Fischbach PS, LeRoy S, et al. Access to the left atrium for delivery of radiofrequency ablation in young patients: retrograde aortic vs. transseptal approach. *Ped Cardiol* 2001;22:204–209.

26. Schaffer MS, Silka MJ, Ross BA, Kugler JD. Inadvertent atrioventricular block during radiofrequency catheter ablation. Results of the Pediatric Radiofrequency Ablation Registry. Pediatric Electrophysiology Society. *Circulation* 1996;94:3214–3220.

27. Kugler JD, Danford DA, Deal BJ, et al. Radiofrequency catheter ablation for tachyarrhythmias in children and adolescents. The Pediatric Electrophysiology Society. *N Engl J Med* 1994;330:1481–1487.

28. Scheinman MM, Wang YS, Van Hare GF, Lesh MD. Electrocardiographic and electrophysiologic characteristics of anterior, midseptal and right anterior free wall accessory pathways. *J Am Coll Cardiol* 1992;20:1220–1229.

29. Mandapati R, Berul CI, Triedman KJ, et al. Radiofrequency catheter ablation os septal accessory pathways in the pediatric age group. *Am J Cardiol* 2003: 92(8);947–950.

5

Atrioventricular Nodal Reentrant Tachycardia

David J. Bradley

Atrioventricular (AV) nodal re-entrant tachycardia (AVNRT) is a form of reentrant supraventricular tachycardia (SVT) whose circuit revolves around the AV nodal tissues. Very similar in clinical presentation to SVT supported by an accessory AV pathway (Chapter 4), it has distinct epidemiological and electrophysiologic features.

SYMPTOMS

The child with SVT complains of episodes of palpitations (heart racing, fluttering, or "beeping"), often accompanied by malaise, pallor, nausea and sweating, reflecting the sympathetic outpouring associated with the rapid heart rate. Shortness of breath may also occur, even with hemodynamically well tolerated SVT; rarely does the patient lose consciousness. Patients may complain of pulsations in the neck—often visible to parents or other observers—due to oscillations in central venous pressure caused by simultaneous contraction of atria and ventricles. Triggers for episodes include physical activity, emotional stress, and also abrupt changes in body position, such as bending over. It appears that initiation of AVNRT is more influenced by increased sympathetic tone than accessory-pathway-supported SVT; it is less common for a child who has SVT at rest to have AV node reentry rather than AV reentry as the cause. In contrast to AVRT, AVNRT is less likely to be incessant, and, therefore, is rarely implicated in tachycardia-induced cardiomyopathy.

INCIDENCE

There are few estimates of the incidence of AVNRT in children. Though it is the most common mechanism of regular SVT in adults, AVNRT is uncommon in small children, comprising up to 3%–13% of infant SVT. Its relative incidence increases across the pediatric age range, equaling, then exceeding that of AVRT during the teenage years. A patient who has the first episode of SVT as an older adolescent is most likely to have AV node reentry as the mechanism; conversely, a child or adolescent who has the first episode of tachycardia in the first year of life but few afterwards, is most likely to have AVRT [i.e., an accessory pathway (AP) supporting the tachycardia].

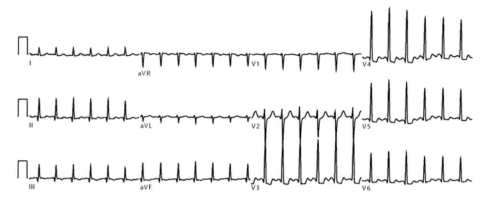

FIGURE 1. 12-Lead electrocardiogram of AVNRT.

ELECTROCARDIOGRAM

The electrocardiogram (ECG) during AVNRT reveals regular, narrow, QRS complexes at a virtually constant rate; rarely, fluctuating conduction velocity (oscillation) in the AV node will render the tachycardia somewhat irregular. During typical "slow-fast" AVNRT, retrograde P-waves are either not visible (buried in the QRS complex) or are located less than 65–70 ms after the QRS onset (Figures 1 and 2). When present, as in atypical "fast-slow" AVNRT, the retrograde P-wave axis is superior, reflecting atrial activation at

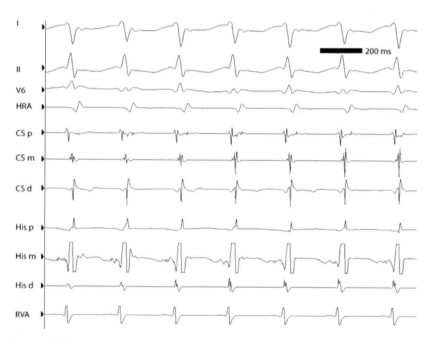

FIGURE 2. I, II, V5 ECG leads and intracardiac electrograms in a 13-year-old girl with AVNRT at approximately 260 bpm. Note the simultaneity between ventricular and atrial activation, characteristic of "typical" AVNRT. HRA = high right atrium; CSp, CSm, CSd = proximal, middle, and distal coronary sinus electrograms respectively. His p, His m His d = proximal, middle, and distal His bundle electrograms respectively; RVA = right ventricular apex.

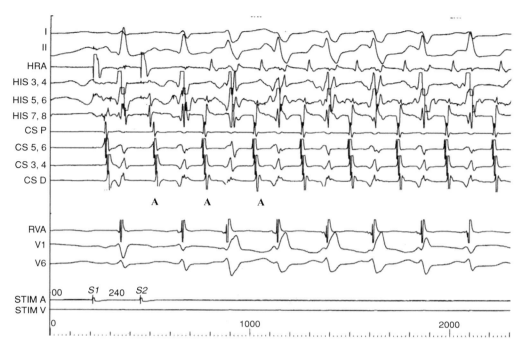

FIGURE 3. "Atypical AVNRT" (slow–slow) initiated by programmed extra-stimulation in a 10-year-old girl. Note the long A2V2 and PR interval indicating slow conduction to the ventricle (3rd QRS complex from the left) and the echo beat (A) beginning the tachycardia. Note that the atrial echos are not simultaneous with the QRS complex as in "typical" AVNRT but are equidistant between the ventricular complexes. Note the rate related right bundle branch bloc in the first 4 beats of the tachycardia followed by 2 narrow QRS complexes. The tachycardia cycle length does not lengthen during RBBB indicating that if there were an accessory pathway it would not be on the right side (Chapter 4).

the septal/tricuspid annulus and usually visible following the QRS complex due to a retrograde P-wave conduction time greater than 80 ms after the QRS complex (Figure 3). As in all forms of SVT, rate-related aberrant right or left bundle branch-block may be seen, producing a wide QRS tachycardia mimicking ventricular tachycardia. Depending on the patient's degree of sympathetic activation and individual AV node characteristics, the rate of AVNRT can range from 120 bpm to 300 bpm, with rates typically 180 bpm to 250 bpm.

Role of Dual AV Node Physiology in AVNRT

The term "dual AV node physiology" denotes the condition in which two functional conducting pathways exist between the atrium and the penetrating His bundle (Figure 4). In the typical example, the antero-superior atrial connections to the AV node comprise a pathway with a faster conduction velocity but a longer refractory period relative to the more slowly conducting postero-inferior atrial fibers, but with a shorter refractory period. As a result, an appropriately timed atrial premature beat (Figure 4A) that falls during the refractory period of the "fast pathway" will be successfully conducted through the AV node via the "slow pathway" (as it is not refractory, Figure 4B). The impulse will enter the His–Purkinje system activating the ventricles, and—due to delay within the slow pathway—will also find the atrial fast pathway fibers no longer refractory and thus available for retrograde excitation (Figure 4C). In AVNRT, the atrium is therefore activated retrogradely at approximately the same time as is the ventricle antegradely (Figures 1 and 2).

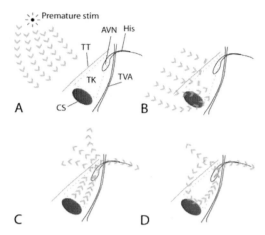

FIGURE 4. Initiation of AVNRT by a premature atrial stimulus. A. Atrial activation proceeds from the source of the premature stimulus. B. The premature atrial activation approaches the AV node, but the anterior inputs ("fast pathway") to the AV node are still refractory, and thus block conduction. C. In contrast, the posterior inputs ("slow pathway") to the AV node have a shorter refractory period and are excitable. Here, conduction persists. The node is activated, leading to activation of the ventricle and (i.e., now the anterior aspect of the atrio-nodal junction is excitable) atrial tissues. D. The re-entrant circuit is established, with activation of atria and ventricles with each "lap" or revolution through the circuit. Abbreviations: AVN = AV node; His = His bundle; Stim = stimulus; TK = triangle of Koch; TT = tendon of Todaro; TVA = tricuspid valve annulus; CS = ostium of the coronary sinus.

The re-entrant circuit continues anterograde through the slow AV nodal pathway and retrograde through the fast, re-activating the atrium and ventricle with each cycle (Figure 4D).

Programmed atrial stimulation is the standard technique to demonstrate these separate conduction pathways. As the coupling interval of the atrial extrastimulus is progressively shortened, the septal A-H interval progressively lengthens, demonstrating decremental conduction (Figure 5). At a critical coupling interval, the "fast" anterior conduction pathway is refractory, blocking conduction. Conduction, however, persists through the slower posterior connections, resulting in an abrupt increase or "jump" (≥ 50 ms associated with only a 10 ms decrease in the premature coupling interval) in the subsequent A-H interval duration or an increase of 50 msc or greater in the H1-H2 interval (Figure 5). This "jump" indicates a shift in conduction from the fast pathway to the slow pathway (i.e., dual AV node physiology) and fulfills two of the three conditions for re-entry (block in one pathway and slow conduction in the other; Chapter 3). Single echoes or multiple beats of tachycardia complete the reentry triad (recovery of excitability in the tissue of origin). Although the above typical "slow–fast" AVNRT (Figures 1 and 2) is the most common pattern seen with dual AV node physiology, variants exist. In the first variant, the so-called "fast–slow" form, the circuit travels in the opposite direction, transcribing a longer RP (slow retrograde conduction) compared to that seen in the typical form and a more normal PR interval (fast antegrade conduction, Figure 3). A third infrequent variant has similar conduction velocity in the two pathways (e.g., "slow–slow") (Figure 6). This form is characterized by a retrograde (negative in II, III, avF) P-wave transcribed in the middle of the tachycardia cycle length. A fourth important characteristic of the re-entrant circuit of AVNRT is its total independence from excitable tissues distal to the circuit, including the bundle of His and ventricles. It is therefore possible to observe AVNRT with 2:1 conduction to the ventricle (Figure 7). It is critical to differentiate this form from atrial flutter by examining for dual AV dual AV node physiology.

Programmed stimulation of the ventricle during electrophysiologic study may demonstrate dual AV node pathway physiology, even when atrial programmed stimulation does not. Because, with ventricular extrastimulation, block occurs retrogradely in the fast pathway with retrograde conduction through the slow pathway, the resultant AVNRT is usually the "fast–slow" form.

Noninvasive Treatment

The decision regarding treatment is largely governed by several historical findings of each individual patient. They include

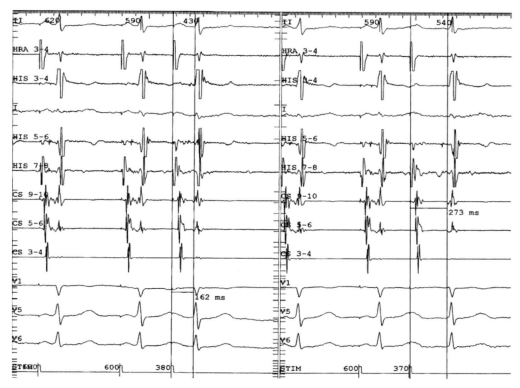

FIGURE 5. Demonstration of dual AV nodal physiology at EP study. In the left-hand panel, atrial programmed extrastimulation is performed using a drive cycle length (S1) of 600 ms and an extrastimulus (S2) of 380 ms. The resultant A2-H2 was 162 ms as measured by the vertical caliper and the H1–H2 interval about 390 ms. In the right-hand panel, the S1S1 is again 600 ms but the S1S2 is now 370 ms. The resultant A2-H2 is now 273 ms (and the H1–H2 interval increased by more than 100 ms) indicating conduction is blocked in the fast pathway but continues antegradely in the slow pathway.

frequency of episodes, duration of episodes, severity of symptoms with episodes (i.e., intensity), response to prior medications, and the patient's and family's understanding and views regarding the various options. Patients who have minimal symptoms with episodes that are short and self-terminating or that respond to vagal maneuvers may pursue no further treatment (Chapter 3). However, for some, the response of the arrhythmia to vagal maneuvers is poor. Athletes may find the unpredictable onset of tachycardia during sports events highly disruptive, whether or not it can be terminated reliably using vagal maneuvers.

Episodes of SVT may be responsive to antiarrhythmic agents (Chapter 21). Because β-blockers are well tolerated and available in convenient dosing (Table 1), they are often the medication of first choice for most patients. This group of medications acts by reducing the number of "triggers" (i.e., premature beats) or by blocking conduction of the atrial extrasystoles through the AV node, and may thereby provide benefit for some patients. Fatigue, malaise, and exercise limitations may occur, particularly with use of older medications in this class, such as propranolol. Newer, more selective, longer acting β_1-blockers such as atenolol or nadolol, which cross the blood-brain barrier to a lesser degree, may be better tolerated. The athlete taking β-blockers, however, often complains of fatigue during vigorous exertion.

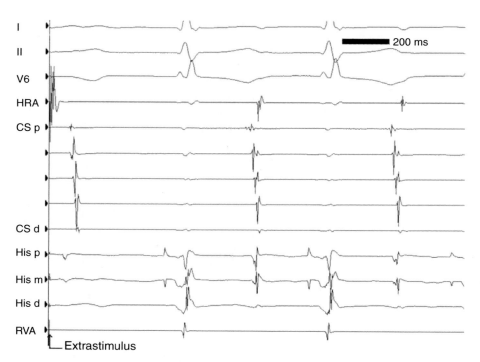

FIGURE 6. Another form of "atypical AVNRT" initiated by programmed extra-stimulation. Note the long RP interval as well as the long PR interval indicating slow conduction both anterograde and retrograde (i.e., "slow–slow").

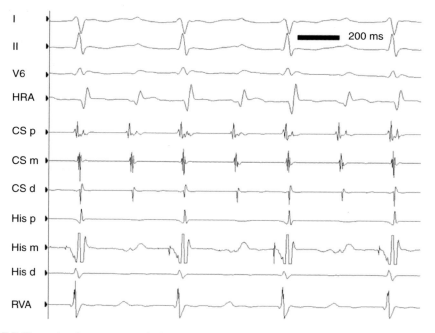

FIGURE 7. The tracing demonstrates typical AVNRT, with a 2:1 A-V relationship between the atria and ventricles.

TABLE 1. Drugs Used in the Treatment of AVNRT

Drug	Class	Dose (mg/kg/24 hrs)	Dose interval (h)	Comment
Digoxin	Glycoside	8–10 mcg	12	Decreasing use, minimal anti-arrhythmic effect
Propranolol	β-blocker	1–4	8	Depression, fatigue may occur
Atenolol	β-blocker	0.5–1.5	12–24	Fewer CNS effects than propranolol
Diltiazem	Ca^{2+} channel blocker	1.5–3.5	6–8	Sustained-release form available
Flecainide	Na^+ channel blocker	1–5	8–12	Structural heart disease a contraindication
Amiodarone	Blocks K^+ and other channels	5	24	Potential end organ toxicity long term

Digoxin, at one time a common first-line therapy for SVT in children, is used much less often. Although relatively inexpensive and readily available, it has little demonstrable direct pharmacologic anti-arrhythmic effect.

Calcium-channel blockers may cause fatigue as a side-effect, but are a satisfactory treatment for some patients with AVNRT, particularly those with unacceptable side effects from β-blockers. The lack of availability of suspensions and the requirement of frequent dosing makes these somewhat more difficult to use.

Class I and III agents such as flecainide (class I) and amiodarone (class III) are often effective in refractory AVNRT and particularly useful when radiofrequency ablation is not an option.

Many children can be safely treated for narrow-complex tachycardia with a normal resting ECG, even when the exact tachycardia mechanism is not known. Infants with SVT no symptoms and only infrequent episodes can be safely managed without pharmacologic agents. ECG distinction between AVNRT and concealed accessory-pathway supported SVT is not always possible, particularly at very fast rates common in children with small hearts, with shorter reentry circuits of SVT. Event monitors can be very useful for the detection of SVT but do not always permit determination of the mechanism.

Although anti-arrhythmic agents remain an option, they no longer have the dominant role for treatment of SVT, especially in older, larger children.

ELECTROPHYSIOLOGY STUDY AND RADIOFREQUENCY ABLATION

Ablation of the SVT substrate, using transcatheter radiofrequency current, or more recently, cryotherapy, while carrying small but serious risks, is regularly performed for the elimination of symptomatic SVT in children (Chapter 22). Many patients and families decide not to begin a daily medication for what is a sporadic symptom, but to proceed to ablation if SVT has become an unacceptable intrusion on their life.

The age at which ablation can safely be performed is declining. Early experience from the Pediatric Electrophysiology Society Registry suggested that patients under 15 kg had a significantly increased risk of serious complications. Some animal models of ablation suggested that radiofrequency lesions grow significantly when performed in young individuals, particularly when they are delivered to the ventricles. However, the clinical experience of large-volume centers and the long-term follow-up of children enrolled in the Prospective Assessment of Pediatric Catheter Ablation (PAPCA) support the use of ablation for highly refractory arrhythmias, even in selected infants and toddlers.

TABLE 2. Components of Electrophysiologic Study for SVT

Study component	Information provided
Adenosine bolus during right ventricular pacing	If V-A block occurs, unlikely to have a V-A conduction pathway other than AV node
Atrial extrastimulus	Anterograde refractory period of AV node pathway(s), atrium, arrhythmia induction
Ventricular extrastimulus	Retrograde refractory period of AV node pathway(s), arrhythmia induction (less commonly)
Incremental pacing—atrial and ventricular	Determination of Wenckebach cycle length, antegrade and retrograde, arrhythmia induction
Isoproterenol infusion	"Unmasking" of dual AV node physiology and arrhythmias not demonstrable in the baseline state, arrhythmia induction
Premature ventricular stimuli during SVT	Determines involvement of possible accessory pathway in supraventricular tachycardia

In contrast to patients with accessory pathways (Chapter 4), our experience with young patients with SVT who present to the electrophysiology laboratory with a top normal PR interval (≥ 0.18 sec) or in a junctional rhythm, as opposed to sinus, are typically found to have AVNRT. Techniques for electrophysiologic study unique to children and young patients are reviewed in Chapter 3. Arrhythmia induction with programmed extrasimulation is similar to the standard methods used in adult patients. Table 2 outlines the specific steps of a SVT protocol. When performing an electrophysiology study of SVT, it is useful to begin with a bolus infusion of intravenous adenosine (200 mcg/kg) while pacing the right ventricle. Adenosine is a potent—and short-acting—blocker of the AV node, but does not block most accessory AV pathways. Ventricular–Atrial (VA) block after such a bolus suggests that no accessory pathway will be found, and that the diagnosis of AVNRT should be pursued. If VA conduction persists after bolus adenosine, the retrograde activation sequence of the atrium will help identify the location of an accessory pathway (Chapter 4).

Next, programmed stimulation is used to define the refractory periods of the cardiac tissues; particularly those of the "fast" and "slow" components of the AV node (see above and Chapter 3).

It is important to note that 30%–50% of normal subjects (i.e., individuals without a history of tachycardia or its associated symptoms) demonstrate dual AV node physiology during electrophysiologic testing. This proportion is smaller in younger children than in adolescents and adults. Interestingly, about one-third of children with AVNRT as their tachycardia mechanism do not demonstrate classical dual AV node physiology in the baseline state. In preparation for ablation, the clinical tachycardia should be induced at EP study. If this cannot be achieved in the baseline state, intravenous isoproterenol 0.02–0.05 mcg/kg/min should be infused, such that the baseline heart rate is increased by at least 25%. A small number of patients in whom dual AV nodal physiology is not found in the baseline state will show such physiology when receiving isoproterenol. The extrastimulus testing should be repeated during isoproterenol infusion to induce tachycardia. In patients with documented SVT on an ECG, dual AV node physiology, but no inducible SVT at electrophysiologic study, proceeding to slow pathway ablation is reasonable to eliminate the most likely substrate.

Intracardiac tracings of typical AVNRT have a characteristic appearance of nearly simultaneous atrial and ventricular activation (Figure 2). After AVNRT is established as the SVT mechanism, radiofrequency (RF) or

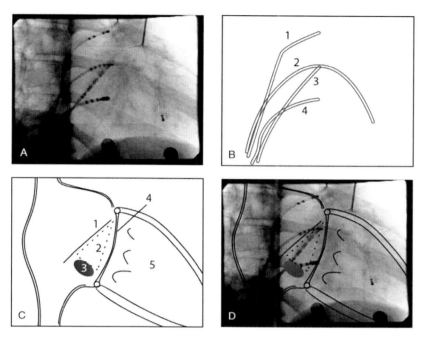

FIGURE 8. Fluoroscopic-anatomic correlation of catheter positions relevant to AVNRT. A. Right anterior oblique (30° angle) fluoroscopic image of the heart with catheters in position. B. Key to catheter positions: (1) High right atrium; (2) Combined His bundle/right ventricular apex catheter; (3) Coronary sinus; (4) Ablation catheter resting in slow pathway region. C. Relevant anatomic structures: (1) Tendon of Todaro/Eustachian ridge; (2) Triangle of Koch; (3) Coronary sinus ostium; (4) Septal tricuspid valve annulus; (5) Body of the right ventricle. D. Anatomic cartoon overlying the fluoroscopic image.

cryotherapy lesions are placed in the slow pathway region, within the triangle of Koch. This triangle comprises the atrial tissues bounded by the coronary sinus ostium and the tendon of Todaro on the upstream side and the tricuspid annulus on the downstream side. It extends anteriorly up the annular edge of the tricuspid valve to its apex at the central fibrous body. The AV node rests just below the apex (Figure 8). To enhance the optimal fluoroscopic localization of the ablation catheter, the signal recorded by the distal bipole is critical to the selection of an appropriate ablation site. Adequate annular position is best identified by a signal that includes both atrial and ventricular components, and in which the ventricular amplitude exceeds the atrial signal by about 3–5 times (Figure 9). The atrial waveform often has a multiphasic morphology ("m" or "w" shaped). Sharp large a and v deflections identify a coronary sinus location of the ablation tip, and are generally inappropriate. It is most sensible to begin ablation in the posterior regions of the slow pathway to maximize the distance from the AV node and His bundle, moving only to more anterior sites if a satisfactory result is not achieved. Successful ablation of the slow pathway can usually be accomplished at a safe distance (≥ 5 mm) from positions in which a His potential is recorded. To stabilize the catheter tip, long sheaths add rigidity to the catheter shaft and are molded to direct the catheter favorably. Additional stability in position is achieved by pausing the patient's ventilator during ablation, usually during expiration. Placement of the catheter in the target area, and especially return to the target area for repeated applications, are greatly assisted by non-radiologic navigation systems,

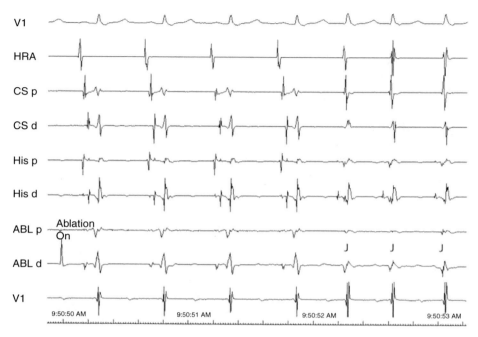

FIGURE 9. Optimal electrograms recorded through the distal closely spaced pair of electrodes on the ablation catheter (ABLd). Note the electrogram recorded through ABL d with small "a" and larger "v". Radiofrequency energy is turned on, and accelerated junctional beats (labeled "J") begin. Note the "JA" association between the a-wave in the CS electrogram and the v-wave in the His electrogram and QRS in V1.

thus reducing radiation exposure, especially important in a young person (Figure 10). This imaging technology allows return to the target site with successive applications of either radiofrequency ablation or cryotherapy without additional imaging with x-ray.

When a satisfactory catheter location has been achieved, RF energy or cryotherapy is delivered from the catheter tip. For RF ablation, a temperature of 48°C–55°C is usually required, delivered for up to 60 seconds at the target site. Temperature-control modes in the RF generators allow for the selection of a maximum desired tissue temperature, and delivery of the power (Watts) required to attain it. With good catheter–tissue contact, a slow-pathway ablation may not require more than 15 W (of the generator's available 50–100 W) to achieve satisfactory ablation. The mark of success of slow pathway RF ablation is the induction of accelerated but relatively (100 bpm) junctional beats upon delivering sufficient heating levels of RF energy. If no junctional acceleration is observed after reaching a sufficient temperature for 10–15 seconds, the ablation is suspended and the catheter is repositioned. Importantly, during RF energy delivery, close attention to the monitoring electrogram must demonstrate "J–A" association [1:1—Junctional (ventricular)—Atrial]. Persistence of the simultaneous relationship between the electrogram and the junction/ventricle electrogram (QRS) suggests persistent integrity of the fast pathway. In smaller children, RF energy may be applied for shorter durations. If accelerated junctional beats demonstrate prolongation or interruption of V–A association, the ablation should be stopped immediately. This may be a sign of injury to the fast AV node pathway, and a harbinger of iatrogenic heart block.

Once one or more satisfactory lesions have been placed, the patient is re-tested for the presence of dual AV node physiology and

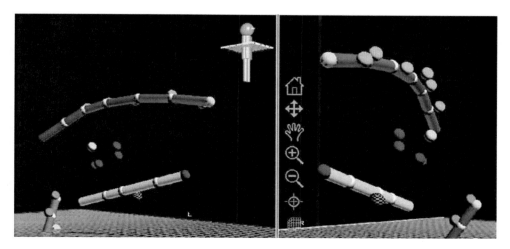

FIGURE 10. Single spot frame of navigation system (LocaLisa®) during cryothermal ablation of slow pathway in 9-year-old boy with AVNRT. The dots indicate His bundle catheter sites (blue), Coronary sinus os (green), and ablation sites (red). Yellow indicates site of increase in PR with delivery of cryothermal energy. Cross-hatched mark is near the inferior margin of tricuspid valve. (Left-hand panel—right anterior oblique projection; Right-hand panel—left anterior oblique projection)

the inducibility of AVNRT. The theoretically most convincing end-point is the elimination of the "jump" in both anterograde and retrograde conduction. This indicates that only a single functional pathway to the AV node remains. However, many electrophysiologists consider the ablation successful when no more than one AVNRT echo beat can be induced with programmed stimulation in the presence of isoproterenol. Recurrence of the clinical symptoms appears to develop at a similar rate with either endpoint. On the other hand, it is well established that more than one inducible beat of tachycardia during post-RF ablation testing has a higher rate of clinical arrhythmia recurrence. Overall, slow pathway modification by RF ablation is acutely (no tachycardia upon the electrophysiology laboratory) successful in 96% of children. Long-term follow-up in children suggests 85%–90% success at 3–5 year follow-up. The result of ablation for AVNRT is referred to as "slow pathway ablation" if there is no slow pathway conduction after the procedure and the PR interval is normal (intact fast pathway); otherwise it is called "slow pathway modification" if slow pathway conduction and echos persist. The term "AV node ablation" is usually reserved for the intentional elimination of AV conduction in patients with intractable atrial tachyarrhythmias (such as atrial fibrillation) with rapid ventricular response.

CRYOTHERAPY: A NEW TRANSCATHETER ENERGY SOURCE

Although cryotherapy (Chapter 22) has been used in the operating room for ablation of cardiac arrhythmias for over 20 years, recent experience with transcatheter cryotherapy offers several features. First, similar to an RF lesion for which an application is usually shorter (≤ 2 minutes), the application of sub-zero temperatures ($-70°C$) to viable tissue for sufficient time (usually 3–4 minutes) leads to permanent complete cell and/or tissue death. However, for temperatures up to $-30°C$, the lesion is reversible and thus functionality (i.e., conduction) can recover, reducing the potential risk of permanent AV block (Figure 11). Second, upon reaching a temperature of $-25°C$ to $-30°C$, the electrode tip adheres to the tissue stabilizing the tip at the desired target site. Although these features suggest certain advantages and early results are encouraging, the final outcome remains to be determined.

The AV Node in Patients with Abnormal Cardiac Anatomy

AVNRT has been identified in patients having a variety of congenital anatomic defects, before and after their surgical repair or palliation. In such patients, the hemodynamic compromise associated with AVNRT episodes can be significant. For example, a child with postoperative Fontan anatomy and mild sinus node dysfunction may have symptoms that can be attributed to the borderline bradycardia; the patient may have an even slower heart rate if taking a β-blocker. Modification of the slow pathway region can be performed successfully even in complex anatomic situations, which may call for a retrograde (trans-aortic valve) approach or limited catheter access due to surgical baffles. When possible, the location of the His bundle should be mapped using more advanced three-dimensional systems, to minimize the chance of injury to the fast AV node pathway.

In patients with complete AV canal defect, the conduction system is displaced posterior and inferiorly, due to incomplete septation. Disparity in mitral and tricuspid annular size may make the triangle of Koch difficult to localize by fluoroscopy. Successful ablation of AVNRT in such subjects requires a methodical positioning of catheters, defining landmarks, and careful selection of appropriate electrograms on the ablation catheter.

Complications of Ablation for AVNRT

Ablation for AVNRT is performed entirely along the septal tricuspid annulus in the normal heart. The most evident serious risk is that of permanently ablating A–V conduction, for which a pacemaker is the usual required treatment. AV block may occur in two ways: either direct thermal injury to the AV node or His bundle by ablating too close to these structures, or by injury to the vascular supply of the AV node. The AV node artery generally arises from the postero-inferior atrium as a branch of the posterior descending coronary artery. Histologic studies have demonstrated the variability both in its position and depth beneath the endocardial surface at which RF ablation is performed. Further, the shape and dimensions of the triangle of Koch in children vary. Complete AV block has been documented with the application of RF energy both immediately anterior and posterior to the coronary sinus ostium, in what would be considered "safe" posterior locations. Although these anatomic features place the AV node at risk, recent experience indicates the likelihood of complete permanent heart block due to RF ablation is low. Nonetheless, it is the practice of experienced operators to limit ablation temperature, power, duration, and number of RF applications in smaller children. For most pediatric patients, recurrent tachycardia is preferable to AV block. There is a small risk of injury to venous structures or cardiac chambers, which differs little in the patient with AVNRT than for other electrophysiologic studies. In general, as the procedure is performed within right-sided chambers, the risk of systemic thromboembolism is very low.

SUMMARY

Atrioventricular nodal reentry tachycardia is the second most common mechanism of paroxysmal supraventricular tachycardia in pediatric patients, and occurs more frequently in older children and adolescents. Although often responsive to vagal maneuvers and sometimes manageable by antiarrhythmic medications, when patients, families, and physicians find the tachyarrhythmia episodes too frequent and too long with too many aggravating symptoms, requiring emergency room visits, ablation can be performed. Slow AV nodal pathway ablation can be achieved with a high degree of success and a relatively low rate of complications using RF energy or cryoablation.

SUGGESTED READING

1. Denes P, Wu D, Dhingra RC, et al. Demonstration of dual A-V nodal pathways in patients with paroxysmal supraventricular tachycardia. *Circulation* 1973;48:549–555.
2. Cohen MI, Wieand TS, Rhodes LA, Vetter VL. Electrophysiologic properties of the atrioventricular node in pediatric patients. *J Am Coll Cardiol* 1997; 29(2):403–407.
3. Crosson JE, Hesslein PS, Thilenius OG, Dunnigan A. AV node reentry tachycardia in infants. *Pacing Clin Electrophysiol* 1995;18(12 part 1):2144–2149.
4. Ko JK, Deal BJ, Strasburger JF, Benson DW. Supraventricular tachycardia mechanisms and their age distribution in pediatric patients. *Am J Cardiol* 1992;69:1028–1032.
5. Ro PS, Rhodes LA. Atrioventricular node re-entry tachycardia in pediatric patients. *Prog Ped Cardiol* 2001;13:3–10.
6. van Hare GF, Chiesa NA, Campbell RM, Kanter RJ, Cecchin F, for the Pediatric Electrophysiology Society. Atrioventricular nodal reentrant tachycardia in children: effect of slow pathway ablation on fast pathway function. *J Cardiovas Electrophys* 2002;13:203–209.

6

Persistent Junctional Reciprocating Tachycardia

Parvin C. Dorostkar

Persistent junctional reciprocating tachycardia (PJRT) is a rare reentrant supraventricular tachycardia comprising less than 1% of supraventricular tachycardias in children.

ANATOMY

Based on an analogy of the known anatomy in patients with Wolff-Parkinson-White syndrome, the hypothetical anatomy of the PJRT circuit is shown in Figure 1. The tachycardia circuit consists of an antegrade limb through the normal atrioventricular (AV) nodal conduction system and a retrograde limb to the atrium via a slowly conducting concealed retrograde accessory pathway with decremental conduction properties.

CLINICAL AND ELECTROCARDIOGRAPHIC FINDINGS

Persistent junctional reciprocating tachycardia usually presents during childhood (<18 years) and in approximately 50% of patients during the first year of life. Using M-mode echocardiography to study the relationship of atrial and ventricular contraction, fetal diagnosis of PJRT has been reported. Persistent junctional reciprocating tachycardia heart rates usually range from 120 to 250 bpm. The ECG (Figure 2) criteria for diagnosis of PJRT are a narrow complex tachycardia, a long R-P interval, and inverted P-waves in leads II, III, AVF, and the left lateral precordial.

The heart rate decreases through a PJRT cycle length increase from 200 to 300 bpm in the first two years of life to 120–150 bpm as the patient matures (Figure 3). Furthermore, the PJRT cycle length increase is principally due to slowing of conduction in the concealed retrograde limb of the reentrant circuit, accounting for 64% of the increase in the tachycardia cycle length (Figure 4). In contrast, the PR interval is relatively stable, accounting for only 36% of the total increase in the tachycardia cycle length (Figure 5).

Although very young patients (<2 years) with PJRT tend to have incessant tachycardia, older subjects with slower retrograde conduction through the accessory pathway also tend to exhibit the incessant form. This apparent contradiction is most likely related to the shorter refractoriness and faster conduction velocity of all cardiac excitable tissue found in

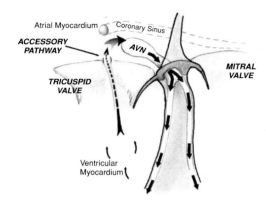

FIGURE 1. Drawing of normal atrioventricular conduction tissue and concealed slowly (only) conducting retrograde accessory pathway supporting PJRT.

young individuals, facilitating, in the presence of the necessary anatomic substrate (i.e., another conducting pathway such as in patients with PJRT), the establishment of a reentrant circuit. As the individual gets older and these properties lengthen, the increasingly slow retrograde conduction of the abnormal accessory pathway compensates for the longer refractoriness of the return tissues (i.e., atrium), allowing for recovery of excitability.

Although PJRT in most patients is incessant, some subjects exhibit a paroxysmal form (Figure 6). One factor that may contribute to this observation is the variation in the cardiac electrophysiologic properties between individuals and even within a single individual at different ages and physiologic state. When a longer refractory period in the atria is coupled with a faster retrograde conduction velocity in the accessory pathway, even for one cycle, the reentry circuit may be interrupted and the tachycardia transiently terminated. Clearly the state of the autonomic nervous system and circulating catecholamines will influence the electrophysiologic properties of the circuit and vary the PJRT rate and pattern. In addition, because of the above noted variation in the electrophysiologic properties of the heart, along with the slowing of heart rate with increasing age, the arrhythmia may not be detected until later in childhood or even adulthood.

Despite the increase in tachycardia cycle length with age, PJRT may continue to be incessant, and thus may lead to tachycardia-mediated cardiomyopathy. The age-related decrease in both the heart rate and the persistence of tachycardia may mask the diagnosis of cardiomyopathy until later in life. Some patients may present with tachycardia-related symptoms or palpitations, as well as decreased exercise tolerance, fatigue, or syncope due to an associated decreased

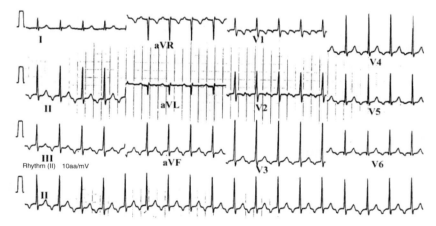

FIGURE 2. 12-Lead electrocardiogram and rhythm (bottom tracing) in a 5-year-old girl with PJRT. Note the negative P-waves in leads II, III, aVF, and the lateral precordial leads along with the long RP interval. Reprinted with permission from Dorostkar, P, et al., Clinical course of PJRT. Journal of the American College of Cardiology Foundation 1999;33(2):3366–375.

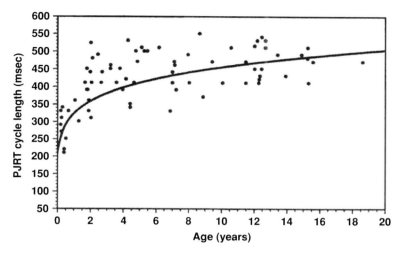

FIGURE 3. Plot of serial tachycardia cycle length versus age years in 9 children with PJRT. Note the trend of an increase in cycle length—slowing of the tachycardia with age. Reprinted with permission from Dorostkar, P, et al., Clinical course of PJRT. Journal of the American College of Cardiology Foundation 1999;33(2):3366–375.

ventricular function. Patients with PJRT may also present with clinical signs of congestive heart failure or echocardiographic findings of impaired ventricular function compatible with cardiomyopathy. Limited series of patients with PJRT have documented diminished left ventricular function in as many as 75% of patients. Signs and symptoms of congestive heart failure may be especially evident in infants with faster PJRT heart rates. The tachycardia will most likely slow as the patient gets older, and some patients have infrequent, spontaneous, and intermittent periods of remission. In some patients, the ventricular shortening fraction or ejection fraction measured by echocardiogram may improve,

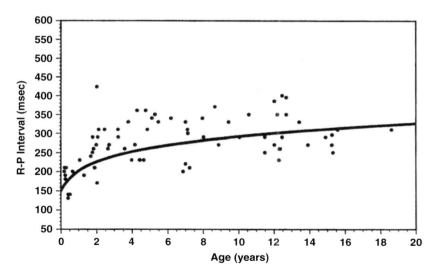

FIGURE 4. Plot of age versus R-P interval demonstrating a slowing in the retrograde conduction limb (increased R-P interval) of the reentrant circuit over time. Reprinted with permission from Dorostkar, P, et al., Clinical course of PJRT. Journal of the American College of Cardiology Foundation 1999;33(2):3366–375.

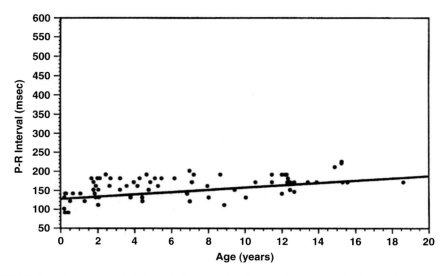

FIGURE 5. Plot of age versus P-R interval demonstrating slow gradual increase in the P-R interval with time. Reprinted with permission from Dorostkar, P, et al., Clinical course of PJRT. Journal of the American College of Cardiology Foundation 1999;33(2):3366–375.

either spontaneously with time as the tachycardia cycle length lengthens with age, as the tachycardia becomes intermittent.

ELECTROPHYSIOLOGY

The tachycardia circuit traverses the normal AV nodal conduction system antegradely and returns to the atrium through a slowly conducting concealed pathway with decremental conduction properties (Figure 1). The incessant nature of the tachycardia is probably due to its characteristically slow, retrograde, and unidirectional (i.e., no antegrade) conducting pathway, allowing the refractory period of the upstream cardiac tissue (atria) to recover so that they always present "an excitable gap" to the reciprocating impulse. Both the conduction velocity and the antegrade effective refractory period of the AV node is typically faster and shorter than the conduction velocity and the retrograde effective refractory period of the accessory pathway, indicating less robust retrograde conduction. The introduction of a ventricular extra-stimulus either from the right ventricular apex or the right ventricular summit during tachycardia while the His bundle is refractory, may either advance, delay, or not affect the next recorded atrial electrogram. A foreshortening of the atrial electrogram interval or termination of the tachycardia, even for only one beat, following the premature ventricular extra stimulus, strongly suggests the presence of an accessory pathway, and a delay in the next recorded atrial electrogram confirms the slowly conducting retrograde accessory pathway, which is typical of PJRT. Rarely, PJRT may co-exist with pre-excitation, as well as other concealed accessory pathways. In older patients (≥ 10 years) with shorter tachycardia

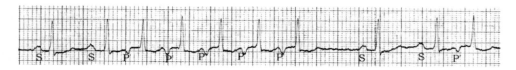

FIGURE 6. Intermittent PJRT. S = sinus impulse; P′ = retrograde P-wave due to reciprocating PJRT impulse.

cycle lengths, retrograde conduction through the accessory pathway is significantly faster in the paroxysmal form of PJRT as compared to the persistent form of PJRT.

TREATMENT

PJRT is often resistant to medical therapy, even with the use of one or multiple antiarrhythmic medications. A few case reports indicate limited efficacy of amiodarone therapy in patients with PJRT; one study reported that control of the incessant nature of the tachycardia was primarily related to the effects of amiodarone on the AV node rather than the accessory pathway associated with PJRT. Because the tachycardia in some patients, aged 59 ± 62 months, may be controlled with variable efficacy with combinations of antiarrhythmic agents, deferment of definitive treatment may be delayed until the patient is older. On the other hand, radiofrequency ablation of the accessory pathway is clearly the treatment of choice when it is judged safe to do so.

ELECTROPHYSIOLOGIC MAPPING AND ABLATION

This slowly conducting accessory pathway of PJRT usually bridges the AV groove in the right posteroseptal area. The most common site of the accessory pathway location was noted to be just superior to the coronary sinus ostium (53%) in one series of 21 patients and right posteroseptal in 25 of 32 patients in another series (76%) (Figure 7).

However, the PJRT accessory pathway has been reported in the right posterior region, mid-septal region, right lateral area, left posterior and left posteroseptal regions, and even left anterior and left lateral locations. Ablation of the earliest atrial activation during tachycardia is associated with successful ablation of the accessory pathway. Even though an accessory pathway potential is usually not noted in the target area, one case report of

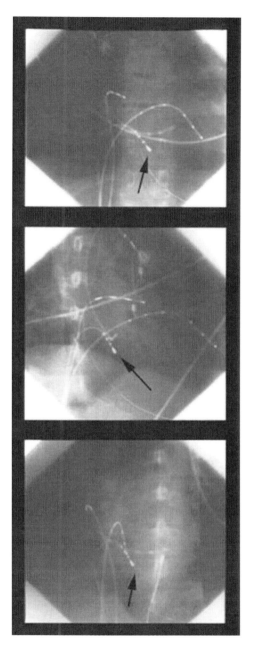

FIGURE 7. Top: Posterior-anterior; middle. Right anterior oblique; bottom left anterior oblique radiographs during an electrophysiology study and radiofrequency ablation in a patient with PJRT. Note: the large electrode-tipped catheter (arrow) with the tip placed in the right posterior septal area.

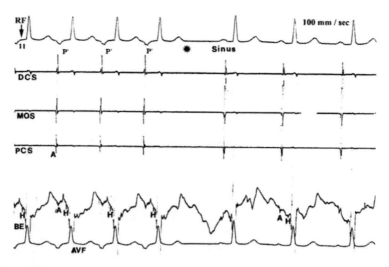

FIGURE 8. Shortly after radiofrequency energy is started (RF arrow), the PJRT is terminated by interruption of the retrograde pathway, indicated by absence (asterisk) of the retrograde P-wave (P').

PJRT documents an accessory pathway potential recording at the site of successful ablation. Another series reports successful ablation in patients as young as 8 weeks of age.

However, because the pathway is usually located in the septal area, the risk of AV block is significant and even somewhat higher in the midseptal area closer to the normal atrioventricular node and His bundle. Although the success of radiofrequency ablation (Figure 8) of the PJRT accessory pathway has been well documented, recurrent pathway conduction is well known. In such patients, repeat electrophysiologic study with ablation is usually successful. The advent of cryoablation may change the ablation strategy in the future (Chapter 22).

intermittent nature of the tachycardia may mask the diagnosis. Thus, the diagnosis may be delayed until tachycardia-related symptoms or palpitations become more apparent. Presentation with heart failure is more common in younger patients. Since the heart rates associated with PJRT will most likely slow with age, radiofrequency ablation may usually be deferred in small children with this tachycardia. Because the tachycardia has a possible spontaneous or intermittent resolution, as well as variable expression of impaired ventricular function, and since it may be effectively and safely treated with ablation, this definitive therapy, whether radiofrequency current or cryotherapy, should be considered only in patients of suitable size and when there are symptoms related to the tachycardia.

CONCLUSION

Persistent junctional reciprocating tachycardia is an arrhythmia that usually presents in infancy or childhood but may not be recognized until adulthood. In older patients, the heart rate may not be sufficiently fast to result in enough symptoms to provoke further examination or evaluation by a physician. Age-related changes in both the rate and the

SUGGESTED READING

1. Coumel P, Cabrol C, Fabiato A, et al. Tachycardie permanent par rhythm reciproque. *Arch Mal Coeur* 1967;60:1830–1864.
2. Scaglione M, Caponi D, Riccardi R, et al. Accessory pathway potential recording in a case of permanent junctional reciprocating tachycardia with decremental conduction localized on the atrial site. *Ital Heart J* 2001;2:147–151.

3. Haissaguerre M, Montserrat P, Warin JF, et al. Catheter ablation of left posteroseptal accessory pathways and of long RP' tachycardias with a right endocardial approach. *Eur Heart J* 1991;12:845–859.
4. Monda V, Scherillo M, Critelli G. Closed chest catheter ablation of an accessory pathway in a patient with permanent junctional reciprocating tachycardia. *J Am Coll Cardiol* 1986;8:740.
5. Perry JC, Garson A, Jr. Complexities of junctional tachycardias. *J Cardiovasc Electrophysiol* 1993;4:224–238.
6. Perticone F, Marsico SA. Familial case of permanent form of junctional reciprocating tachycardia: possible role of the HLA system. *Clin Cardiol* 1988;11:345–348.
7. Medeiros CM, Lucchese FA. Permanent form of junctional reciprocating tachycardia with only even-numbered beats. *J Electrocardiol* 1989;22:249–256.
8. Critelli G. Permanent junctional reciprocating tachycardia. *Pacing Clin Electrophysiol* 1996;19:256–257.
9. Mantovan R, Viani S, Stritoni P. [Permanent junctional reciprocating tachycardia (Coumel type): an unusual location of a retrograde accessory pathway]. *G Ital Cardiol* 1999;29:315–320.
10. Klein GJ, Kostuk WJ, Ko P, Gulamhusein S. Permanent junctional reciprocating tachycardia in an asymptomatic adult: further evidence for an accessory ventriculoatrial nodal structure. *Am Heart J* 1981;102:282–286.
11. McGuire MA, Lau KC, Davis LM, et al. Permanent junctional reciprocating tachycardia misdiagnosed as 'cardiomyopathy'. *Aust N Z J Med* 1991;21:239–241.
12. Critelli G. Recognizing and managing permanent junctional reciprocating tachycardia in the catheter ablation era. *J Cardiovasc Electrophysiol* 1997;8:226–236.
13. Zalzstein E, Zucker N, Sofer S, et al. Successful radiofrequency ablation in a 3-month-old baby with permanent junctional reciprocating tachycardia: a new era in the treatment of incessant life-threatening arrhythmias in infants. *Am J Perinatol* 1995;12:82–83.
14. Arribas F, Lopez-Gil M, Nunez A, Cosio FG. Wolff-Parkinson-White syndrome presenting as the permanent form of junctional reciprocating tachycardia. *J Cardiovasc Electrophysiol* 1995;6:132–136.
15. Gaita F, Haissaguerre M, Giustetto C, et al. Catheter ablation of permanent junctional reciprocating tachycardia with radiofrequency current. *J Am Coll Cardiol* 1995;25:648–654.
16. Dorostkar PC, Silka MJ, Morady F, Dick M, 2nd. Clinical course of persistent junctional reciprocating tachycardia. *J Am Coll Cardiol* 1999;33:366–375.
17. Chien WW, Cohen TJ, Lee MA, et al. Electrophysiological findings and long-term follow-up of patients with the permanent form of junctional reciprocating tachycardia treated by catheter ablation. *Circulation* 1992;85:1329–1336.
18. Jaeggi E, Fouron JC, Fournier A, et al. Ventriculo-atrial time interval measured on M-mode echocardiography: a determining element in diagnosis, treatment, and prognosis of fetal supraventricular tachycardia. *Heart* 1998;79:582–587.
19. Coumel P. Junctional reciprocating tachycardias. The permanent and paroxysmal forms of A-V nodal reciprocating tachycardias. *J Electrocardiol* 1975;8:79–90.
20. Josephson ME SS. *Preexcitation Syndrome, in: Clinical Cardiac Electrophysiology: Techniques and Interpretations*. Lea & Febiger: Philadelphia, 1979.
21. Bartolome FB, Sanchez Fernandez-Bernal C, Torres Feced V. Anterograde decremental conduction by left free wall accessory pathway in the permanent form of junctional reciprocating tachycardia. *Rev Esp Cardiol* 2000;53:878–880.
22. Yagi T, Ito M, Odakura H, Namekawa A, et al. Electrophysiologic comparison between incessant and paroxysmal tachycardia in patients with permanent form of junctional reciprocating tachycardia. *Am J Cardiol* 1996;78:697–700.
23. Giani P, Maggioni AP, Volpi A, et al. Blood levels and electrophysiological effects of intravenous amiodarone in patients with junctional reciprocating tachycardia. Preliminary observations. *Acta Cardiol* 1984;39:9–17.
24. Chen RP, Ignaszewski AP, Robertson MA. Successful treatment of supraventricular tachycardia-induced cardiomyopathy with amiodarone: case report and review of literature. *Can J Cardiol* 1995;11:918–922.
25. Drago F, Silvetti MS, Mazza A, et al. Permanent junctional reciprocating tachycardia in infants and children: effectiveness of medical and non-medical treatment. *Ital Heart J* 2001;2:456–461.
26. Elbaz M, Fourcade J, Carrie D, et al. Atrial insertion of accessory pathways in permanent reciprocating junctional tachycardia. *Arch Mal Coeur Vaiss* 1995;88:1399–1405.
27. Jaeggi E, Lau KC, Cooper SG. Successful radiofrequency ablation in an infant with drug-resistant permanent junctional reciprocating tachycardia. *Cardiol Young* 1999;9:621–623.
28. Aguinaga L, Primo J, Anguera I, et al. Long-term follow-up in patients with the permanent form of junctional reciprocating tachycardia treated with radiofrequency ablation. *Pacing Clin Electrophysiol* 1998;21:2073–2078.

7

Sinoatrial Reentrant Tachycardia

Macdonald Dick II

Sinoatrial reentrant tachycardia, a rare form of supraventricular tachycardia in children, involves reentrant in and around the sinoatrial node. It is postulated that the sinus impulse, as it emerges from the sinoatrial node, enters an area of slow conduction within the atria. It then recycles through peri-sinoatrial nodal tissue to reenter the site of origin. Clinical criteria for establishing this diagnosis are outlined in Table 1 and Figures 1 and 2. This form of supraventricular tachycardia accounts for considerably less than 1% of SVT in children.

When considering the diagnosis of this tachycardia, one must exclude other physiologic or pathologic states that can cause a sustained acceleration of sinus rhythm. It is very unusual for the sinus rate to exceed 220 bpm under any physiologic or pathologic state; the maximal heart rate of an elite athlete can attain is usually 220 bpm under maximal sustained exercise. A good formula for the maximal attainable heart rate is: 208 bpm minus 0.7 times the subject's age.

Hyperthyroidism, febrile illnesses, and hypovolemic states can increase the heart rate (Figure 3) above what the apparent metabolic needs of the patient are at the time. Hyperthyroidism may present with an accelerated sustained heart rate usually about 120 bpm in the absence of other signs or symptoms. Febrile illnesses or hypovolemic states rarely exceed the maximal heart rate appropriate for the patient's age. Inappropriate sinus tachycardia, defined as a sinus tachycardia without discernible cause, is usually seen in anxious older children and adolescents who are experiencing increased stress in their lives. A 24-hour ECG Holter monitor will uncover normal heart rate variability without the findings of abrupt onset or termination of a tachycardia. The post orthostatic tachycardia syndrome (POTS) is a form of dysautonomic regulation resulting in an abrupt accelerated sinus rhythm when the person arises from the supine or sitting position to the standing position. It is a form of neurocardiogenic syncope (Chapter 16).

If an underlying condition is present, treatment is specific for that disorder. The sinoatrial reentrant tachycardia can be converted with either adenosine, intracardiac, or transesophageal overdrive pacing DC cardioversion. Chronic treatment is rarely necessary in children. Radiofrequency ablation has been reported to be successful in adults. Inappropriate sinus tachycardia and POTS can be addressed with beta blockade if simple measures such as fluid and electrolyte replacement are unsuccessful. In contrast to sinoatrial node reentrant tachycardia, inappropriate sinus tachycardia does not respond to adenosine. Although often vexing to the patient,

TABLE 1. Criteria for Diagnosis of Sinoatrial Reentrant Tachycardia

Initiation (and termination) by appropriately timed premature atrial (or sinus) stimuli, either spontaneous or through programmed extrasimulation.

Sinus-like P-waves (positive P waves in limb leads 1, 2, 3, and negative P-wave in avR) during the tachycardia (Figure 1).

Normal PR interval (identical to that during sinus rhythm) with a long RP interval during the tachycardia (Figure 2).

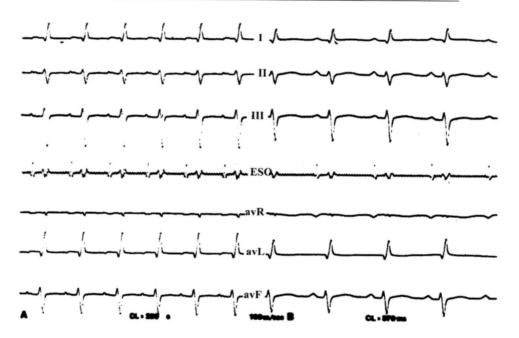

FIGURE 1. Six-limb lead electrocardiogram (top three and bottom three tracings) with an esophageal bipolar lead recording (middle tracing), all at paper speed of 100 mm/sec, in a 2-month-old boy with tricuspid atresia. Panel A on the left demonstrates supraventricular tachycardia at 270 bpm. Panel B demonstrates sinus rhythm at 160 bpm. Note the PR interval and P-wave morphology and axis during the tachycardia are virtually identical to the PR and P-wave morphology and axis observed during sinus rhythm, suggesting the origin of the tachycardia to be around the sino-atrial (sinus) node.

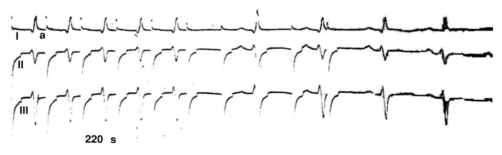

FIGURE 2. Transesophageal atrial overdrive pacing (stimulus rate -s- at a pacing cycle length of 220 ms) in the patient from Figure 1. Successful cardioversion to sinus rhythm is achieved, supporting the diagnosis of a reentrant mechanism in the region of the sinus node (inferred from Figure 1), i.e., sinoartial reentrant tachycardia.

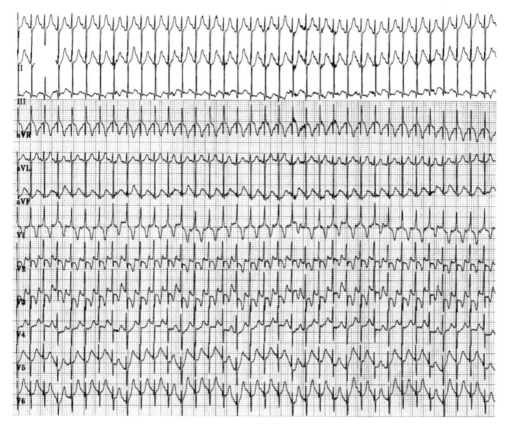

FIGURE 3. Sinus tachycardia at 250 bpm in a 14-month-old girl with a urinary tract infection and fever of 40°C.

especially an adolescent, inappropriate sinus tachycardia appears to be benign. Modification of the sinoatrial node with radiofrequency in children is rarely necessary.

SUGGESTED READING

1. Blaufox AD, Numan M, Knick BJ, Saul JP. Sinoatrial node reentrant tachycardia in infants with congenital heart disease. *Am J Cardiol* 2001;88:1050–1054.
2. Sanders WE Jr, Sorrentino RA, Greenfield RA, et al. Catheter ablation of sinoatrial node reentrant tachycardia. *J Am Col Cardiol* 1994;23:926–934.
3. Still A-M, Huikuri HV, Airaksinen KE, et al. Impaired negative chronotropic response to adenosine in patients with inappropriate sinus tachycardia. *J Cardiovasc Electrophysiol* 2002;13:557–562.
4. Fischer DJ, Gross DM, Garson A Jr. Rapid sinus tachycardia. Differentiation from supraventricular tachycardia. *Am J Disease Children* 1983;137:164–166.
5. Kanjwal MY, Kosinski DJ, Grubb BP. Treatment of postural orthostatic tachycardia syndrome and inappropriate sinus tachycardia. *Curr Cardiol Rep* 2003 5:402–406.
6. Shen WK. Modification and ablation for inappropriate sinus tachycardia: current status. *Card Electrophysiol Rev* 200;6:349–355.

8

Intra-atrial Reentrant Tachycardia—Atrial Flutter

Ian H. Law and Macdonald Dick II

Intra-atrial reentrant tachycardia (IART) is a form of reentrant supraventricular tachycardia in which the reentry circuit is contained exclusively within the atria. This reentrant circuit relies upon anatomical or functional barriers within the atria; these circuits are often exposed and amplified by diseased atrial myocardium.

There are a number of types of IART. Atrial flutter is common in older adults, especially those with heart disease. In contrast, it is uncommon in infants and even less common in children. Other forms of IART include atypical atrial flutter and incisional IART. In the patient with surgically corrected or palliated congenital heart disease, incisional IART is an increasingly more prevalent and important arrhythmia.

This chapter will focus on the diagnosis and management of both typical atrial flutter in the child and incisional IART in the patient following congenital heart disease surgery.

TYPICAL ATRIAL FLUTTER

Typical atrial flutter comprises less than 1% of arrhythmias in children with a normally structured heart, the majority in infancy. At the same time, only 6%–7% of older children with atrial flutter or IART, have a normal structural heart.

Mechanism

Experimental animal studies and clinical observations have shown that in order for typical atrial flutter to develop, suitable substrates are required. Prerequisites include: (1) a corridor or isthmus bounded by structural barriers such as the tricuspid valve, the coronary sinus, the superior or inferior vena cava, or a "scar"; (2) a functional change in the myocardium resulting in prolonged intra-atrial or inter-atrial conduction and increasing atrial refractoriness; and (3) eccentric premature depolarization away from the normal preferentially atrial conduction pathway. Functional barriers, such as the crista terminalis with its anisotropic properties, may play an important role in the initiation and maintenance of atrial flutter.

ECG Characteristics

Typical (type I) atrial flutter has characteristic inverted "sawtooth" P-waves in leads II, III, and AVF, implying inferior to superior

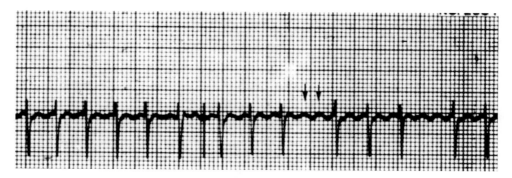

FIGURE 1. Atrial flutter in a 12-hour-old infant. Note the variable ventricular rate and the flutter waves (arrows, cycle length = 140 ms).

atrial activation. In infants, these rates are very fast, ranging from 350 to 500 bpm (Figure 1). Atrial rates in older children and adults are usually around 300 bpm (range: 240–350 bpm). There is usually atrioventricular (AV) conduction, yielding a pulse rate of 150 bpm; in clinical practice, a fixed heart rate of 120 to 150 bpm strongly suggests the presence of atrial flutter with 2:1 AV block (Figure 2). A second form, atypical atrial flutter, has positive P-waves in leads II, III, and AVF, and usually has slower atrial rates around 200 bpm. The upright P-waves imply superior to inferior atrial activation.

Treatment

There are several methods of treating this arrhythmia including observation, antiarrhythmic medications, atrial overdrive pacing (Figure 3), and DC cardioversion. The initial goal of medical management in the young is termination of the arrhythmia. Rate control is less often employed because of: the infrequent recurrence in the child with a normal structural heart; the difficulty in slowing the ventricular rate with cardiac glycosides, beta receptor, or calcium channel blockers (Chapter 21) due to relatively rapid

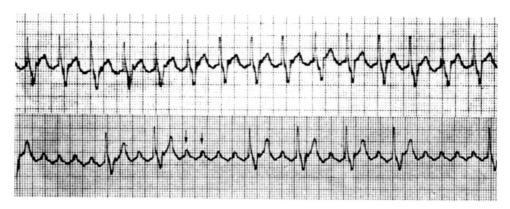

FIGURE 2. Tracings from a 24-hour Holter monitor in a 15-year-old boy with Ebstein's anomaly of the tricuspid valve. Top tracing: The fixed ventricular rate of 130 bpm strongly suggests atrial flutter with 2:1 conduction, which is confirmed in the lower tracing when greater AV block occurs and the flutter waves become more striking and clear (arrows).

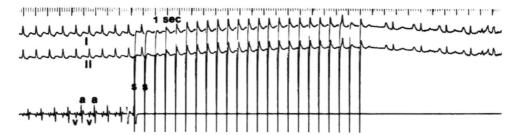

FIGURE 3. Transesophageal overdrive pacing and conversion of SVT in a 2-month-old boy. The pacing cycle length is slightly shorter than the tachycardia cycle length and engages both the orthodromic and antidromic limbs of the reentrant circuit where the impulse traversing both limb collides and self extinguishes, terminating the flutter followed by emergence of the sinus impulse.

conduction in the young AV node; the desire to avoid anti-coagulation in the young; the risk of a negative inotropic effect, especially in the very young; and the success of overdrive pacing or DC cardioversion.

In the newborn and infant, spontaneous conversion is common; however, if the arrhythmia lasts longer than 24 hours or if there is clinical deterioration, either cardioversion or overdrive pacing is warranted. Because the atrial flutter is usually composed of a single highly organized reentrant circuit, minimal biphasic energy (≤ 0.05–1.0 joules/kg) delivered through patches placed in the anterior/posterior position is all that is necessary.

Transesophageal atrial overdrive pacing (Figure 3) offers the advantage of rapid termination with targeted electrical energy (i.e., to the atria not the entire myocardium). As the atrial flutter rates are often 400 bpm or greater in the newborn and infant, very rapid atrial pacing is required. Transesophageal pacing stimuli are delivered at rates 20%–25% faster than the flutter rate until the flutter circuit is entrained (captured). When both the orthodromic limb and the antidromic limb of the reentrant circuit are sufficiently engaged by the pacing impulse (wavefront) so as to collide and self-extinguish within the circuit, the pacing is terminated and sinus (or a slower atrial) rhythm will escape. These steps are repeated with faster pacing rates until there is either termination of the flutter or loss of atrial capture. Transesophageal atrial pacing requires higher pacing outputs (about 10–15 mA at 4–6 ms pulse duration) to capture the atria; these outputs are well tolerated for brief periods.

In the older child, if initially controlling the rate followed by rhythm termination is the primary goal, sotalol and amiodarone, both primarily potassium channel blockers (Chapter 21), have been shown to be successful ($\geq 60\%$ of the time). For the older patient who has recurrence and does not respond to pharmacological management, radiofrequency ablation (infra vide) is an option and has been shown in adult subjects to have an greater success (80% vs. 50%) than pharmacological management.

INCISIONAL INTRA-ATRIAL REENTRANT TACHYCARDIA

Incidence

In contrast to typical atrial flutter, incisional intra-atrial reentrant tachycardia (IART) is an increasingly more prevalent and important arrhythmia in the young cardiac patient following heart surgery. The vast majority of young patients with IART have heart disease (about 84%), most of them following surgery. The incidence of incisional IART is related to both the underlying congenital heart defect and the type of surgical repair. In general, the more complex the defect and the surgery (number of incisions and suture lines within the atria), the higher is the incidence. Thus, in patients between 12 months and 25 years of age with post operative IART, the most common anatomic diagnoses include: D-transposition of the great arteries following an atrial switch operation (Mustard or Senning; about 20%), complex congenital heart disease with single ventricle following Fontan palliation (about 18%), atrial septal defects post repair (about 8%), surgically corrected tetralogy of Fallot and variants (about 8%), dilated cardiomyopathy (about 6%), and repaired AV septal defects (5%). With the development and widespread success of the arterial switch operation replacing the atrial switch operation, the incidence of IART in patients with transposition has markedly decreased. In contrast, as the number of candidates for single ventricle palliation (Fontan sequence) has increased, the prevalence of incisional IART in the pediatric and young adult population remains in the 10% to 40% range. In addition, the incidence of incisional IART appears to increase with age and to be directly related to age at operation, the type of operation (most prevalent in the direct atriopulmonary connection), and to the presence of sinus node dysfunction. Thus, a substantial number of Fontan patients (16%) may develop incisional IART in their lifetime.

ECG Characteristics

In contrast to typical atrial flutter, incisional IART usually has a slower atrial rate—often less than 300 bpm (cycle length $\geq$ 200 msec). Atrial rates of 185–270 bpm (cycle length: 325–220 msec) are seen in Mustard patients and rates of 220–325 bpm (cycle length: 230–470 msec) in Fontan patients. The P-wave morphology is variable, often smaller and fractionated, and at times can be difficult to detect, particularly by single channel recording found in most transtelephonic transmitters. A multiple lead ECG tracing is often necessary to detect the small P-waves. A fixed ventricular rate of 100 to 150 bpm in patients following extensive atrial surgery or abnormality should suggest IART with 2:1 conduction to the ventricles (Figures 4 and 5). Furthermore, as the atrial rate slows, the possibility of 1:1 AV conduction increases, resulting in faster ventricular rates and an increased risk of hemodynamic compromise. Sodium channel blockers may accentuate this risk by

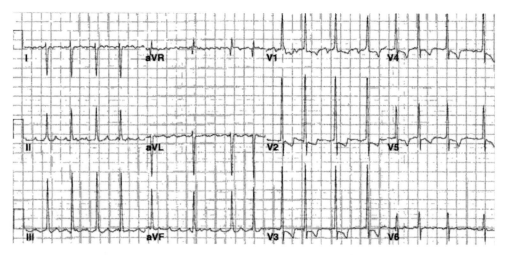

FIGURE 4. 12-Lead electrocardiogram in a 14-year-old girl with transposition of the great arteries 13 years following the Mustard operation and intra-atrial reentry tachycardia. Note the very small flutter waves, especially in the limb and lateral precordial leads.

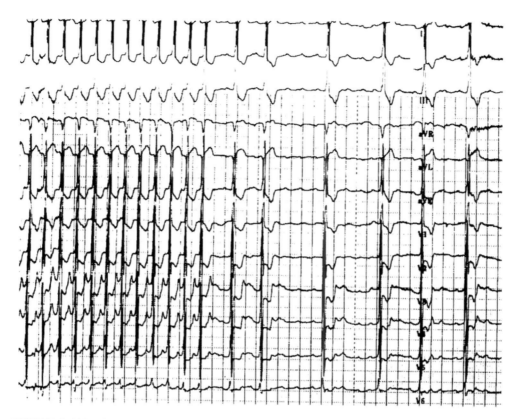

FIGURE 5. 12-Lead electrocardiogram rhythm strip obtained on a 6-year-old Fontan patient during intra-atrial reentrant tachycardia. During 1:1 AV conduction the P-waves are difficult to visualize. Following adenosine infusion, the small, low amplitude P-waves are seen.

slowing the atrial rate while exerting, through a vagally-mediated reflex, a slight but variable increase in AV node conduction, thus possibly facilitating a faster ventricular rate.

Etiology

The primary reason for the differences between the typical form of atrial flutter and incisional IART is the nature of the barriers around which the reentry impulse circulates. In typical atrial flutter, the anatomical barriers are the tricuspid valve annulus, coronary sinus, inferior and superior vena cavae, and crista terminalis. In contrast, incisional IART may involve not only these anatomically fixed structural barriers, but also incisions, suture lines, patches, and baffles within the atria, as well as areas of fibrosis. The atria of patients with heart disease are also often subjected to increased pressure and volume loads, resulting in progressive fibrosis and changes in the electrophysiological properties. A number of experimental animal studies have demonstrated the critical importance of the natural anatomical barriers in typical atrial flutter. More recent studies have experimentally demonstrated the role of surgical incisions and suture lines in both facilitating IART, especially the contribution of the crista terminalis, as well as potentially preventing inducibility of IART. Clinical intracardiac electrophysiologic studies have confirmed the vulnerability of patients for IART after the Senning, Mustard, and Fontan procedures. This susceptibility is most likely due to (1) surgically created atrial discontinuities; (2) focal atrial scarring and conduction block associated with suture lines; (3) atrial fibrosis and thickening caused by pericardial inflammation and abnormal wall stress; (4) abnormal atrial size and anatomy, and (5) increased atrial refractoriness associated with sinus node dysfunction and prolonged duration of atrial activation and may even be present before any of the three operations, especially in Fontan patients.

A major vulnerable site for IART in Mustard or Senning patients is the triangle of Koch (location of AV nodal reentry tachycardia), which is often divided by the baffle so that some or all of the target triangle is separated from the systemic venous pathway and not easily accessible (requiring a trans-baffle approach to the target area). Thus the critical isthmus may either be near the coronary sinus on the systemic venous side or near the tricuspid annulus on the pulmonary venous side or may involve both sites. A second site is the junction between the superior vena cava and systemic venous atrium.

In post Fontan patients with IART, intra-atrial mapping studies have demonstrated several conduction corridors or isthmuses between the atriotomy incision and the crista terminalis, around the atrial septal defect patches or between the inferior vena cava and tricuspid valve.

Electrophysiology Study and Evaluation

Effective electrophysiology study and successful ablation of IART depends on an understanding of the complexity of the congenital heart disease, previous surgical intervention, and available vascular access. Typical atrial flutter requires the basic electrophysiologic pacing and recording equipment, along with a high right atrial catheter and a multi-pole (duo-decapolar) catheter placed around the tricuspid valve—the latter to determine the course and rotation [counter clockwise looking from the right ventricular apex (the usual rotation) or clockwise] of the flutter circuit. In addition, a multi-polar catheter in the coronary sinus (to monitor left atrial activation) or a temporary active fixation lead can serve as a stable electrical reference and pacing site. A His bundle catheter is usually not necessary, although it may be helpful in demarcating the area of the AV node. Often in patients with complex anatomy, the His bundle as well as the ventricular chambers may not be accessible. As the anatomy becomes more complex, the use of three-dimensional mapping and advanced computer-based navigation systems as well as other multi-pole catheters placed between the superior vena cava and inferior vena cava in the abnormal atria is critical to understanding the potential circuits and to intervening with ablation techniques.

In addition, studies for incisional IART usually require more time, personnel, as well as deeper lesion formation. Nonetheless, radiofrequency ablation of both the typical form of atrial flutter and incisional IART utilizes the same principle: creation of a line of block across a critical isthmus of the reentrant circuit.

Goals for electrophysiologic evaluation include: (1) evaluation of possible arrhythmia-related symptoms (palpitations, tachycardia, syncope) in patients with post-surgical congenital heart disease; (2) evaluation of anti-arrhythmic therapy; and (3) mapping of intra-atrial reentrant circuits prior to transcatheter ablation. It is important to note the following: (1) the clinical IART may not be induced in the laboratory; (2) the tachycardia induced may not be the clinical tachycardia; (3) multiple reentry circuits (i.e., tachycardias) may be induced in the catheterization laboratory; and (4) the clinical IART may exhibit different cycle lengths. Careful sorting out of the clinical IART from non-clinical laboratory artifact is essential. A full understanding of the patient's specific cardiac malformation and surgical interventions is indispensable. All surgical notes should be

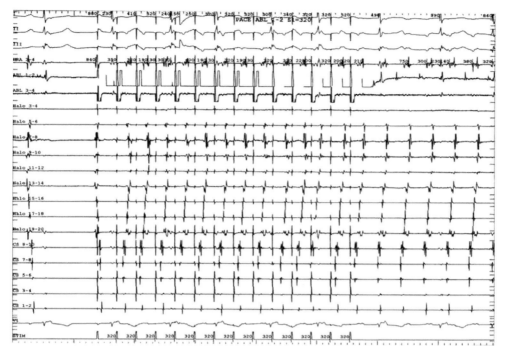

FIGURE 6. Initiation of clinical IART by burst atrial pacing in a young woman following repair of the tetralogy of Fallot.

reviewed for a detailed description of the atrial anatomy, as well as a precise description of the surgical intervention. Attention needs to be directed toward placement of atriotomy incisions, atrial septal defect patches, by-pass cannula insertion sites, atrial appendage anatomy, intra-atrial baffle placement, and coronary sinus position in relation to the intra-atrial baffle. A heart catheterization with appropriate contrast injections can supply anatomic (and hemodynamic) features missing from the surgical record such as the configuration of intra-atrial baffles, the drainage of the coronary sinus, and the size and hemodynamic load of the atria. On occasion, patients will present to the electrophysiology laboratory in IART, permitting immediate evaluation and mapping of the clinical arrhythmia.

When the patients presents in sinus rhythm, a basic electrophysiology study is performed, including assessment of sinus node and AV node function (Chapter 3). This information can prove valuable when considering antiarrhythmic therapy or determining whether an IART is capable of rapid conduction through the AV node. The atrial anatomy can also be determined by endocardial mapping, paying close attention to anatomical landmarks such as the vena cavae, coronary sinus, atriotomy scars, anastomotic sites, the His bundle area, and baffles or patches. Intra-atrial reentrant tachycardia is then induced by either incremental rapid atrial pacing (Figure 6) or programmed atrial extrasimulation with 1–3 extrastimuli. Isoproterenol continuous infusion may be required (0.02–0.05 mcg/kg/min). Upon induction of the tachycardia, a brief hemodynamic assessment should be performed to assure adequate blood pressure and perfusion. When hypotension is encountered, intravenous phenylephrine can be helpful in increasing the blood pressure and reflexively slowing conduction through the AV node. Discontinuation of isoproterenol or a beta-blocking may also be used to prevent rapid AV conduction. The induced tachycardia can be confirmed to be a truly IART by verifying that the AV node and ventricle are not critical to the reentrant circuit, such as observing variable AV conduction and by entraining the tachycardia with atrial pacing.

Entrainment Mapping

Entrainment mapping (Chapter 3) is performed by pacing at 10 to 30 msec less than the spontaneous IART cycle length for at least 10 beats at twice the diastolic threshold. Pacing too rapidly may terminate the tachycardia or initiate a different reentrant tachycardia. Pacing too slowly may not achieve constant capture. Upon consistent capture, the presence of either manifest or concealed entrainment is determined. Manifest entrainment occurs when there is progressive fusion of the paced P-wave (atrial activation) with the tachycardia P-wave (different atrial activation) on the 12-lead electrocardiogram, indicating that the pacing site is remote from the reentrant circuit. Concealed entrainment occurs when there is no surface fusion noted and a delay is seen between the stimulus artifact and the paced P-wave, indicating that the pacing site is within the reentrant circuit. When concealed entrainment is obtained at different pacing cycle lengths, the site of pacing is felt to be within the protected isthmus of conduction. Another criterion of concealed entrainment or whether one is "inside" or "outside" the atrial reentrant circuit is by measuring the post-pacing interval—the interval between the last paced atrial capture and the first spontaneous atrial activation on the same intracardiac electrode pair used for pacing (the mapping pair of electrodes). If the post-pacing interval is equal to or only 10 to 20 msec longer (up to 50–60 msec in slow IART) than the tachycardia cycle length, the pacing site is said to be "inside" the reentry circuit. If the post-pacing interval is greater than 20 msec longer ($\geq$50–60 msec in slow IART) than the tachycardia cycle length, the pacing site is "outside" the atrial flutter circuit (Figure 7). Further support for pacing "inside" the reentrant circuit is present when the atrial activation sequence recorded through multiple closely spaced bipolar pairs of intracardiac electrodes is identical during both pacing and the tachycardia. Finally, an isthmus or narrow corridor of slow conduction is identified by examining the timing of the local electrogram relative to the onset of the surface electrocardiogram P-wave. Optimally, the local electrogram should occur 20–40 msec prior to the onset of the P-wave.

Three-Dimensional Mapping

Due to the complexity of incisional IART circuits, and the possibility of multiple circuits, three-dimensional mapping systems are a very valuable adjunct to traditional mapping techniques. As the available systems each have limitations, it is important to confirm any findings with conventional techniques, such as entrainment or endocardial (relative timing of the local electrogram with the onset of the P-wave) mapping.

Basket Contact Mapping. Basket contact mapping is performed by expanding a multi-electrode open lumen basket array within the atrium. Once the array is fully expanded, the points contacting the atrial myocardium provide simultaneous atrial activation allowing for instantaneous mapping. Reference markers are placed on the basket allowing for correlation of the recorded electrical activity to the fluoroscopic anatomic location. The advantage of this system is that it is relatively inexpensive and has a relatively short learning curve. The disadvantages are its limited spatial resolution and its reliance on electrode contact for signal acquisition. Often the electrodes are greater than 1 cm in distance and, therefore, mapping is required between the electrodes for precise location of an isthmus. There are a variety of basket sizes for different size atria; usually, a suitable one can usually be found for normal atrial anatomy.

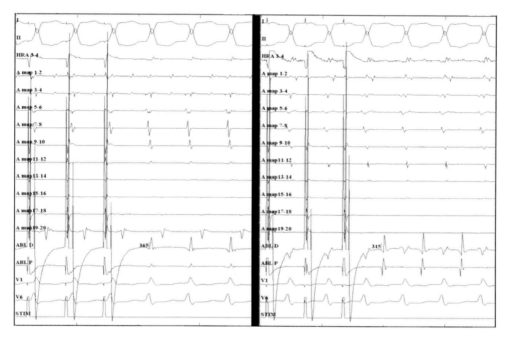

FIGURE 7. Surface electrocardiograms and intracardiac electrograms obtained on a 19-year-old Fontan patient while performing entrainment mapping during IART (atrial cycle length = 310 msec). Pacing through the ablation electrode pair 1-2 at a slightly faster cycle length (300 msec) reveals consistent atrial capture. In the left panel, the post-pacing cycle length (365 msec) is significantly longer than the tachycardia cycle length, which indicates that the site of atrial pacing is not within the reentrant circuit. In the right panel, the post-pacing cycle length (315 msec) is nearly identical to the tachycardia cycle length, which indicates that the pacing site is within the reentrant circuit. HRA = high right atrial reference catheter; A map = atrial mapping catheter—1-2 = distal electrode pair; ABL = ablation catheter; STIM = stimulation channel marker.

Unfortunately, the baskets are suboptimal in very dilated, irregularly shaped atria as seen in those patients with repaired congenital heart disease and IART. This limitation can be minimized by mapping only limited areas of the atrium at any given time, sacrificing the remaining spatial data. Finally, the basket limits movement of other mapping catheters within the atrium.

Contact Three-Dimensional Mapping. Contact three-dimensional mapping consists of an electro-anatomical mapping system (CARTOMERGE Image Integration Module by Biosense Webster, Inc.) using a special catheter with a sensor in the tip, operating within a low intensity magnetic field. The sensor on the tip of the catheter is detected within the magnetic field and its 3-D position within the chamber is linked to local electrical activity in a point by point manner. This system requires a specialized mapping catheter, a reference pad placed on the patient's back, and a stable electrical reference in the chamber of interest. The atria can be mapped during sinus rhythm and during the tachycardia, identifying areas of atrial activation and inactivation (scars and anatomical landmarks that serve as barriers) and the reentrant circuit. This system has the advantage of providing much greater anatomical detail than the basket contact system. Because low amplitude signals suggest abnormal tissue or scar, voltage maps can construct a 3-D image of these areas. The anatomic resolution is less than 1 mm, and after the atrial geometry is cre-

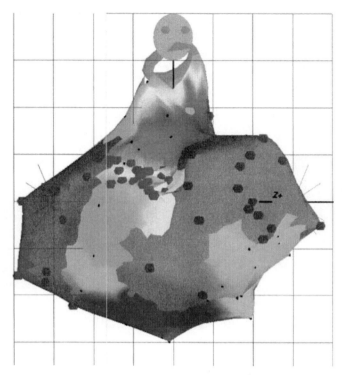

FIGURE 8. Biosense Webster CARTOMERGE electro-anatomic map (right anterior oblique projection—note the direction of the face and eyes of mannequin) of the right atrium during sinus rhythm in a 19-year-old man with pulmonary atresia with intact ventricular septum, 15 years after the Fontan operation. He also had intra-atrial reentry tachycardia. Note the two large scars (gray areas of no voltage) comprising the barriers through which the sinus impulse (the IART wavefront) traveled. The brown dots indicate where radiofrequency ablation lesions were placed to interrupt the circuit successfully. The activation sequence is color-coded: earliest = red, followed by yellow, green, blue, purple—indicating a head to feet direction of the sinus impulse.

ated, the catheter can be manipulated without fluoroscopic guidance.

The major disadvantage of the system is that it requires a stable rhythm to obtain meaningful data. In patients who have multiple reentrant circuits, a new map must be created each time a different cycle length is induced. This disadvantage can be minimized by finding common areas critical to each of the circuits. Another approach is to map in sinus rhythm, identify the barriers, and target these areas for ablation lesions (Figure 8). A second disadvantage is limited catheter selection. However, new catheters with a variety of curves, including a cooled-tip catheter allowing for deeper lesion formation, are now available.

Non-contact Endocardial Mapping. The non-contact mapping technology (Endocardial Solutions, Inc.) utilizes a specialized 64-electrode balloon array to simultaneously record far field electrical activity, compute virtual electrograms, and project them on a 3-D recreation of the chamber of interest using a reversed Laplace transformation. The 3-D model is created by tracing the endocardial surface and computing the distance from the axis of the balloon by triangulation from specialized electrodes on either side of the balloon array and the balloon array itself. After the atrial geometry has been created, non-contact endocardial mapping allows for instantaneous identification of an IART circuit and, as opposed to contact mapping, has the

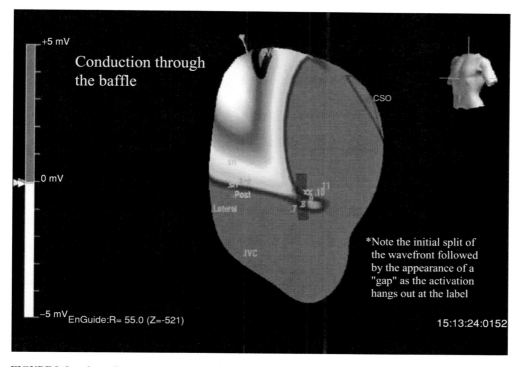

FIGURE 9. Spot frame from a non-contact mapping system in a 16-year-old with IART. The map is of the right atrium in the right anterior oblique projection. Note the "gap" as the wavefront moves caudally across the right free wall and slips through a gap on the baffle. Additional ablation lesions placed at that area eliminated conduction through the gap and the tachycardia.

advantage of identifying a new circuit without creating a new geometry. This can be very advantageous in patients having multiple reentrant circuits with different cycle lengths. Optimally, a common isthmus critical to the different circuits is found. In a similar manner, areas of scar tissue and other anatomical barriers can be mapped during sinus rhythm or atrial pacing to localize a potential isthmus that can then be targeted for ablation. Following creation of an ablation line, pacing on either side of the line while mapping can rapidly identify "gaps" that require additional energy delivery (Figure 9).

As with the contact method system, the catheter can be guided without the use of fluoroscopy, reducing radiation exposure. The non-contact method system also allows for rapid atrial mapping, which shortens the amount of time a hemodynamically unstable patient must remain in the arrhythmia. As opposed to the electro-anatomical contact mapping system, non-contact mapping does not require the use of specialized ablation catheters, allowing freedom of choice when selecting the type of catheter and energy source for ablation (radiofrequency or cryothermal).

A disadvantage of the non-contact method system is the concern of thrombus formation on the deployed balloon array; therefore, activated clotting times should be maintained between 250 and 300 seconds. A second disadvantage is that the balloon array has a relatively large profile and requires a 9 French sheath for introduction. In addition, when the balloon array is fully expanded, it occupies a large amount of space within the right atrium, which can limit ablation catheter manipulation, and possibly impede flow in and out of the atrium. While the resolution of the non-contact method system is quite good, resolution beyond 4–5 cm from the center of the balloon diminishes, limiting its use in the very dilated atrium. A final disadvantage of the non-contact method system is that far field ventricular activity can be interpreted as atrial activity. This can be particularly problematic in patients who have 1:1 AV conduction where the atrial activation may occur during ventricular depolarization or repolarization, leading to false identification of the atrial activation. This problem can be minimized by administering adenosine during the IART, allowing numerous cycles of the atrial activity to be mapped during AV block.

Treatment

Observation. Treatment of IART is dependent on several factors including: associated heart disease, frequency and duration of episodes, associated hemodynamic compromise, patient/parental preference, and failure of previous management strategies. Clearly, common atrial flutter in the infant with no structural heart disease has an excellent prognosis and is unlikely to require long-term management. For patients who have had a single, well tolerated, self-limited episode of incisional IART, observation as the initial strategy after cardioversion is an option. These patients are advised on the importance of early physician notification of recurrence, as sustained tachycardia increases the risk of both hemodynamic compromise and thrombus formation. While the risk of thrombus formation is not known, prophylactic anti-platelet therapy is recommended. As the risk of thrombus formation increases with tachycardia duration, patients on aspirin therapy should have cardioversion attempted within the first 24 hours of IART. An echocardiogram, often transesophageal, is useful to exclude a preexisting thrombus before cardioversion. If one is seen, warfarin therapy should be instituted for 3 weeks prior to cardioversion if the clinical state permits.

Medical Management. As incisional IART tends to be a progressive disease, most patients will require antiarrhythmic therapy at some point (Chapter 21). Unfortunately, as this can be a difficult arrhythmia to control, often more than one antiarrhythmic agent is needed—as many as 2–3 antiarrhythmic drugs/patient may yield only "poor results." In patients who have infrequent episodes of IART or have had significant hemodynamic compromise due to rapid AV conduction, rate control should be considered in addition to anti-platelet or anti-coagulation therapy. Beta-blocker and calcium channel blockers (dilatizam) are first-line agents. Beta-blockers may suppress sinus node function in an at-risk population, as well as possibly depressing ventricular function. Calcium channel blockers can exert a negative inotropic effect, but usually are well-tolerated.

For patients with frequent and/or poorly tolerated episodes of IART, rate and rhythm control is desirable. Anticoagulation therapy should also be considered for at least 3–6 months until appropriate arrhythmia suppression is obtained. The mainstays of rhythm control agents include sodium channel blockers and potassium channel blockers as individual agents or in combination. In part, the antiarrhythmic mechanism of action of sodium channel blockers is to slow myocardial conduction velocity (Chapter 21). While this can help in preventing reentrant atrial arrhythmias, it can also adversely slow an IART, converting what once was a hemodynamically well-tolerated rapid atrial tachycardia with 2:1 AV conduction to a slower reentrant tachycardia with 1:1 AV conduction and a paradoxically faster ventricular rate. When a sodium channel blocker is used, consideration should be given to the addition of a rate control agent to protect against this adverse effect.

Potassium channel blockers, sotalol and amiodarone, which prolong the myocardial effective refractory period, are valuable agents in preventing IART. They also have the beneficial effect of slowing AV conduction. Two of the more commonly used potassium channel blockers include sotalol and amiodarone. Both have proven efficacy, however sotalol has a higher risk of proarrhythmia while amiodarone has a greater risk of systemic adverse reactions (Chapter 21).

Radiofrequency Ablation. Radiofrequency ablation, although an effective tool in the management of incisional IART, is not as acutely successful (75%, with a recurrence as high as 50%) as ablation for typical atrial flutter. This lower success rate and higher recurrence rate is most likely due to the thicker, diseased atrial myocardium, a greater number of anatomical barriers and circuits, and a high prevalence of concurrent sinus node disease. The goal of ablation is to create a line of block within the critical isthmus of the intra-atrial circuit. Most often this requires a long linear lesion line; infrequently, a point lesion can be created within a narrow isthmus. In patients who have numerous intra-atrial length tachycardia circuits, a transcatheter atrial Maze procedure, creating a number of lines of block between several anatomical barriers, can not only address the current clinical arrhythmia but also areas of potential future reentrant circuits.

Technique. The key to success in ablating IART is extensive preparation. As patients with congenital heart disease can have multiple inducible circuits, it is essential to have a 12-lead ECG of the clinical tachycardia, or even more advantageous, have the patient present to the laboratory in the clinical arrhythmia. Each induced arrhythmia should be compared to the clinical arrhythmia, matching P-wave morphology and cycle length as closely as possible.

As discussed above, mapping of the arrhythmia can be accomplished by several techniques. Three-dimensional mapping is helpful in identifying the intra-atrial circuit but should not replace other techniques, such as entrainment mapping. There is a direct correlation between the thoroughness of mapping and the success of the ablation procedure. The sinus node and AV node in patients with complex congenital heart disease may be markedly displaced and therefore placed at risk of inadvertent injury. In addition, the sinus node cannot be mapped during the IART and therefore the anatomical position must either be assumed, or mapped in sinus rhythm. The phrenic nerve lies along the lateral wall of the right and left atrium and can be injured in the process of performing long linear lesions in this area. High output pacing on the lateral atrial wall prior to ablation, observing for phrenic nerve stimulation, can help prevent collateral damage to this structure.

Other areas of interest to be identified when mapping the atrium include the vena cava and coronary sinus orifices, the AV valve annulus, the crista terminalis, intra-atrial baffles (e.g., Mustard operation), atrial septal defect and other patch material, anastomotic sites (e.g., atriopulmonary connection Fontan operation), and atriotomy scars. Split potentials can often be seen along linear scars or the crista terminalis. Fractionated atrial electrograms can be found in areas of diseased myocardium, and are often within areas of slow conduction such as an isthmus. Voltage can detect scarred areas (low signal amplitude) as well as areas of large electrograms with rapid onset, which are indicative of thick atrial tissue. Local atrial electrograms that precede the onset of the P-wave by only 20–50 msec are suggestive of an exit area of the tachycardia in the isthmus and therefore are an area to target for ablation.

After the IART circuit has been characterized, attention can be directed toward performing the radiofrequency ablation. In typical atrial flutter, the usual site for radiofrequency ablation is the isthmus between the tricuspid valve annulus and the inferior vena. As this is the most prevalent isthmus for all forms of IART, and a relatively easily accessible area, a prophylactic line of block in this area is also prudent when ablating intra-atrial reentrant circuits in patients with complex anatomy. A successful lesion should reduce the amplitude of the atrial electrogram

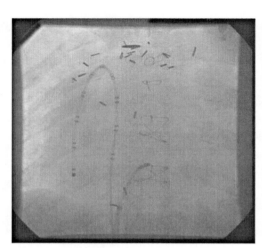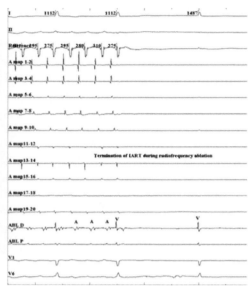

FIGURE 10. AP cine image (left panel) and surface electrocardiogram and intracardiac electrograms obtained on a 22-year-old Fontan patient while applying radiofrequency energy during IART. The radiofrequency energy terminates the tachycardia resulting in a prolonged period of atrial asystole and a junctional escape rhythm. A Map = atrial mapping catheter—1-2 = distal electrode pair; ABL = ablation catheter; A = atrial electrogram; V = ventricular electrogram.

by 90% and often terminates the tachycardia (Figure 10). Criteria for an effective "line of block" include the inability to induce the atrial reentrant tachycardia and the presence of bi-directional conduction block. Bidirectional conduction block is confirmed by pacing within the proximal coronary sinus and observing prolonged conduction to the low lateral right atrium as conduction proceeds counterclockwise around the tricuspid valve, and similarly, delayed conduction to the proximal coronary sinus when from pacing in the low lateral right atrium, with clockwise conduction around the tricuspid valve to the coronary sinus (viewing the tricuspid orifice from the right ventricular apex). Three-dimensional mapping can assist in identifying areas of breakthrough conduction along the line of block, allowing targeted ablation lesions to close the gap (Figure 9).

In patients with incisional IART, the isthmus block between the tricuspid valve and inferior vena cava is seldom sufficient to treat this arrhythmia. Electrophysiologic investigations have demonstrated that the tachycardia in atrial switch and Fontan patients is often AV node reentrant tachycardia. The optimal site to target for radiofrequency ablation is one that is narrow, has low voltage electrograms, exhibits concealed entrainment, has local electrograms that precede the P-wave by 20 or more milliseconds, and the line of block terminates the tachycardia. When these criteria cannot be met, a next best step is to find a narrow area between two barriers where a linear lesion can be made. In patients that have multiple inducible intra-atrial reentrant circuits, a shared isthmus can sometimes be found (dual figure of eight circuits). On occasion, when IART cannot be induced or sustained in the laboratory, substrate mapping (i.e., identifying barriers or scars and possible corridors) can serve as a surrogate to mapping during the intra-atrial reentrant circuit, using similar criteria to identify the area for ablation.

In patients with multiple inducible intra-atrial reentrant circuits and no clear isthmus, an endocardial "Maze" procedure can be performed. This is accomplished by creating lines of block through potential areas of reentrant conduction. Typical lesion sets include a line from the superior vena cava to the inferior vena cava, the tricuspid valve to the inferior vena cava with or without incorporating the coronary sinus os, the atrial septal defect patch/scar to the line adjoining the vena cavae, and the right atrial appendage to the line adjoining the vena cavae. Other lesion lines may be required to disrupt potential circuits around anatomical boundaries depending on the individual electro-anatomic characteristics.

After identifying the critical isthmus and/or other areas to be targeted for disruption of the IART circuit(s), lesions can be created. To be effective, a non-interrupted transmural lesion is required. Unfortunately, in patients with incisional IART, there are many obstacles that stand in the way of lesion formation. The endocardium is often thick, reducing the likelihood of a transmural lesion, and fibrotic, limiting the transmission of the radiofrequency energy. The low blood flow within the dilated atria also limits the blood cooling effects on the catheter tip.

To overcome these obstacles, several techniques can be undertaken. Usually, a larger tip catheter (8 mm) is required. In addition, a high output radiofrequency generator may be required in high flow areas (70 watts or greater). However, if the blood flow velocity in the area of lesion formation is low, temperature regulated low energy output is usually more effective in delivering deeper lesions. To assist in keeping the radiofrequency ablation tip cool in low flow areas, several techniques have been tried. The most straightforward is to place the ablation catheter within a long sheath and infuse saline during energy delivery; however, the desired results are achieved inconsistently and ablation catheter manipulation is limited by the long sheath. A second solution is to use an irrigated tip catheter that infuses saline out of the same catheter tip that is delivering energy; however, this technique has also been less reliable. Both of these techniques also require fluid boluses, which may be undesirable.

An intracatheter cooling system has been developed that does not require fluid boluses to the patient. Fluid flow within the catheter tip enables more stable temperatures, diminishing the possibility of coagulum formation at the tip and allowing for deeper energy penetration. While favorable results have been reported, this technology requires specialized catheters and infusion equipment, adding expense and increasing the complexity of the procedure. Also, when incorporating techniques that assist in deeper radiofrequency lesion formation, the potential for adverse effects such as myocardial perforation and damage to surrounding tissue such as the phrenic nerve is increased.

With the advent of cryothermal ablation systems, lesion formation within low flow areas may be more effective. Initial work in the cryoablation of typical atrial flutter has proven effective but has emphasized the need for larger tip catheters for deeper lesion formation. Cryoablation has the added advantage of less collateral damage, diminishing the concern of injury to surrounding structures such as the coronary arteries and phrenic nerve. The ability to "cryomap" at lower temperatures ($-20°C$ to $-30°C$) also reduces the probability of damaging the sinus or AV nodes.

In rare circumstances when antiarrhythmic and atrial ablative therapy is unsuccessful, consideration may be given to AV node ablation (Figure 11), especially in patients who have rapid AV conduction. Accessing the AV node can also be challenging in complex congenital heart disease as it is often anatomically displaced or protected by intra-atrial patches/baffles. Such is the case in patients who have undergone certain variants of the atrial switch or Fontan operation. Another disadvantage of this therapy is that the patient becomes pacemaker dependent, and while many patients are candidates for transvenous pacemakers, those with single ventricle physiology or complex systemic venous anatomy require epicardial lead placement.

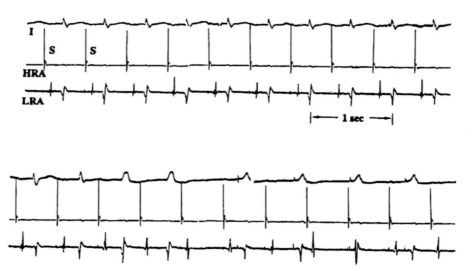

FIGURE 11. Tracing from an 18-year-old girl with transposition of the great arteries, Mustard operation, and intractable atrial arrhythmias. The tracing is recorded during delivery of radiofrequency energy (beginning at the second from the last beat on the top tracing) and with the ablation electrode positioned where the His bundle electrogram was recorded. The top and bottom tracings are continuous. Note AV block beginning with the third A electrogram in the bottom tracing, progressing to complete heart block and with a temporary transvenous pacemaker escape rhythm. Reprinted with permission from Russell, M.W et. al. Catheter interruption of atrioventricular conduction using radiofrequency energy in a patient with transposition of the great arteries. Pacing Clin Electrophysiol 1995;18: 113–116.

Surgical Management. Patients with incisional IART may have associated cardiac structural abnormalities resulting in significant hemodynamic compromise that requires surgical correction. In these patients, concurrent surgery for both arrhythmia management and correction of the structural abnormalities is a possible approach, such as concurrent arrhythmia surgery performed during Fontan conversion to a total cavopulmonary anastomosis Fontan. The arrhythmia surgery consists of an isthmus cryoablation and a right-sided maze (or a Cox III maze for patients with atrial fibrillation).

Given the encouraging results of surgical management, consideration for this surgical strategy is an excellent option after failed medical and/or catheter based management. Because of the inherit risks associated with an open heart procedure, surgical arrhythmia management is best reserved for those patients who require surgical intervention for structural problems or failed transcatheter ablation.

Pacemaker Management. Pacemakers have long been used for the treatment of sinus and AV node disease that is associated with congenital heart disease (Chapter 17). The addition of atrial anti-tachycardia pacemakers to the treatment options for IART is relatively new. Initially, only a single chamber (atrial) device was available. This device, while effective, could not be used to treat patients with AV node disease and was limited by its ability to only sense atrial activity. An early concern was that rapid atrial pacing might induce ventricular arrhythmias. The more recent atrial anti-tachycardia device is a dual-chamber pacemaker, allowing for discrimination between atrial and ventricular arrhythmias and determination of whether there is 1:1 AV conduction. This newer pacemaker system not only can treat the atrial arrhythmias with overdrive pacing protocols but also has pacing prevention algorithms that provide rate stabilization, minimizing the triggers for atrial arrhythmias. An anti-

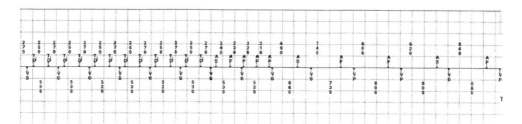

FIGURE 12. Pacemaker strip from a 25-year-old woman palliated with a Senning operation. The rhythm at the beginning of the pacemaker strip is suggestive of IART having 2:1 atrioventricular conduction. The pacemaker appropriately detects the arrhythmia (TD notations), and successfully atrial overdrive paces (AP) the patient into a sinus (AS), then an atrial paced rhythm.

tachycardia pacing protocol can convert up to 50% of atrial tachycardias and may be even more effective in patients who have had the Mustard or Senning operations.

The overall success of this atrial anti-tachycardia pacemaker in the congenital heart disease population is similar to that of the adult population. The atrial anti-tachycardia pacemaker is particularly suited for patients with indications for atrial or dual chamber pacing and IART (Figure 12). The applicability of this device for the treatment of IART is limited by its dependence on 2:1 or greater AV conduction for the classification and treatment of the atrial arrhythmia. When there is 1:1 AV conduction, the overdrive pacing protocol is disabled. Modifications of future devices may include the ability to manually override the disabled overdrive pacing therapies in patients with confirmed IART and a low risk of rapid AV conduction. Other advances may improve the atrial arrhythmia detection algorithm.

Prevention

Traditionally, the treatment of IART has been reactive. As our understanding of this atrial arrhythmia has grown, a greater effort has been put forth in the area of prevention. The idea of modifying surgical procedures to prevent arrhythmias is not new; in fact, modifications of the atrial switch and Fontan operations were in part driven by the desire to prevent sinus node disease and IART.

Unfortunately, these past attempts have not lived up to expectations.

A novel strategy is now being investigated. Based upon the experimental data from animal studies, a prospective study is underway that investigates the utility of placing a prophylactic strategic incision during the lateral tunnel Fontan completion. The strategic incision interrupts a potential isthmus site. At present, enrollment has been completed and no deleterious effects have been found by incorporating the incision into the operation. Long-term follow-up will be required to determine the efficacy of this preventive strategy.

Prognosis

The prognosis of typical atrial flutter in the pediatric population is excellent; the majority of infants do not have recurrence beyond their first episode. In contrast, incisional IART, when untreated, is progressive and can be associated with significant morbidity and mortality. In one study of atrial flutter in the young, the congenital heart disease population had a mortality rate of 5% in patients with medically controlled IART, and 20% in medically uncontrolled patients. Antiarrhythmic agents can be effective in both rate and rhythm control but often multiple medications are required, and proarrhythmic effects and other adverse reactions limit usage and decrease compliance.

Ablative therapy for typical atrial flutter, when needed, has a high success rate.

Ablation for IART averages 75% short-term success rates and up to 50% recurrence. Advances in mapping and ablation technology may greatly improve the long-term success rates. Arrhythmia surgery is advocated strongly by several groups, but due to its inherent increased risk, this procedure, in our experience, is limited to patients who require an open heart procedure or who fail transcatheter techniques.

Atrial anti-tachycardia pacemaker therapy for patients with IART is particularly suitable if other indications (sick sinus syndrome) for pacing are present. While the reported arrhythmia treatment success rate in the repaired/palliated congenital heart disease population is less than desired (54%), atrial rate stabilization algorithms may decrease the atrial arrhythmia burden. Also, patients with the atrial switch have shown a greater success rate and therefore may be more suitable for this treatment.

Ideally, prevention of the IART is the goal. Addressing the reported risk factors, modifying surgical approaches (i.e., minimizing injury to the sinus node), and incorporating strategic incisions through well-described isthmus regions have the potential for reducing the incidence of IART. Long-term follow-up of current studies will address this issue.

SUGGESTED READING

1. Mendelsohn A, Dick M, Serwer GA. Natural history of isolated atrial flutter in infancy. *J Pediatr* 1991;119(3):386–391.
2. Garson A Jr, et al. Atrial flutter in the young: a collaborative study of 380 cases. *J Amer Coll Cardiol* 1985;6(4):871–878.
3. Watson RM and Josephson ME. Atrial flutter. I. Electrophysiologic substrates and modes of initiation and termination. *Am J Cardiol* 1980;45(4):732–741.
4. Pfammatter JP, Paul T. New antiarrhythmic drug in pediatric use: sotalol. [Comment.] *Pediatr Cardiol* 1997;18(1):28–34.
5. Perry JC, et al. Pediatric use of intravenous amiodarone: efficacy and safety in critically ill patients from a multicenter protocol. *J Amer Coll Cardiol* 1996. 27(5):1246–1250.
6. Campbell RM, Dick M, Jenkins JM, et al. Atrial overdrive pacing for conversion of atrial flutter in children. *Pediatrics* 1985;75:730–740.
7. Rowland TW, et al. Idiopathic atrial flutter in infancy: a review of eight cases. *Pediatrics* 1978;61(1):52–56.
8. Casey FA, et al. Neonatal atrial flutter: significant early morbidity and excellent long-term prognosis. *Am Heart J* 1997;133(3):302–306.
9. Natale A, et al. Prospective randomized comparison of antiarrhythmic therapy versus first-line radiofrequency ablation in patients with atrial flutter. *J Amer Coll Cardiol* 2000;35(7):1898–1904.
10. Kanter RJ, et al. Radiofrequency catheter ablation of supraventricular tachycardia substrates after Mustard and Senning operations for d-transposition of the great arteries. *J Amer Coll Cardiol* 2000;35(2):428–441.
11. Van Hare GF, et al. Mapping and radiofrequency ablation of intra-atrial reentrant tachycardia after the Senning or Mustard procedure for transposition of the great arteries. *Am J Cardiol* 1996;77(11):985–991.
12. Triedman JK, et al. Intra-atrial reentrant tachycardia after palliation of congenital heart disease: characterization of multiple macroreentrant circuits using fluoroscopically based three-dimensional endocardial mapping. *J Cardiovasc Electrophysiol* 1997;8(3):259–270.
13. Rosenblueth A, Garcia Ramos A, Studies on flutter and fibrillation. Part II The influence of artificial obstacles on experimental auricular flutter. *Am Heart J* 1947;33:677–684.
14. Frame LH, et al. Circus movement in the canine atrium around the tricuspid ring during experimental atrial flutter and during reentry in vitro. *Circulation* 1987;76(5):1155–1175.
15. Kurer CC, Tanner CS, Vetter VL, Electrophysiologic findings after Fontan repair of functional single ventricle. *J Amer Coll Cardiol* 1991;17(1):174–181.
16. Law IH, et al. Inducibility of intra-atrial reentrant tachycardia after the first two stages of the Fontan sequence. *J Amer Coll Cardiol* 2001;37(1):231–237.
17. Gandhi SK, et al. Lateral tunnel suture line variation reduces atrial flutter after the modified Fontan operation. *Ann Thorac Surg* 1996;61(5):1299–1309.
18. Durongpisitkul K, et al. Predictors of early- and late-onset supraventricular tachyarrhythmias after Fontan operation. *Circulation* 1998;98(11):1099–1107.
19. Fishberger SB, et al. Factors that influence the development of atrial flutter after the Fontan

operation. *J Thorac & Cardiovasc Surg* 1997; 113(1):80–86.
20. Olgin JE, et al. Role of right atrial endocardial structures as barriers to conduction during human type I atrial flutter. Activation and entrainment mapping guided by intracardiac echocardiography. *Circulation* 1995;92(7):1839–1848.
21. Balaji S, et al. Management of atrial flutter after the Fontan procedure. *J Amer Coll Cardiol* 1994;23(5):1209–1215.
22. Baker BM, et al. Catheter ablation of clinical intra-atrial reentrant tachycardias resulting from previous atrial surgery: localizing and transecting the critical isthmus. *J Amer Coll Cardiol* 1996;28(2):411–417.
23. Triedman JK, Alexander ME, Love BA, et al. Influence of patient factors and ablative technologies on outcomes of radiofrequency ablation of intra-atrial re-entrant tachycardia in patients with congenital heart disease. J Am Coll Cardiol 2002;39(11):1827–1835.
24. Triedman JK, Bergau DM, Saul JP, et al. Efficacy of radiofrequency ablation for control of intraatrial reentrant tachycardia in patients with congenital heart disease. *J Am Coll Cardiol* 1997;30(4):1032–1038.
25. Kalman JM, et al. Ablation of 'incisional' reentrant atrial tachycardia complicating surgery for congenital heart disease. Use of entrainment to define a critical isthmus of conduction. *Circulation* 1996;93(3):502–512.
26. Van Hare GF, et al. Electrophysiologic study and radiofrequency ablation in patients with intracardiac tumors and accessory pathways: is the tumor the pathway? *J Cardiovasc Electrophysiol* 1996;7(12):1204–1210.
27. Mavroudis C, et al. Fontan conversion to cavopulmonary connection and arrhythmia circuit cryoablation. *J Thorac & Cardiovasc Surg* 1998;115(3):547–556.
28. Russell MW, Dorostkar PC, Dick M 2nd, et al. Catheter interruption of atrioventricular conduction using radiofrequency energy in a patient with transposition of the great arteries. *Pacing Clin Electrophysiol* 1995;18(1 Pt 1):113–116.
29. Urcelay G, Dick M 2nd, Bove EL, et al. Intraoperative mapping and radiofrequency ablation of the His bundle in a patient with complex congenital heart disease and intractable atrial arrhythmias following the Fontan operation. *Pacing Clin Electrophysiol* 1993;16(7 Pt 1):1437–1440.
30. Theodoro DA, et al. Right-sided maze procedure for right atrial arrhythmias in congenital heart disease. *Ann Thorac Surg* 1998;65(1):149–153; discussion 153–154.
31. Stephenson EA, et al. Efficacy of atrial antitachycardia pacing using the Medtronic AT500 pacemaker in patients with congenital heart disease. *Am J Cardiol* 2003;92(7):871–876.
32. Mavroudis C, et al. Total cavopulmonary conversion and maze procedure for patients with failure of the Fontan operation. *J Thorac & Cardiovasc Surg* 2001;122(5):863–871.
33. Darbar D, Olgin JE, Miller JM, Friedman PA. Localization of the origin of arrhythmias for ablation: from electrocardiology to advanced endocardial mapping systems. *J Cardiovasc Electrophysiol* 2001;12(11):1309–1325.

9

Atrial Fibrillation

Peter S. Fischbach

Atrial fibrillation is a supraventricular tachycardia characterized by a disorganized activation and contraction of the atrium. The characteristic "irregularly irregular" rhythm is generated by the rapid irregular bombardment of the atrioventricular (AV) node with electrical impulses emanating from the atrial myocardium with variable conduction of the atrial impulses through the AV node (Figure 1). The ventricular response is dependent on the ability of the AV node to transmit electrical signals (i.e., its refractory state). This is determined by its inherent electrophysiologic properties as well as autonomic tone.

INCIDENCE

Atrial fibrillation is the most common rhythm disorder seen in man, with prevalence in the general population estimated at 0.4%. There is a significant age-related discrepancy in the distribution of the arrhythmia with 6% of the population greater than 80 years of age suffering from atrial fibrillation. Atrial fibrillation remains an extremely rare arrhythmia in the pediatric population, although there have been a few reports of familial forms of atrial fibrillation (Chapter 18).

ASSOCIATED DISEASE

Atrial fibrillation commonly occurs as a co-morbid condition with other cardiovascular abnormalities. It is frequently associated with congestive heart failure, mitral valve stenosis and insufficiency, hypertension, and hyperthyroidism. Atrial fibrillation is also associated with Wolff-Parkinson-White syndrome. Atrial fibrillation occurring in conjunction with Wolff-Parkinson-White syndrome is a potentially life-threatening situation due to the lack of decremental conduction through the accessory pathway. Some accessory pathways are capable of rapidly conducting the atrial signals to the ventricle leading to ventricular fibrillation (Chapter 20, Figure 3). Following elimination of the accessory pathway, either surgically or via transcatheter ablation, atrial fibrillation resolves in a large portion of these patients. Atrial fibrillation also may be associated with several other forms of supraventricular tachycardia in young patients. Our experience has been that young patients presenting with atrial fibrillation will frequently have other underlying mechanisms of SVT such as AVNRT, AVRT, ectopic atrial tachycardia, or atrial flutter. Ablation of these other tachycardia substrates results in the resolution of atrial fibrillation.

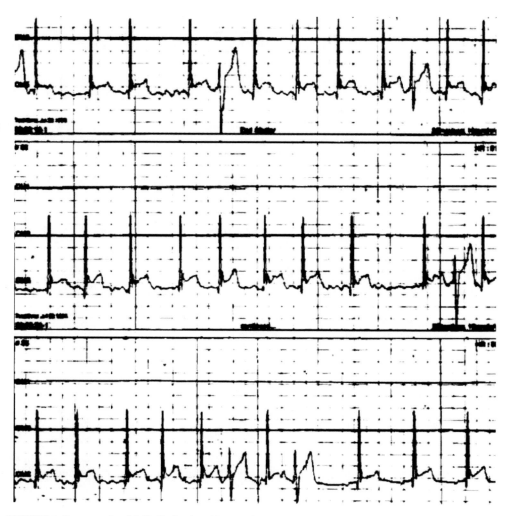

FIGURE 1. Paroxysmal atrial fibrillation in a 17-year-old boy. Note the low amplitude fast waveforms (atrial fibrillatory waves) between the "irregularly irregular" QRS intervals. There are a few different QRS morphologies due to aberrant conduction. The atrial fibrillation spontaneously terminates (bottom tracing) as it frequently does in the paroxysmal form in patients without structural heart disease.

MECHANISM

The pathophysiologic mechanisms underlying the initiation and perpetuation of atrial fibrillation remain under investigation (Figure 2). It is possible that the mechanisms that support the initiation of atrial fibrillation are different from those that sustain the arrhythmia. The three most likely mechanisms include the multiple wavelet theory, single-circuit reentry, and multiple-circuit reentry. In the first model, multiple reentrant waves continuously circulate through the atrium using migrating central cores of refractory tissue to rotate around. The multiple wavelets circulate throughout the atrium following ever-changing pathways created by myocardium recovering from refractoriness. A critical mass of atrial tissue in this model is necessary to support a minimum number of wavelets in order to sustain atrial fibrillation. This has been cited as a rational for the low

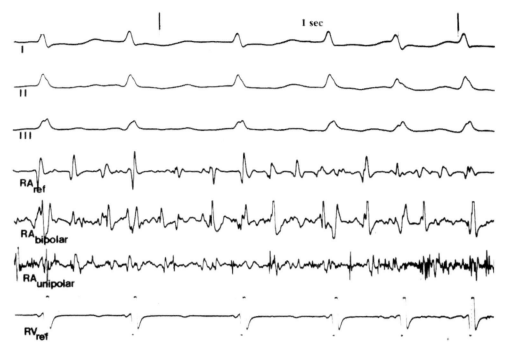

FIGURE 2. Three electrocardiographic leads and 4 intracardiac electrograms [3 in the right atrium (RA) and 1 from the right ventricular apex (RV)] illustrating the multi-phasic variable amplitude of the atrial fibrillation wavefronts.

incidence of atrial fibrillation in infants and children, as well as in small mammals (Chapters 2 and 3). The second proposed mechanism for atrial fibrillation is the single circuit reentrant model. In this model, a single "mother rotor" serves as the hub with multiple accessory circuits emanating from it. The third potential mechanism is atrial fibrillation arising from a rapidly discharging ectopic focus with fibrillatory conduction. This mechanism has come into favor with recent findings of excitatory loci occurring within the pulmonary veins. Recent clinical experience has demonstrated that focal radiofrequency ablation within the pulmonary veins as well as pulmonary vein isolation has successfully treated atrial fibrillation.

Sustained atrial fibrillation produces profound changes within the atrial myocardium, creating a substrate more conducive to supporting sustained atrial fibrillation. The electrical and structural remodeling of the atrium includes increasing fibrosis, as well as alterations in the expression of gap junctions and ion channels. These changes alter the mechanical and electrical properties of the atrial tissue by slowing the conduction velocity and shortening the refractory period in the atrium, leading to tissue that is more likely to sustain atrial fibrillation.

THERAPY

Atrial fibrillation has proven to be an extremely difficult arrhythmia to manage with pharmacological therapy, achieving far less than optimal results. Multiple large clinical trials have been performed using nearly all of the currently available antiarrhythmic drugs, each of them with disappointing results. Large trials have also been performed comparing rate control (inhibition of conduction through the AV node) with rhythm control (restoration of sinus rhythm) for the ideal therapy. These studies have demonstrated no significant difference between these treatment options using total mortality, congestive heart failure, and re-hospitalization as end points. Much of the difficulty with pharmacological therapy lies in the ever-present risk of ventricular proarrhythmia.

There are several non-pharmacological treatment strategies that have recently emerged. The Cox maze procedure creates multiple surgical incisions that are then repaired in the atrium in an attempt to channel the electrical signal between the sinus node and the AV-node while minimizing the formation of a reentrant loop. The surgical maze procedure has proven to have a reasonably high success rate, albeit with considerable associated surgical morbidity. An attempt to recreate this treatment principle using radiofrequency ablation has been attempted using long linear lesions, but with relatively poor results. The understanding of the role of rapidly discharging ectopic foci residing within the pulmonary veins has refocused the trans-catheter therapies. Application of radiofrequency energy or cryotherapy (the later particularly in the operating room) are now directed at either electrically isolating the pulmonary veins from the left atrium or at directly ablating the ectopic focus within the veins. A recent large study has demonstrated rhythm control (return to normal sinus rhythm) using radiofrequency ablation techniques that is superior to either rate control or rhythm control using pharmacological agents. The difference between rhythm control using radiofrequency ablation or pharmacological agents likely rests in the proarrhythmia associated with pharmacological therapy.

Finally, pacemakers and implanted defibrillators have been used in an attempt to treat atrial fibrillation or to prevent its onset. To date, no significant improvement has been achieved relative to pharmacological therapies.

In conclusion, while atrial fibrillation is a common arrhythmia in older patients, it remains rare in pediatric and young adult patients. When encountered in pediatric patients, it often is the paroxysmal form with spontaneous remission and infrequent episodes. In addition, it can occur in the setting of other forms of supraventricular tachycardia, which are amenable to radiofrequency ablation therapy, resulting in the resolution of the atrial fibrillation. For older patients with symptomatic or hemodynamic consequences, new ablation strategies aimed at eliminating or isolating the triggers within the pulmonary veins appear promising. Due to the infrequency of this arrhythmia in the young, transcatheter treatment targeted directly to atrial fibrillation is rarely indicated or necessary.

SUGGESTED READING

1. Allessie M, Ausma J, Schotten U. Electrical, contractile and structural remodeling during atrial fibrillation. *Cardiovasc Res* 2002;54:230–246.
2. Nattel S. Therapeutic implications of atrial fibrillation mechanisms: can mechanistic insights be used to improve AF management. *Cardiovasc Res* 2002;54:347–360.
3. Wijffels MC, Kirchhof CJ, Dorland R, Allessie M. Atrial fibrillation begets atrial fibrillation: a study in awake chronically instrumented goats. *Circulation* 1995;92:1954–1968.
4. Scheinman MM, Morady F. Non-pharmacologic approaches to atrial fibrillation. *Circulation* 2001;103:2120–2125.
5. Jais P, Haissaguerre M, Shah DC, et al. A focal source of atrial fibrillation treated by discrete radiofrequency ablation. *Circulation* 1997;95:572–576.
6. Hohnloser SH, Kuck KH, Lilienthal J. Rhythm or rate control in atrial fibrillation—Pharmacological Intervention in Atrial Fibrillation (PIAF): a randomised trial. *Lancet* 2000;356:1789–1794.
7. The Atrial Fibrillation Follow-up Investigation of Rhythm Management (AFFIRM) Investigators. A comparison of rate control and rhythm control in patients with atrial fibrillation. *N Engl J Med* 2002;347:1825–1833.
8. Pappone C, Rosanio S, Augello G, et al. Mortality, morbidity, and quality of life after circumferential pulmonary vein ablation for atrial fibrillation: outcomes from a controlled nonrandomized long-term study. *J Am Coll Cardiol* 2003;42:185–197.

10

Atrial Ectopic Tachycardias/ Atrial Automatic Tachycardia

Burt Bromberg

Disorders of impulse formation account for approximately 10% of the tachycardias in infants and children. These include automatic atrial tachycardia (AAT), also known as ectopic atrial tachycardia (AET), originating from a single focus in the atrium, multifocal atrial tachycardia (MAT, or chaotic atrial tachycardia) thought to arise from multiple foci within the atria (Chapter 11), congenital junctional ectopic tachycardia (JET), and accelerated junctional rhythm, both originating from tissue in the region of the atrioventricular (AV) node. They are characterized electrophysiologically by a point origin, in contrast to tachycardias supported by macrorentrant circuits where larger segments of tissue with abnormal conduction perpetuate the arrhythmia.

The precise cellular mechanism of disorders of impulse formation can only be inferred in patients with abnormal automaticity. Both abnormal transmembrane potassium (pacemaker) and calcium (triggered activity) currents have been proposed, based on both cellular laboratory observations and indirect clinical evidence.

AUTOMATIC ATRIAL TACHYCARDIA/ATRIAL ECTOPIC TACHYCARDIA

Clinical Presentation

Automatic atrial tachycardia (AAT), or ectopic atrial tachycardia (AET), is the most common manifestation of abnormal automaticity in otherwise healthy children. AAT typically occurs in children with no structural heart disease. Infants and toddlers may have a longstanding tachycardia that is not recognized until they develop symptoms and signs of congestive heart failure. In contrast, older patients usually present with palpitations; a few will present with tachycardia-related cardiomyopathy.

The heart rate in AAT tends to be slower than in those with reentrant supraventricular tachycardia (SVT), typically in the range of 120 to 160 bpm. Both a paroxysmal form, similar in onset and cessation to the reentrant forms of SVT, and a more chronic form have been described. It is the chronic form, often at relatively slow rates (<150 bpm), that can lead to a tachycardia-induced cardiomyopathy. Occasionally, these tachycardias resolve spontaneously. AAT is rarely seen in children, particularly in infants and toddlers, with congenital heart disease; when it does occur, it is most often post-operatively or following interventional catheter procedures. Most of these will resolve following a period of medical management.

Diagnosis

Most AATs can be diagnosed from the ECG. Each QRS complex of the tachycardia is preceded by a P-wave, although at faster rates this may be difficult to ascertain since the P-wave becomes buried in the preceding T-wave (Figure 1). The ectopic P-wave has a different morphology from the sinus beat (Figure 2), but this may not always be a reliable indicator if the ectopic focus is near the sinus node.

Noninvasive testing may clarify the diagnosis (Table 1). Holter monitoring may demonstrate a characteristic gradual acceleration of the heart rate in AAT, the so-called "warm-up" period. The first beat of the tachycardia typically occurs late in the cycle, and the P-wave will have the same morphology as subsequent P-waves. Participation of the ventricle in the tachycardia is not required, excluding an accessory pathway in these patients. During periods of reduced adrenergic tone and heightened parasympathetic tone (sleep, carotid sinus massage), the P-waves of an AAT may not conduct to the ventricles, resulting in varying degrees of atrioventricular block. On occasion this block can be induced with adenosine (Figure 3); on the other hand, some AATs may be transiently suppressed with adenosine, limiting the usefulness of this test.

Automatic atrial tachycardia also has distinguishing features observed at electrophysiology study. First, it cannot be terminated by overdrive pacing. If ventricu-

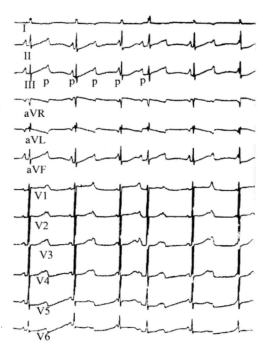

FIGURE 1. 12-Lead ECG from a 9-month-old boy with atrial automatic tachycardia at 150 bpm. There is variable AV conduction. The P-wave is large in many leads and has a leftward inferior (positive in leads I, II, aVF) but anterior (positive in lead V1) axis suggesting a high left atrial origin near the right upper pulmonary vein.

lar participation can be excluded during either spontaneous or adenosine-induced AV node Wenckebach, the remaining diagnostic possibility for a narrow QRS tachycardia with P-waves preceding each QRS complex is atrial flutter (intra-atrial reentrant tachycardia). While macrorentrant tachycardia,

TABLE 1. Criteria for Automatic Atrial Tachycardia

Criteria for Automatic Atrial Tachycardia
P-wave preceding each QRS
P-wave of different morphology from sinus
Persistence of tachycardia in presence of AV block
"Warm-up" period
Catecholamine sensitive

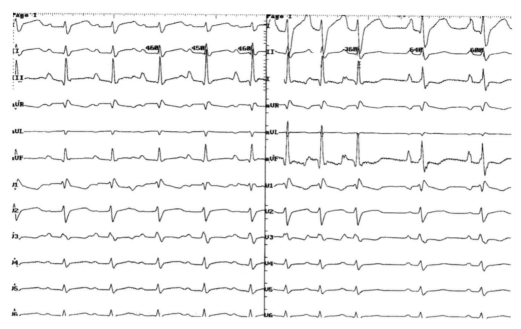

FIGURE 2. Automatic atrial tachycardia originating from the atrial mid-septum. Left panel: At a slower cycle length, the P-wave of the tachycardia preceding the QRS is easily discernable, but not obviously distinguishable in shape from a sinus beat. Right panel: During radiofrequency catheter ablation, the ectopic focus heats up, the tachycardia cycle length shortens, and the P-wave cannot be clearly identified within the preceding T-wave. After the third beat, the tachycardia terminates. Cycle lengths are displayed in lead II. Successful ablation of the ectopic focus is often preceded by an acceleration of the tachycardia, followed by a sudden prolongation of the cycle length of the normal sinus impulse. The morphology of the ensuing sinus beats is only subtly different from that of the automatic tachycardia (see leads I, aVR, V1-2).

including atrial flutter, can be interrupted with appropriate atrial overdrive (burst) pacing, an automatic tachycardia is only momentarily suppressed before the ectopic focus gradually warms up and returns. Another diagnostic finding is the sequence of atrial-atrial-ventricular electrograms following cessation of ventricular burst pacing, elicited in almost 80% of patients with a paroxysmal form of this arrhythmia.

Treatment

Treatment presents two options: antiarrhythmic medications or radiofrequency ablation. There are no controlled, randomized series to evaluate drug therapy; available data are difficult to interpret because of selection bias in retrospective studies. Digoxin, propranolol, and verapamil are ineffective. Class III drugs like amiodarone and sotalol, and IC drugs (propafenone and flecainide) have been more successful in small retrospective series. The addition of a beta-blocker to flecainide may be helpful, and in refractory cases, flecainide in combination with amiodarone may be considered.

The acute success rate of radiofrequency ablation for AAT approaches 95% to 100%. Experience in children demonstrates that a majority of the foci are in the left atrium, particularly near the pulmonary veins and in the atrial appendage. In contrast, the anatomic distribution in adult patients is predominantly on the right side, particularly near the SVC/RA junction, right atrial appendage, and

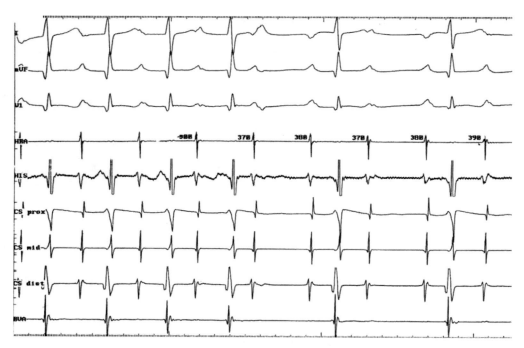

FIGURE 3. Atrioventricular block with adenosine during AAT. The atrial tachycardia persists at a cycle length of 370 to 390 msec despite AV block induced by adenosine. In addition to demonstrating that the ventricular is not a participant in the tachycardia, the P-wave, which had been buried within the preceding T-wave, becomes clearly discernable.

the coronary sinus os. When the focus(i) in children is on the right side, they cluster in and around the right atrial appendage. The technical aspects of successful ablation include identification of the earliest intra-atrial electrogram preceding the onset of the surface P-wave by at least 20 msec, pace-mapping to assure that P-waves originating from the pacing site are identical to those of the AAT and computer-based intracardiac mapping techniques (Figure 4).

In view of the possibility of spontaneous resolution, as well as the availability of drugs with perhaps modest efficacy and a low incidence of adverse effects, a trial with flecainide or amiodarone for symptomatic patients with AAT is reasonable. However, for patients who have poor control of the tachycardia, especially with evidence of diminished ventricular function, or who remain dependent on pharmacologic therapy after 1–2 years, radiofrequency catheter ablation is indicated.

JUNCTIONAL TACHYCARDIAS

Junctional Ectopic Tachycardia

Junctional ectopic tachycardia (JET) is an abnormal automatic arrhythmia that originates from the region of the AV node or His bundle and may be seen in a variety of settings. There are three distinct clinical entities in children: the first and most common presents in the post-operative patient (JET); the second presents in infancy or early childhood (congenital JET); and the third presents later in childhood and adolescence (accelerated junctional tachycardia). Although the mechanism is similar in these three forms, each entity has a sufficiently unique clinical history and

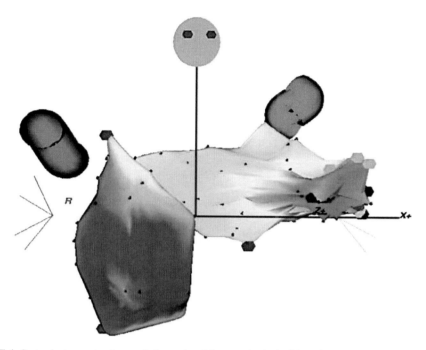

FIGURE 4. Carto electro-anatomic map (left anterior oblique projection) of AAT in a 16-year-old boy arising from the tip of the left atrial appendage (red-orange), moving through the left atrium (yellow-green) to the right atrium (blue-purple). Radiofrequency ablation was successful at this site (red). The gray and red tubular structures represent the right and left upper pulmonary vein respectively.

treatment program that warrants clinical distinction.

Post-operative Junctional Ectopic Tachycardia

Natural History. Post-operative JET is the most common form of an abnormal, automatic rhythm in children. However, because of its association with hospitalized post-operative congenital heart disease patients, its usually transient course, and its decreased frequency over the past two decades, in part due to change from the atrial to the arterial switch procedures in patients with D-transposition of the great arteries, its importance is often overlooked. Its greatest impact today is among patients undergoing the Fontan sequence [Norwood, hemi-Fontan (or Glenn) and Fontan procedures]. The exact mechanism for this arrhythmia is unknown; however, the fluid and electrolyte shifts of the early post-operative state, trauma, stretch, or ischemia in the region of the AV node and/or His bundle are likely candidates.

The heart rate of post-operative JET may be extremely rapid, ranging from 150 to 240 bpm (Figure 5), potentially compromising cardiac output. The atria and ventricles are activated nearly simultaneously from impulses originating within the AV node and/or His bundle causing the contraction of the atria against closed atrioventricular valves (Figure 6). This lack of atrioventricular synchrony compromises the cardiac output by decreasing ventricular filling. Uncontrolled, this tachycardia may further lead to a fall in blood pressure, initiating a downward spiral characterized by increased endogenous catecholamines and increased inotropic support to maintain adequate blood pressure and renal perfusion. As a result of vasoconstriction, the

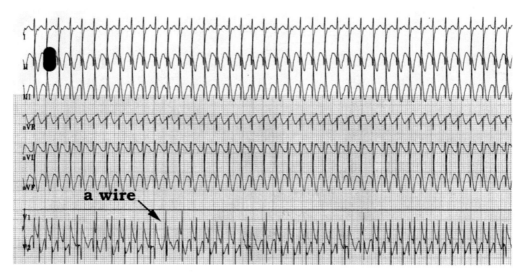

FIGURE 5. Post-operative JET in 4-month-old girl following a repair of an atrioventricular septal defect (complete). The heart rate is approximately 260 bpm. Note the atrioventricular dissociation clearly demonstrated by recording of the atrial electrogram through the transthoracic temporary atrial epicardial electrode "a wire"; the "a" rate is approximately 160 bpm; there is variable retrograde conduction from (or capture of the atrium by) the ectopic junctional focus to the atrium. She responded to amiodarone and had an otherwise normal postoperative course. This illustration underscores the importance of post-operative transthoracic wires in diagnosing arrhythmias.

patient's skin may feel cool, but the core temperature rises because of an inability to dissipate heat. This increase in both endogenous and exogenous adrenergic input as well as the elevation in core temperature exacerbates the tachycardia. Thus, it is critical to interrupt this cycle of escalating heart rate, increasing vasoconstriction, and falling cardiac output. Although post-operative JET is often self-limited, slowing to a tolerated heart rate within 48 hours, rapid rates ($\geq 180-190$ bpm) associated with decreasing blood pressure require

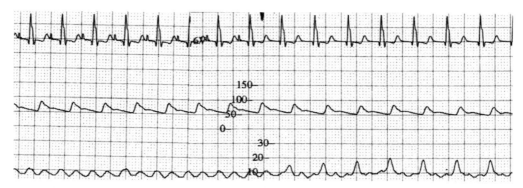

FIGURE 6. ECG tracing, blood pressure (middle) and right atrial recording (bottom) in a 2-year-old post-Fontan patient in the intensive care unit. In the left-hand part of the tracing there is sinus tachycardia (CL = 410 msec) with a blood pressure of 85/50 mmHg, and a right atrial pressure of peak 10 mmHg. In the right hand part of the tracings there is junctional rhythm (CL = 430 msec) with "canon" waves of 15–20 mmHg and a blood pressure of 75/50 mmHg, due to loss of atrioventricular synchrony and thus atrial "kick."

intervention. The slower rates often do not need treatment or need only atrial pacing at a slightly higher rate than the junctional rate to provide atrial-ventricular synchrony. They rarely persist beyond 5–7 days.

Diagnosis. This tachycardia should be suspected in any post-operative patient within 48 hours following cardiopulmonary bypass who exhibits a narrow QRS tachycardia in which P-waves are either not discernable or are dissociated from the QRS. If retrograde conduction is intact, the VA interval will be between -20 msec and $+40$ msec and the right atrial pressure tracing will exhibit a characteristic "cannon" A-wave. Sometimes retrograde conduction to the atrium is impaired, resulting in A-V dissociation (Figure 6). This latter finding indicates retrograde block, excluding the atrium as a participant in the tachycardia. The diagnosis of JET can be confirmed with an atrial electrogram recorded through temporary transthoracic epicardial leads or a transesophageal lead (Figure 5).

Treatment. A number of complementary strategies are available. A junctional rate below 170 bpm is usually well tolerated. A rate between 170 and 190 bpm often calls for intervention and a rate above 190 bpm usually needs to be urgently addressed. First, the goal of management of the post-operative hemodynamic state is to maximize cardiac output, and minimize factors exacerbating the tachycardia by carefully titrating fluid and volume replacement, inotropic agents (dopamine, epinephrine, and dobutamine), body temperature, and pain control. Second, at the slower junctional rates (<180 bpm), atrial overdrive pacing (typically 10 bpm higher than the junctional rate) can often restore atrioventricular synchrony and improve ventricular filling and cardiac output. High atrial pacing thresholds can be overcome with a high output stimulator such as the kind used for transesophageal pacing. The improved ventricular filling offsets the shorter ventricular filling time imposed by the faster rate. As the cardiac output improves, the inotropic support can be reduced and the intrinsic sympathetic drive falls, often reducing the junctional rate. As the junctional rate falls, the atrial overdrive pacing can be weaned, allowing for a longer time of ventricular filling and a gradual return to a more normal hemodynamic state.

A third useful strategy is mild systemic hypothermia, usually around a core temperature of 35°C. Below 32°C, ventricular function may be impaired and the slowed conduction velocity could theoretically predispose to other arrhythmias.

The fourth approach involves antiarrhythmic therapy. Digoxin, used in the past, has little place in this clinical setting. Prior to the availability of intravenous amiodarone, a procainamide drip in conjunction with mild hypothermia was widely used with good success. Propafenone, another class I drug (Chapter 21), lost favor when proarrhythmia effects were attributed to it. Since FDA approval of intravenous amiodarone, it has become the drug of choice. In a multi-center study, postoperative JET was controlled in 12/13 patients. The onset is rapid, typically within 30–60 minutes, and is generally well tolerated, with minimal proarrhythmia or depressed contractility. However, intravenous amiodarone may produce hypotension during the initial bolus and has been associated with a "gasping syndrome" (product report) leading to the manufacturer's warning that continuous infusion should be used in infants only with full bedside cardio-respiratory monitoring.

As a last alternative, ECMO can be used for control of the postoperative junctional tachycardia. The tachycardia is suppressed almost immediately once the patient is placed on extracorporeal support. Fortunately, the need for ECMO has become rare with the availability of intravenous amiodarone.

Congenital Junctional Ectopic Tachycardia

Natural History. Congenital JET (Figure 7) is a rare, but potentially life-threatening

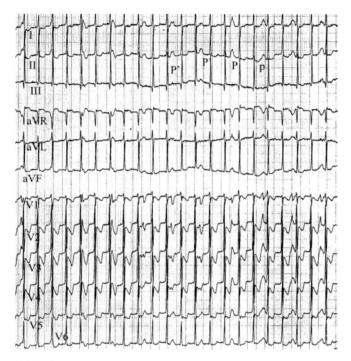

FIGURE 7. 12-Lead ECG of congenital JET in a 9-month-old girl. Note the narrow QRS tachycardia at 170 bpm and AV dissociation (large P-waves in lead II. Note brief period of 2:1 retrograde conduction to the atria (P'). This girl was treated with amiodarone for 12 months and the arrhythmia subsided. The amiodarone was discontinued and JET has not returned after 9 years.

tachycardia. Most patients present before 6 months of age. A large multi-center report showed an overall mortality of 35%. Affected infants usually present with nonspecific symptoms of congestive heart failure. However, some may initially appear compensated, only to subsequently develop a rapid, malignant tachycardia. Interestingly, while 7 of the 9 deaths occurred in patients with obvious low output secondary to an intractable tachycardia, 2 deaths occurred in patients who were apparently well-controlled. One was noted to have a bradyarrhythmia immediately prior to cardiac arrest. Histologic abnormalities of the AV node and His bundle identified at autopsy include entrapment, distortion, and division of the AV node within the central fibrous body, left-sided AV node, fibrosis, inflammation, and focal degeneration. This has led to speculation that in addition to uncontrolled tachycardia and a consequent tachycardia-induced cardiomyopathy, another mechanism of sudden death in these patients may be complete heart block. Furthermore, a familial pattern observed in half of the patients implicates programmed degeneration of the AV node and His bundle in some cases.

Diagnosis. The diagnosis should be suspected from the ECG, in which a narrow QRS tachycardia may show AV dissociation; typically the P-waves on the 1 lead ECG are large and wide (Figure 7) suggesting inherent anatomic electrophysiologic abnormalities in the atrial segments of the heart. Confirmation of AV dissociation should be obtained with an atrial electrogram, assessed by either a transesophageal or an intra-atrial lead (Figure 8). The heart rate typically is <200 bpm, although it may be as slow as 130

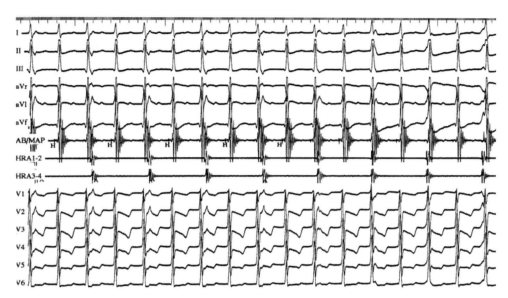

FIGURE 8. Same patient as in Figure 7. There is a His bundle (H) tracing (ABMAP) and intra-atrial electrograms (HRA 1–2 and 3–4) recorded through closely spaced bipolar electrodes, along with 12-lead ECG. The tracing shows a ventricular rate of 100 bpm, each V preceded by an H potential. There is atrioventricular dissociation with an atrial rate of about 50 bpm.

to 150 bpm. At the slower rates, there may be a 1:1 ventricular–atrial relationship, which could obscure the diagnosis. The response to adenosine is either to block V-A conduction or transient conversion to sinus rhythm, followed by resumption of the tachycardia.

Treatment. Because of the rarity of congenital JET, and thus a small sample size and lack of controls, treatment is empiric. Currently, amiodarone is the treatment of choice, yielding, in a multi-center study, improvement in 85% of patients treated with high dose amiodarone and survival of almost 75% of patients for an average follow-up of 6.1 years. Drug therapy was discontinued in over half of the survivors, who remained in sinus rhythm an average of 5.7 years later, although Holter monitoring continued to show an underlying slower junctional tachycardia (60–110 bpm) during times of decreased sinus rate.

In the absence of spontaneous resolution, ablation therapy is the definitive treatment. Prior to the availability of intravenous amiodarone in the United States, intentional AV node ablation with radiofrequency ablation followed by permanent ventricular pacing was an option; it is now rarely used. With increased experience, radiofrequency ablation, and more recently, cryothermal ablation, of the junctional tachycardia can be successfully delivered with preservation of atrioventricular conduction. The Pediatric Radiofrequency Ablation Registry, along with other reports, describes successful ablation in infants (<1.5 years of age).

Accelerated Junctional Tachycardia

Within this subset of junctional tachycardias are two different clinical entities: a paroxysmal form and a non-paroxysmal form (accelerated junctional escape). The paroxysmal form is seen in otherwise healthy adults, as well as older children and adolescents with no structural heart disease. The mechanism appears to be one of enhanced automaticity in pacemaker currents, typically originating from the AV node in the region of the slow pathway insertion (Figures 9 and 10).

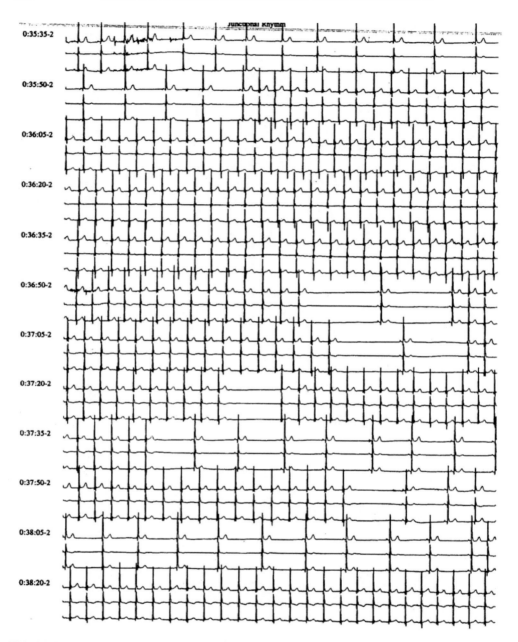

FIGURE 9. Holter tracing of a 7-year-old boy with accelerated junctional tachycardia. Note the junctional tachycardia at approximately 100–120 bpm and the pauses with suppression of sinus escape. His tachycardia was successfully ablated using cryothermal energy.

The non-paroxysmal junctional form is most commonly found in adults with acute myocardial ischemia, digoxin toxicity, chronic obstructive pulmonary disease, rheumatic carditis, hyperkalemia, hypercalcemia, or after heart surgery, where the mechanism may be triggered activity. It is infrequently observed in adolescents and

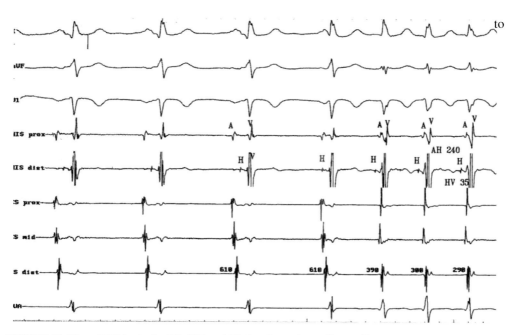

FIGURE 10. Three ECG leads (I, aVR, V1) and His bundle tracings (4th and 5th from the top) and three coronary sinus electrograms with right ventricular (bottom training) electrogram. The tracing initially demonstrates sinus rhythm, with a PR interval of 120 msec. The junctional rhythm suddenly accelerates capturing the atrium retrogradely.

young adults with no apparent predisposing conditions.

Natural History. Unlike the incessant junctional tachycardias in infancy or following surgery, these junctional tachycardias are slower and much better tolerated. Clinically, the paroxysmal form may have sudden onset and cessation, and persists seconds to hours or it may have a gradual onset, accelerating to overtake the sinus node (Figure 11). The heart rates range widely, from 110 to 250 bpm. At times of increased adrenergic stimulation (e.g., fever, excitement, exertion), the tachycardia is often sustained and more rapid. While the rapid prolonged episodes may precipitate evaluation, Holter monitoring frequently demonstrates slower junctional rates of which the patient is often unaware. Similarly, patients with the incessant but slower non-paroxysmal form may only become symptomatic with adrenergic stimulation, when the heart rate can accelerate to 130–150 bpm. These tachycardias are usually benign, although they may persist for decades.

Diagnosis. On the surface ECG, P-waves may be dissociated from the QRS complex (Figure 10), or more commonly, not evident because of retrograde conduction. Electrophysiologic studies demonstrate a His bundle preceding the QRS, with a short VA time and long AV time that has the appearance of typical AV node reentry tachycardia (Figure 11). Occasionally, there will be atrial dissociation from the tachycardia, reducing the likelihood of AV node reentry tachycardia (Figure 12). However, AV node reentry tachycardia can reliably be excluded by pacing at a cycle length shorter than that of the tachycardia; the AV interval shortens, eliminating block in a fast pathway, and eliminating conduction down a slow pathway as a possibility (Figures 13 and 14). The response to overdrive pacing is variable. A transient

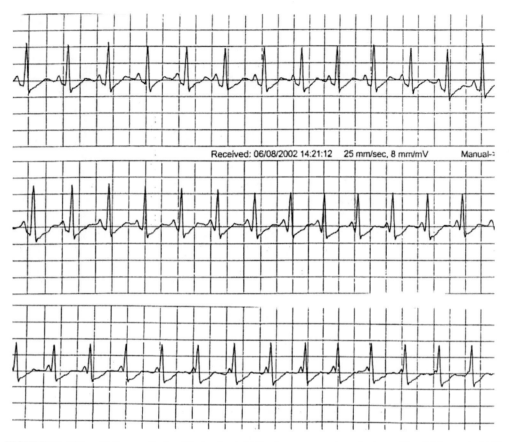

FIGURE 11. JET on Holter monitoring. The tracing initially demonstrates sinus rhythm, with a PR interval of 120 msec. However, the PR interval gradually shortens as the junctional focus accelerates. Eventually the P-wave is buried within the QRS complex, clearly implying a tachycardia that originates independently within the AV node or His bundle.

interruption of the tachycardia both to atrial burst pacing and adenosine may be observed.

Treatment. The decision to treat is based largely on the intensity, duration, and frequency of the symptoms that are related to the tachyarrhythmia. Some patients can be followed without initiating any therapy. If patients are troubled by the palpitations, beta blockers are an appropriate choice. Other medications such as flecainide or amiodarone might be effective. However, given the potential toxicity of amiodarone and the risk of proarrhythmia associated with flecainide (albeit low in a child with a structurally and functionally normal heart), radiofrequency catheter ablation or, more recently, cryotherapy should be considered if the tachycardia persists. In a recent report, 10/13 children >1.5 years of age with an accelerated junctional tachycardia had been successfully treated with catheter radiofrequency ablation. Mapping techniques do not reliably identify the precise origin of the tachycardia focus in the region of the AV and His bundle. However, electro-anatomic mapping or navigation systems (Figure 15) may provide a useful "topographical map" of the various structures in the AV node region (His bundle, tricuspid valve orifice, tendon of Todaro and

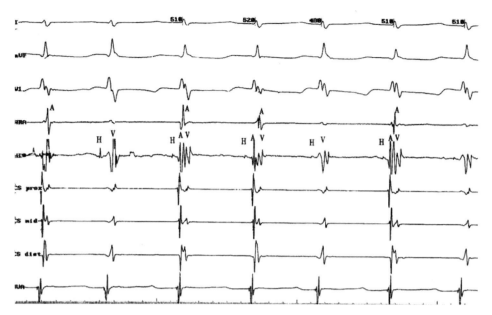

FIGURE 12. JET with V-A dissociation. V-A dissociation without interruption of the tachycardia is most apparent in the HRA and CSp tracings. HRA: high right atrium; His: His bundle; CSp: coronary sinus proximal; CSd: coronary sinus distal; RVA: right ventricular apex.

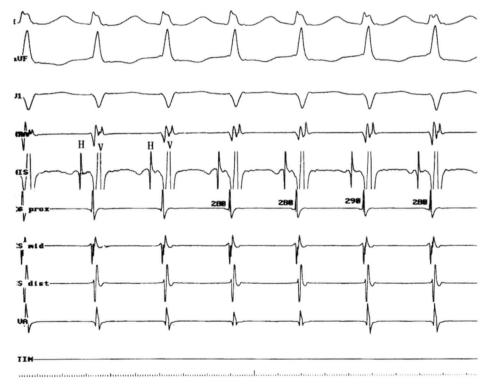

FIGURE 13. Entrainment during junctional ectopic tachycardia. The AH interval is 240 msec during the junctional ectopic tachycardia, which has a cycle length of 290 msec. The long AH interval and short or negative VA interval is consistent with typical AV node reentry tachycardia (AVNRT), where antegrade conduction proceeds down the slow pathway.

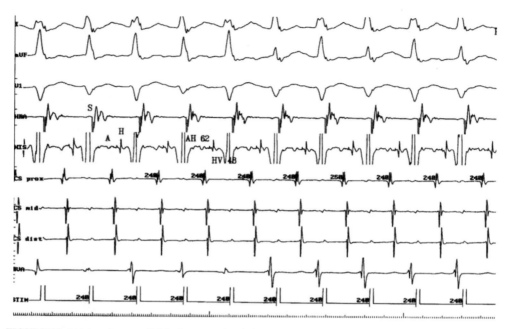

FIGURE 14. Atrial pacing at a slightly faster rate (cycle length 240 msec) reduces the AH interval to 62 msec. This eliminates slow pathway conduction as a consideration, thereby excluding AVNRT as a possible mechanism.

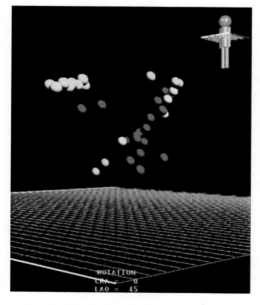

FIGURE 15. Navigation system (LocaLisa®) in LAO projection. Blue dots indicate location of His bundle catheter, Green dots = coronary sinus os. Red dots = cryothermal ablation application sites. Yellow dots = sites of ablation inducing prolongation of the PR interval. Delivery of energy was terminated when the PR interval lengthened.

coronary sinus os) that facilitates repeated placement of the ablation catheter tip on the target site, as well as potentially reduce fluoroscopy time. It is useful to note that ablation in the region of the slow pathway is safe and may encompass the focus of the tachycardia. Catheter cryoablation may reduce the risk further as the cryo-effect can be monitored at lower but reversible temperatures prior to delivering the permanently ablative $-70°C$ temperature.

SUGGESTED READING

1. Huikuri HV, Poutiainen A, Makikallio TH, et al. Dynamic behavior and autonomic regulation of ectopic atrial pacemakers. *Circulation* 1999;100:1416–1322.
2. Josephson ME, Spear JF, Harken AH, et al. Surgical excision of automatic atrial tachycardia:anatomic and electrophysiologic correlates. *Am Hrt J* 1982;104:1076–1085.
3. Rosen MR, Gelband H, Merker C, Hoffman BF. Mechanisms of digitalis toxicity: effects of ouabain on phase four of canine Purkinje fiber transmembrane potentials. *Circulation* 1973;47:681–689.
4. Rosen MR, Fisch C, Hoffman BF, et al. Can accelerated atrioventricular junctional escape rhythms be explained by delayed after depolarizations? *Am J Cardiol* 1980;45:1272–1284.
5. Gillette PC, Garson Jr. A. Electrophysiologic and pharmacologic characteristic of automatic ectopic atrial tachycardia. *Circulation* 1977;56:571–575.
6. Gillette PC, Wampler DG, Garson Jr A, Zinner A, Ott D, Cooley D. Treatment of atrial automatic tachycardia by ablation procedures. *J Am Coll Cardiol* 1985;6:405–409.
7. Koike K, Hesslein PS, Finlay CD, et al. Atrial automatic tachycardia in children. *Am J Card* 1988;61:1127–1130.
8. Bauersfeld U, Gow RM, Hamilton RM, Izukawa T. Treatment of atrial ectopic tachycardia in infants <6 months old. *Am Hrt J* 1995;129:1145–1148.
9. Josephson ME. *Clinical Cardiac Electrophysiology*. Second edition. Lea and Febiger, Philadelphia/London. 1993.
10. Knight BP, Zivin A, Souza J, et al. A technique for the rapid diagnosis of atrial tachycardia in the electrophysiology laboratory. *J Am Coll Cardiol* 1999;33:775–781.
11. Fenrich A Jr., Perry JC, Friedman RA. Flecainide and amiodarone: combined therapy for refractory tachyarrhythmias in infancy. *J Am Coll Cardiol* 1995; 25:1195–1198.
12. Fish FA, Gillette PC, Benson DW Jr. Proarrhythmia, cardiac arrest and death in young patients receiving encainide and flecainide. The Pediatric Electrophysiology Group. *J Am Coll Cardiol* 1991;18:356–365.
13. Walsh EP, Saul JP, Hulse JE, et al. Transcatheter ablation of ectopic atrial tachycardia in young patients using radiofrequency current. *Circulation* 1992;86:1138–1146.
14. Tracy CM, Swartz JF, Fletcher RD, et al. Radiofrequency catheter ablation of ectopic atrial tachycardia using paced activation sequence mapping. *J Am Coll Cardiol* 1993;21:910–917.
15. Kay GN, Chong F, Epstein AE, et al. Radiofrequency ablation for treatment of primary atrial tachycardias. *J Am Coll Cardiol* 1993;21:910–919.
16. Bisset GS, Seigel SF, Gaum WE, Kaplan S. Chaotic atrial tachycardia in childhood. *Am Hrt J* 1981;101:268–272.
17. Salim MA, Case CL, Gillette PC. Chaotic atrial tachycardia in children. *Am Hrt J* 1995;129:831–833.
18. Dodo H, Gow RM, Hamilton RM, Freedom RM. Chaotic atrial rhythm in children. *Am Hrt J* 1995;129:990–995.
19. Yeager SB, Hougen TJ, Levy AM. Sudden death in infants with chaotic atrial rhythm. *AJDC* 1984;138:689–692.
20. Fish FA, Mehta AV, Johns JA. Characteristics and management of chaotic atrial tachycardia of infancy. *Am J Cardiol* 1996; 78:1052–1055.
21. Walsh EP, Saul JP, Sholter GF, et al. Evaluation of a staged treatment protocol for rapid automatic junctional tachycardia after operation for congenital heart disease. *J Am Coll Cardiol* 1997;29:1046–1053.
22. Dodge-Khatami A, Miller OI, Anderson RH, et al. Surgical substrates of postoperative junctional ectopic tachycardia in congenital heart defects. *J Thorac Cardiovasc Surg* 2002;123(4):624–630.
23. Rosales AM, Walsh EP, Wessel DL, Triedman JK. Postoperative ectopic atrial tachycardia in children with congenital heart disease. *Am J Cardiol* 2001;88(10):1169–1172.
24. Bronzetti G, Formigari R, Giardini A, et al. Intravenous flecainide for the treatment of junctional ectopic tachycardia after surgery for congenital heart disease. *Ann Thorac Surg* 2003;76(1):148–151.
25. Simmers TA, Sreeram N, Wittkampf FH, Derksen R. Radiofrequency catheter ablation of junctional ectopic tachycardia with preservation of atrioventricular conduction. *Pacing Clin Electrophysiol* 2003;26(5):1284–1288.
26. Hoffman TM, Bush DM, Wernovsky G, et al. Postoperative junctional ectopic tachycardia in

27. Laird WP, Snyder CS, Kertesz NJ, et al. Use of intravenous amiodarone for postoperative junctional ectopic tachycardia in children. *Pediatr Cardiol* 2003;24(2):133–137.
28. Janousek J, Paul T, Reimer A, Kallfelz HC. Usefulness of propafenone for supraventricular arrhythmias in infants and children. *Am J Cardiol* 1993;72:294–300.
29. Perry JC, Fenrich AL, Hulse JE, et al. Pediatric use of intravenous amiodarone: efficacy and safety in critically ill patients from a multicenter protocol. *J Am Coll Cardiol* 1996; 27:1246–1250.
30. Villain E, Vetter VL, Garcia JM, et al. Evolving concepts in the management of congenital junctional ectopic tachycardia. *Circulation* 1990;81:1544–1549.
31. Bharati S, Moskowitz WB, Scheinman M, et al. Junctional tachycardia: anatomic substrate and its significance in ablative procedures. *J Am Coll Cardiol* 1991;18:179–186.
32. Gillette PC, Garson Jr A, Porter CJ, et al. Junctional automatic ectopic tachycardia: new proposed treatment by transcatheter His bundle ablation. *Am Hrt J* 1983;106:619–623.
33. Dubin AM, Cuneo BF, Strasburger JF, et al. Congenital junctional ectopic tachycardia and congenital complete atrioventricular block: a shared etiology? *Heart Rhythm* 2005;2(3):313–315.
34. Plumpton K, Justo R, Haas N. Amiodarone for postoperative junctional ectopic tachycardia. *Cardiol Young* 2005;15(1):13–18.
35. Sarubbi B, Musto B, Ducceschi V, et al. Congenital junctional ectopic tachycardia in children and adolescents: a 20-year experience based study. *Heart* 2002;88(2):188–190.
36. Bae EJ, Kang SJ, Noh CI, et al. A case of congenital junctional ectopic tachycardia: diagnosis and successful radiofrequency catheter ablation in infancy. *Pacing Clin Electrophysiol* 2005;28(3):254–257.
37. Blaufox AD, Felix GL, Saul JP, and Participating Members of the Pediatric Catheter Ablation Registry. Radiofrequency catheter ablation in infants <18 months old. *Circulation* 2001;104:2803–2808.
38. Ruder MA, Davis JC, Eldar M, et al. Clinical and electrophysiologic characterization of automatic junctional tachycardia in adults. *Circulation* 1986;73:930–937.
39. Kugler JD, Danford DA, Deal BJ, et al. Radiofrequency catheter ablation for tachyarrhythmias in children and adolescents. *NEJM* 1994;330:1481–1487.

11

Multifocal Atrial Tachycardia

David J. Bradley

Multifocal atrial tachycardia (MAT, also known as chaotic atrial rhythm) is an unusual arrhythmia in children. The defining characteristics—P-waves of three or more distinct morphologies with irregular P–P intervals in the context of a rapid rate, greater than 100 beats per minute—are the same as those in adults (Figure 1).

Clinical Context

Whereas the adults with MAT are nearly uniformly affected with severe chronic lung disease, pediatric patients with MAT are a heterogeneous group including some of the healthiest and the most severely ill. In all large pediatric series, a significant association (30%–60%) is observed between MAT and respiratory illness, including both infectious processes, such as bronchiolitis or croup, and non-infectious conditions such as respiratory distress syndrome of prematurity and bronchomalacia. Resolution of the arrhythmia does not always exactly parallel recovery from the respiratory illness or treatment of it. Nor does oxygen saturation *per se* appear linked to the presence of MAT.

MAT in the context of severe, life-threatening illness, including myocarditis, complex cyanotic heart disease and birth asphyxia has been described. Patients with structural heart disease may comprise up to 33% of pediatric MAT. The routine cardiac monitoring of cardiac patients while in the hospital for management of the cardiac malformation probably leads to overrepresentation of this patient subgroup. No particular cardiac lesions have emerged as associated with the incidence of MAT, although coarctation of the aorta, hypertrophic cardiomyopathy, atrial septal defect, tetralogy of Fallot, and complex single ventricle lesions have been reported. Thus, a thorough evaluation of the patient with newly diagnosed MAT is necessary to identify any coexisting conditions that may impact the patient's prognosis.

The most common presentation of MAT, in approximately half of cases, is as an incidental finding in an otherwise healthy infant. The markedly rapid and irregular rhythm may be an unexpected observation in a presumed "well" child. A history of recent viral upper respiratory symptoms may be elicited, but in many patients there is none. Such children may feed and grow normally, without any evidence of abnormality or congestive heart failure.

As in other forms of supraventricular tachycardia, sustained or recurrent, rapid MAT has the potential to affect cardiac

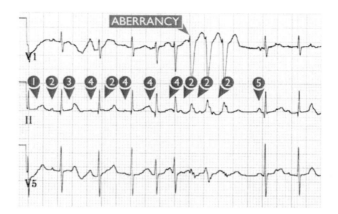

FIGURE 1. Three electrocardiograph leads (I, II, III) in a 3-month-old boy with MAT. Note the five different P-wave morphologies (arrows). Note also the aberrant conduction (complexes 8–10 from the left). Reprinted with permission from Bradley DJ et al. Clinical course of MAT in infants and children. *J Am Coll Cardiol* 2001:38: 401–408, the American College of Cardiology Foundation.

function, sometimes severely. In our published series, 27% (4 of 15) patients who had echocardiograms demonstrated either dilation of cardiac chambers or abnormal indices of function. Longitudinal follow-up of these patients has revealed that the cardiac function normalizes once the arrhythmia resolves or is brought under control.

Most children diagnosed with MAT are infants, and those over five years of age are rare. In our experience, 1 (5%) was over a year old at presentation, and three demonstrated heart rate irregularity on fetal monitors.

Incidence

MAT accounts for a very small proportion of arrhythmias treated in major pediatric cardiology centers. As an estimate of its incidence in healthy newborns, in a survey of the heart rhythms of 3,383 infants by ambulatory monitoring, two (0.02%) were found to have MAT.

Clinical Course

The course of MAT, regardless of treatment, is self-limited. Half of patients will have no residual evidence of MAT five months after diagnosis (Figure 2). In our long-term follow-up, recurrence has not been observed.

Electrocardiographic Features

MAT is typically extremely irregular and can be very rapid, with atrial rates as high as 400 beats per minute. Care must be taken to identify different P-wave forms on the ECG, although the effort to find them is often prompted by MAT's signature ECG appearance, which includes scattered,

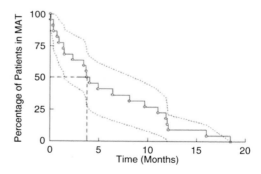

FIGURE 2. Resolution of MAT. Fifty percent of patients (indicated by broken line) were free of arrhythmia five months after diagnosis. The dotted line indicates a 95% confidence interval. Reprinted with permission from Bradley DJ et al. Clinical course of MAT in infants and children. *J Am Coll Cardiol* 2001:38:401–408, the American College of Cardiology Foundation.

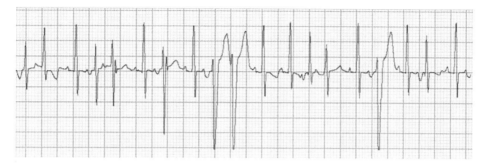

FIGURE 3. ECG tracing of MAT in a 5-month-old boy with three multiple P-wave morphologies with the aberrant conduction.

aberrantly conducted beats and pauses, due both to sinus node suppression after a run of MAT beats, as well as blocked conduction of premature atrial impulses (Figures 1 and 3). Prolonged ECG recordings demonstrate periods of sinus rhythm alternating with MAT in most patients.

Mechanism

The multiple P waveforms on the electrocardiogram that characterize MAT suggest that the mechanism of the arrhythmia is possibly triggered activity at several ectopic sites. This mechanism has not, however, been demonstrated. If an ensemble of ectopic foci gives rise to the arrhythmia, it is unusual that they would fire together in bursts, rather than each creating spontaneous, unifocal atrial runs at separate times.

An alternative mechanistic explanation is variable propagation through the atrium of rapid impulses from a single focus. Supportive of this concept is one report of successful radiofrequency ablation of MAT in an infant at a single atrial site. Additionally, at least two reported patients had the diagnosis of (unifocal) ectopic atrial tachycardia prior to their diagnosis of MAT.

The poor response of MAT to direct current cardioversion and to overdrive pacing, however, along with the observed cycle length irregularity, is strong evidence that the mechanism is not re-entrant.

Antiarrhythmic Therapy

The poor response of MAT to standard, "first-line" antiarrhythmic agents compels the clinician to make the assessment whether treatment is indeed necessary. Though correction of magnesium deficit, administration of beta blockers (such as metoprolol) and calcium channel blockers (verapamil) are reported effective in adults, no consistent response to any of these in pediatric patients has been described. The reported treatment strategies are varied, however, and no well-designed trials have been carried out. With regard to beta-blockade, the few reported patients with MAT treated with propranolol have shown no clear response. Probably due to its known, albeit rare, potential to cause hemodynamic collapse in infants, verapamil has not been described for MAT in children. Flecainide was of reported benefit in two patients. Amiodarone however, appears to be the current treatment of choice for pediatric MAT, when treatment is necessary.

Patient Management

The asymptomatic infant with newly diagnosed MAT who has no evidence of inter-current illness, cardiac anomaly or dysfunction may be observed regularly as an outpatient without further investigation. Follow-up should include ECG and ambulatory (Holter) monitoring, as MAT may occur sporadically and not be detected on ECG alone. The patient should be followed until there is no further evidence of MAT. An excellent outcome can be anticipated.

The patient with any suspected cardiac pathology should have anatomy and function defined by echocardiogram. Concurrent illness should be treated. Periods of MAT can be expected to alternate with sinus rhythm on cardiac monitor. If the patient's reduction in cardiac function is significant or complicates other conditions, medical therapy with amiodarone should be begun. Outcome in such patients will depend on recovery of function and resolution of the associated condition. Once the rhythm has normalized, recurrence is not expected.

SUGGESTED READING

1. Shine KI, Kastor JA, Yurchak PM. Multifocal atrial tachycardia. Clinical and electrocardiographic features in 32 patients. *N Eng J Med* 1968;279(7): 344–349.
2. Bradley DJ, et al. The clinical course of multifocal atrial tachycardia in infants and children. *J Am Coll Cardiol* 2001;38(2):401–408.
3. Yeager SB, Hougen TJ, Levy AM. Sudden death in infants with chaotic atrial rhythm. *Am J Dis Child* 1984;138(7):689–692.
4. Southall DP, et al. Frequency and outcome of disorders of cardiac rhythm and conduction in a population of newborn infants. *Pediatrics* 1981;68(1):58–66.
5. Houyel L, Fournier A, Davignon A. Successful treatment of chaotic atrial tachycardia with oral flecainide. *Int J Cardiol* 1990;27(1): 27–29.

12

Ventricular Tachycardia

Craig Byrum

The diagnosis of ventricular tachycardia (VT) is often assigned in children with hesitation. One explanation is that pediatric cardiologists and pediatricians understand that VT is a potentially lethal tachyarrhythmia that is seen mostly in a adult patients with ischemic heart disease. Even when young patients have scars on their chest from previous congenital heart disease surgery, the tendency is to consider first other diagnoses in patients with wide QRS complex tachycardia. In fact, most young patients with VT have a structurally normal heart. Interestingly, the tachycardias of ventricular origin are commonly benign in children. Thus, the clinical presentation in a child with VT is a function not only of the rate of the tachycardia but also of the underlying health of the myocardium, the health of the cardiac valves, and the reactivity of the systemic vascular bed in supporting blood pressure during the episode. Accordingly, the clinical response to the patient can be tailored to the individual clinical presentation; it may vary from simple conservative management without any intervention to emergent cardioversion leading ultimately to invasive management.

DEFINITION

The diagnosis of VT implies an arrhythmia that arises from the ventricular myocardium or from the specialized conduction tissue distal to the His bundle. One may define VT by a variety of inter-related frames of reference or criteria (Table 1). It is customary to define VT by its duration: a salvo of VT is a few beats in a row of ventricular origin beats. Non-sustained VT refers to a longer salvo, which self terminates before 30 seconds and does not result in hemodynamic deterioration. "Sustained VT" refers to an episode extending beyond 30 seconds. Repetitive VT presents as one salvo or episode of non-sustained ventricular tachycardia occurring right after another interrupted by a few sinus beats. In this form, the beats of ventricular origin may actually outnumber those of sinus origin, and yet symptoms of palpitations may be absent. The term incessant VT refers to lengthy sustained VT episodes that dominate the cardiac rhythm and most often present in infancy. The infant is frequently asymptomatic and does not require (or respond to) medications.

TABLE 1. Factors to Consider in Clinically Defining VT

Asymptomatic	Symptomatic
Non-sustained	Sustained
Monomorphic	Polymorphic
No heart disease	Structural or functional myocardial disease
Focal myocardial process	Diffuse myocardial process
Specific drug response (e.g., calcium blocker)	No specific drug response
Exercise suppressible	Exercise provocable
Automatic or triggered automatic	Reentry or multiple reentrant
No underlying electric disease	Underlying electrical disease

A second criterion for VT diagnosis is defined by the electrocardiographic morphology. Monomorphic VT refers to one dominant QRS form during VT. In contrast, polymorphic VT refers to a beat-to-beat change in the QRS shape and duration within the episode. A specific form of polymorphic VT is "bidirectional tachycardia" where the QRS axis typically shifts across the baseline. Polymorphic VT is more predisposed to degeneration into ventricular fibrillation.

A third characteristic is the relationship between the underlying cardiac structure and function. VT may be seen in the presence of a structurally normal heart. On the other hand, VT may arise in the setting of acquired myocardial dysfunction, which can be either acute (ischemic, toxic, or inflammatory) or chronic (neoplastic, dysplastic, or myopathic) and the myocardial processes may be focal or diffuse. Examples are Kawasaki disease, tricyclic poisoning, myocarditis, focal cardiac tumors, fatty replacement areas in arrhythmogenic right ventricular dysplasia, and dilated or hypertrophic cardiomyopathy. Lastly, ventricular tachycardia is well known to occur in certain types of congenital heart disease, particularly in the presence of post-operative scars, fibrosis, patches, conduits, incisions, and suture lines.

A fourth feature of VT is its relationships to physical effort. Some ventricular tachycardias are exercise provoked where others are suppressed by exercise and increasing sinus rates. Exercise-induced VT may be found in individuals with a history of exercise-provoked palpitations or syncope.

A fifth feature used to classify VT is its underlying electrophysiologic mechanism. Understanding the mechanism may be very important in deciding between medical versus invasive management. Three known mechanisms occur: reentry, automaticity, and triggered automaticity. Reentrant arrhythmias will be seen in settings where the heart has sustained damage locally through scar formation after surgery or more diffusely from myocardial ischemia due to coronary artery anomalies, or to coronary perfusion abnormalities related to chronic abnormal loading conditions. Whereas most reentry causing VT occurs within the working ventricular myocardium, reentry utilizing the bundle branch and Purkinje system may rarely be seen (almost unique to the dilated cardiomyopathy patient). Automatic tachycardia, on the other hand, appears to arise more often in the setting of a structurally normal heart. Stretch on the myocardium may induce automaticity in patients with abnormal preload with or without scars.

A sixth characteristic VT in the young is its response to a specific drug, albeit in a small number of patients. A number of right ventricular and left ventricular tachycardias may be suppressed with adenosine, beta blockers, and one specific left ventricular type of VT with calcium channel blockers. Some VT arrhythmias are mediated by cyclic-AMP and others are provoked by catecholamines.

Finally ventricular tachycardia is associated with primary electrical heart disease, generally affecting repolarization of the heart. The most well known of these is the congenital long QT syndrome. Some of these ventricular tachycardias are catecholamine-provoked whereas others come about after pauses in cardiac rhythm. Recently, the Brugada syndrome has been identified as an important cause of sudden death in certain subpopulations (Chapter 18). Nonetheless, it is important to note that a number of young patients present with VT and/or sudden death episodes in whom no underlying structural, functional or electrical abnormality can be identified.

CLINICAL PRESENTATION

A variety of clinical presentations may be seen with VT. Those patients who are asymptomatic often have no underlying structural heart disease and have either repetitive non-sustained VT or sustained but rather slow VT (≤150 bpm). Those patients with non-sustained VT may be detected by an irregular heartbeat during routine primary care. There are some patients who have a sensation of racing heart during the tachycardia; those patients may be experiencing either non-sustained or sustained VT. Accelerated idioventricular rhythm comprises a group of asymptomatic VT patients; this VT is benign and typically does not exceed 120/minute. It rarely requires treatment or leads to an unfavorable outcome. On the other hand, VT at faster rates over time may lead to a tachycardia-mediated cardiomyopathy. In the patient presenting with clinical congestive heart failure in a structurally normal heart, the issue between the cause and effect may be difficult to resolve. Patients with VT secondary to a dilated cardiomyopathy may need an implantable cardiovertor-defibrillator (ICD) with or without antiarrhythmic therapy and eventually perhaps cardiac transplantation. A patient whose chronic tachycardia has caused the cardiomyopathy and clinical heart failure will benefit from primary treatment of the tachycardia either through medication or catheter ablation, which often reverses the cardiac dysfunction and symptoms.

The most dramatic but probably the least common clinical presentation of VT in the young is syncope or sudden death. These events are unusual in a structurally normal heart save for those with underlying electrical heart disease or those with infiltrative or hypertrophic diseases of the myocardium. Most of the patients with this presentation will have had surgery for congenital heart disease or have an acutely or chronically dysfunctional myocardium after an infectious or inflammatory insult.

ELECTROCARDIOGRAPHIC FINDINGS

Certain ECG findings should always be sought for in the diagnosis of VT. The QRS duration during VT is longer than the normal sinus rhythm QRS complex for age. It is thus useful to consider the duration of the QRS as a relative measure. For example, a VT in an infant may show a QRS duration of only 70 or 80 msec, well within normal limits for an older child but is wider than the normal 40 or 50 msec seen during sinus rhythm in an infant. Beats of ventricular origin that comprise the VT arrhythmia show both initial and terminal conduction delay because of the slower cell-to-cell conduction occurring through most of the arrhythmia cycle. This finding may be difficult to appreciate in younger patients especially with faster heart rates but may be unmasked by increasing the gain (20 mm/mvolt gain—double standardization) and the sweep speed on the electrocardiogram (50 mm/sec paper sweep—double speed). A second element of the electrocardiographic diagnosis is atrioventricular (AV) dissociation (Figure 1), due to the independent activities of the VT mechanism and the sinus node. The demonstration of AV dissociation may require very careful inspection of continuous tracings. The diagnosis may be very difficult to make at high VT rates since the slower sinus origin

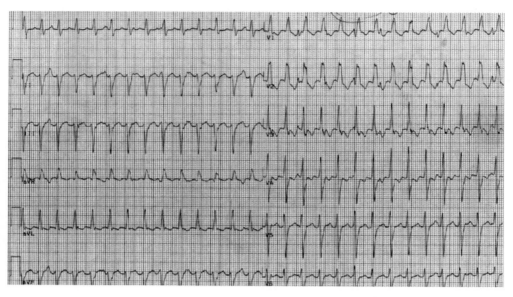

FIGURE 1. Electrocardiogram in a 13-year-old presenting with findings of congestive heart failure. This wide complex tachycardia exhibits a right bundle branch block morphology and left superior axis and shows a ventricular rate of 165. AV dissociation is nicely demonstrated in lead V1 where P-waves are seen marching through at a slower rate of 113 per minute. This tachycardia proved to be highly sensitive to calcium channel blocker therapy. After resolution of congestive heart failure symptoms, radiofrequency ablation along the inferior aspect of the left ventricular septum rendered the VT non-inducible after activation and pace mapping using 2 catheters coming in from transseptal and retrograde aortic approaches. This patient has been arrhythmia free for 10 years and with normal left ventricular function.

P-waves may be buried either in the long QRS or the abnormal ST-T wave complex. If AV dissociation is present, a long rhythm strip may reveal fusion and capture beats in which the sinus node exerts at least 50% influence over ventricular excitation on at least one beat. The likelihood of observing these beats is inversely proportional to the VT rate. An exception to the AV dissociation rule is the young patient with VT who has a structurally normal heart. In these patients, it is not uncommon to observe 1:1 ventriculo-atrial (VA) conduction across the healthy AV node (Figure 2) or, less frequently, ventriculo–atrial periodic conduction block. When Wenkebach VA conduction is observed during VT, the arrhythmia is most likely of ventricular origin (Figure 3). When there is 1:1 VA conduction, the diagnosis of VT is not assured. Adenosine given to such a patient will break the tachycardia if it is an AV nodal dependent supraventricular tachycardia. In the VT patient, transient VA block induced by the adenosine reveals the persistent wide QRS complex to be independent of the AV node and atria.

Often one can use vector analysis to predict the direction of the VT in the frontal plane as well as in the horizontal plane, and thereby derive a hypothesis as to the origin of the tachycardia (left versus right; superior versus inferior; anterior versus posterior—Figure 4).

DIFFERENTIAL DIAGNOSIS

When confronted with a wide QRS tachycardia in a young patient, the clinician to should consider the diagnosis of VT first. After failing to meet the criteria for VT as outlined above, other considerations may include sinus tachycardia or atrial tachycardia with either preexcitation or preexistent bundle branch block. The other considerations would be orthodromic atrioventricular reentry

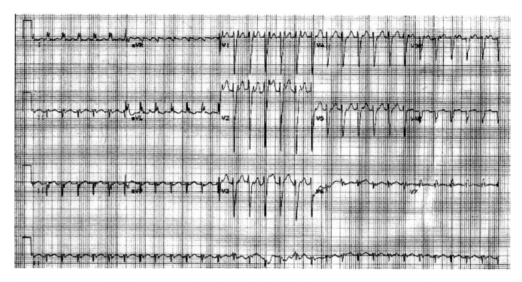

FIGURE 2. Top panel: Wide complex tachycardia in a 3-day-old with no symptoms who had a fetal tachycardia just before delivery. The QRS is wide for age at 100 msecs and exhibits a left superior axis and is negative across the precordium. Origin from the lower anterior freewall of the ventricular myocardium is suspected. The terminal portion of the QRS is notable for showing 1:1 VA conduction during the VT rate of 180 per minute in this patient with a structurally normal heart. Bottom panel: Holter monitor strip on same patient on oral beta blocker trial showing lack of VA conduction and fusion complexes (beat 3 and others) and sinus captures (beat 6 and others).

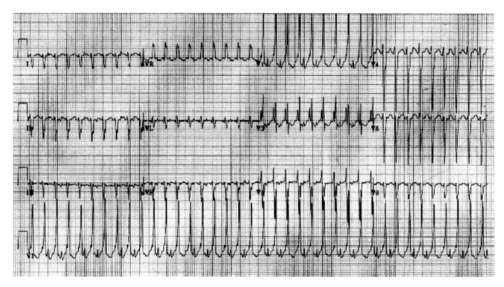

FIGURE 3. Ventricular tachycardia presenting in a 22-day-old with mild LV dysfunction. The tachycardia exhibits a right superior axis and right bundle branch block pattern. The QRS duration is only 90 msec which, for a neonate, is a prolonged QRS. Note the presence of retrograde VA conduction in a 3:2 Wenkebach pattern (bottom tracing). The arrhythmia responded to intravenous verapamil given by slow push after a preparatory intravenous dose of calcium. The infant was maintained in continuous remission on oral verapamil and then went into spontaneous remission off medication with no VT at 18 months of age.

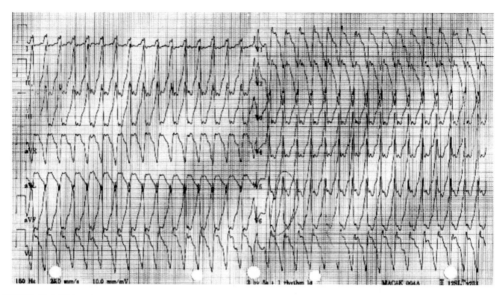

FIGURE 4. This VT presented in a 31-year-old followed late after repair of tetralogy of Fallot. He underwent repair with a transannular patch at 6 years of age. Persistent right ventricular volume overload necessitated placement of a pulmonary valve at 29 years of age. This VT was seen two years later. The left inferior axis and left bundle branch block morphology predicted origin from the right ventricular outflow tract. Following an ablation procedure in that area guided by non-contact mapping, the ventricular tachycardia was non-inducible.

tachycardia with bundle branch aberrancy; or antidromic atrioventricular reentry utilizing either typical or atypical (Mahaim) accessory pathways. Also AV node reentry with either bundle branch block aberrancy or bystander ventricular preexcitation would be a consideration. In each of these instances, AV association would be expected. Rarely, preexcited atrial fibrillation in a WPW patient could be confused with VT but marked irregularity of both the QRS rate and morphology are clues to the diagnosis (Chapters 4, 5 and 20).

CLINICAL INVESTIGATION

Important points from the history in a patient presenting with ventricular tachycardia include noting the timing—whether the symptoms are paroxysmal or chronic. Associated symptoms during tachycardia may include dizziness, particularly at the onset of tachycardia, vague chest discomfort, some dyspnea, weakness, and headache. The history, as in those patients with supraventricular tachycardia, will typically indicate a sudden cessation of the arrhythmia. A history of acquired or congenital heart disease and treatment is important. A history of substance abuse, exposure to toxins or contact with viruses known to be causal in the diagnosis of myopericarditis is significant. The family history should focus on syncope, seizures, sudden cardiac death, or specific familial cardiac arrhythmias.

Findings from the examination reflect the inefficiency of cardiac contraction and the AV dissociation. Pulses may be weak. The observation of cannon waves in the neck reflecting episodic atrial contraction against a closed AV valve would be helpful, but are difficult to discern in an infant. Diastolic filling sounds in the heart may reflect underlying cardiomyopathy. Pathologic murmurs or chest scars signal the presence of congenital heart disease (or previous cardiac surgery). Evidence for electrolyte abnormalities, inflammation, cardiac ischemia, and thyroid abnormality may be helpful.

Echocardiography with Doppler for those without a history (or surgery) of congenital heart disease should examine for focal or diffuse structural or functional myocardial abnormalities such as cardiac tumors, right ventricular dysplasia with fatty replacement, or abnormal hypertrophy, as well as myocardial dysfunction resulting from abnormalities of preload or contractility.

Magnetic resonance imaging may be useful but not diagnostic in detecting fatty replacement in the patient with arrhythmogenic right ventricular dysplasia,. In addition, it offers the advantage of viewing the ventricles in three dimensions and may give a better view of both anatomy and function. Patients who have other focal myocardial disease below the detection threshold of echocardiography may be discovered by this imaging technique.

Treadmill exercise testing is of limited use except for assessing the effectiveness of antiarrhythmic or ablative therapy in those patients who have exercise-provokable VT. In addition, it may be diagnostic in a patient whose symptoms are consistently exercised-provoked. Likewise, Holter monitoring may be useful for assessing lack of induction of VT by spontaneous premature ventricular extrasystoles, as well as a measure of the density of single ventricular premature beats or non-sustained VT episodes.

Transcatheter electrophysiologic investigation is usually reserved for the VT patient in whom an intervention is anticipated. As part of the electrophysiologic study, hemodynamic data and perhaps angiography will be helpful to outline the current hemodynamic state of the heart and circulation. In the current era, electrophysiologic investigation may be indicated for those patients with a high density of premature ventricular beats which may serve as triggers for VT and symptoms suspicious for tachyarrhythmia. It may also be indicated for those patients with underlying structural heart disease, especially those with post–operative congenital heart disease who have non-sustained VT, to unmask potentially life-threatening inducible sustained VT. In individual circumstances, electrophysiologic study may be performed to target the VT focus or reentry circuit for ablation. A final group of patients may need electrophysiologic investigation to check for effectiveness of orally administered anti-arrhythmic therapy. Patients with underlying structural or functional abnormalities of the heart and VT who have presented with life–threatening symptoms should promptly be considered for implantation of an ICD.

TREATMENT

The acute medical response to a patient with VT is a function of the hemodynamic compromise. The first line response in the patient with wide QRS complex tachycardia, congestive heart failure or shock is direct current cardioversion. If patients present without hemodynamic compromise but solely with palpitations, tailoring a deliberate response to the history, clinical situation, and electrocardiographic findings is preferred. Because some patients with VT in a structurally normal heart may respond to adenosine, a trial of this agent may be helpful. VT that arises from the posterior inferior aspect of the left ventricular septum and exhibits a right bundle branch block and superior axis morphology (Figure 1) may be calcium channel dependent and respond to slow intravenous push of verapamil (Figure 3). In infants in this category, a preparatory slow intravenous push of calcium at usual resuscitative dose will help avoid any lowering of blood pressure if the verapamil is given too quickly. In patients with other VT morphologies who are clinically stable, a trial of procainamide slow push over 20 minutes (10–15 mg/kg) is acceptable. The drug would be effective for any VT that is procainamide responsive but also would likely convert most other wide QRS tachycardias that are dependent on an accessory pathway. Intravenous amiodarone may also be safely used in most patients. Any acute medical management of a VT is best performed in a setting where cardioversion/defibrillation can be accomplished if necessary.

After initial management and conversion to sinus rhythm, chronic therapy is considered. Because improvement in function may reduce the frequency of VT and the associated symptoms, any underlying functional hemodynamic abnormality of the heart should be corrected before embarking on therapy specific for the VT. An example is reduction of right ventricular volume overload by pulmonary valve replacement in a patient with free pulmonic insufficiency after repair of tetralogy of Fallot.

In patients with minimal symptoms and no underlying structural or functional myocardial disease, conservative management is appropriate, anticipating the high likelihood of spontaneous resolution of the arrhythmia (Figure 5). In older patients without heart disease, conservative management may be safe for those with repetitive non-sustained VT in the absence of symptoms and in the absence of any evidence for ventricular dysfunction (Figure 6).

The second option for treatment is antiarrhythmic therapy. In certain situations this can be guided by the morphology of the tachycardia and the clinical situation. Beta blockers may be very effective for patients who have no underlying heart disease and modestly symptomatic exercise-provocable monomorphic VT from the right or left ventricular outflow tract. Likewise, chronic oral verapamil therapy is well tolerated and effective for patients with calcium sensitive left VT. Digitalis has no direct role in the treatment of ventricular arrhythmias but may offer indirect benefit by improving myocardial function. Similarly the use of low dose beta blockers or afterload reduction therapy in patients with VT associated with heart failure may reduce the arrhythmia burden.

Type IA and Type IB antiarrhythmic agents such as procainamide, quinidine, disopyramide, diphenylhydantoin and tocainide play a diminishing role in the treatment of VT primarily due to their high side effect profile as well as the availability of more effective antiarrhythmic medications. Thus, the use of Type IA or Type IB agents should be individualized to the patient who cannot benefit from catheter ablative therapy or who has benefited from an individual medication demonstrated at electrophysiologic study and clinical follow-up to suppress the tachycardia.

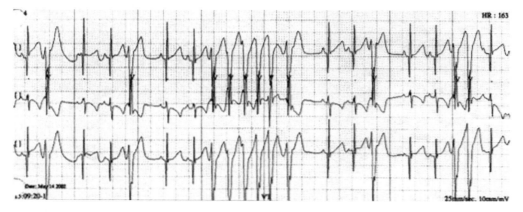

FIGURE 5. Three-channel Holter tracings showing high density ventricular ectopic activity occurring in singles but also repetitively as VT salvos up to 6 in a row in a 4-month-old with no symptoms and with no abnormality on Doppler echo assessment. The VT rate is 255 per minute average. This arrhythmia subsided spontaneously two months later. A trial of oral propranolol had no effect on the arrhythmia and it spontaneously resolved off medication within 2 months.

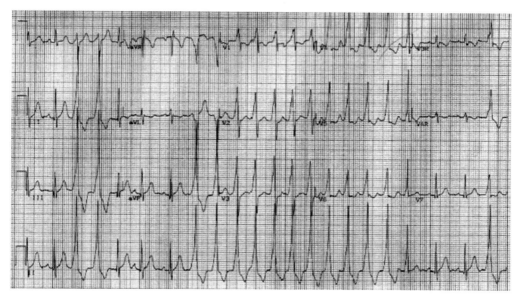

FIGURE 6. Twelve-lead electrocardiogram in an asymptomatic 9-year-old without underlying structural heart disease shows a salvo of 11 in a row terminating with a fusion complex. The left bundle branch block morphology and inferior axis is typical for a high septal origin VT. The positive QRS complex in lead V1 may point to a left anterior septal location for this tachycardia. It completely suppressed with exercise early into a Bruce protocol at a sinus rate of 150 and then slowing re-emerged in recovery as sinus rates dropped. The patient is followed conservatively without medication.

Type IC agents have been shown to be effective in the treatment of supraventricular tachycardia in the young. In an increasing experience in VT patients, they have a low side effect profile and can be used safely in patients with and without structural heart disease except for individuals with coronary artery ischemia. Because pro-arrhythmia seems to be an early phenomenon seen soon after initiating medication, these medications should be started in a well-monitored setting if there is any underlying structural heart disease. Caution should be exercised in patients with ventricular dysfunction, particularly after heart surgery.

Type III agents may also be highly effective in the suppression of VT. Sotalol has a beta blockade effect in addition to its Type III action and maybe doubly beneficial in certain situations. While amiodarone may be highly effective, the benefit needs to be weighed against the risk of long-term side effects when a young patient is placed on chronic therapy (Chapter 21).

SURGICAL MANAGEMENT

In the current era of effective catheter ablation, effective anti-arrhythmic medication, and an increasing experience in the young with the use of ICDs that can be placed without thoracotomy (Chapter 17), the role of surgical management of VT has markedly decreased. Available techniques would include, with careful intra-operative mapping, excision or cryoablation of discrete areas of automatic activity or reentry circuits. An approach of this type could be justified in a patient in whom the myocardial lesion was too big or too epicardial for catheter ablative techniques or who required cardiac surgery. Alternatively, with the location of VT origin confirmed at EP study before surgery, wide excision of larger lesions such as aneurysmal tissue that houses a VT could be contemplated. Emerging experience with pericardial approaches to ablation suggests that even these unusual cases may ultimately be able to be handled without the need for surgery.

Catheter Ablation

Catheter ablation of VT is most successful in patients with a structurally normal heart and focal origin of their tachycardia. The procedure is indicated for patients who have symptomatic sustained or repetitive non-sustained type monomorphic VT; for patients with exercise provocable VT from the right ventricular outflow tract who are not responsive to beta blockade; and for those patients with calcium sensitive left ventricular septal tachycardia who have breakthrough on calcium blocker therapy. Sustained monomorphic VT from the right ventricular outflow tract in a few patients may originate from the left side of the outflow septum requiring a retrograde aortic and/or a transseptal—transmitral approach. Another group of patients in whom catheter ablation can be contemplated would be those with arrhythmogenic right ventricular dysplasia. Caution must be exercised however since the dysplasia is often multi-focal and since ablation of the fatty dysplastic tissue is probably more risky.

Patients with underlying structural heart disease are more likely to have reentry circuits. The simplest situation is the patient with a single reentry circuit. The first requirement for the reentry circuit is an electrically active corridor of myocardium available for conduction between two electrically inert boundaries. An example of two inert boundaries that defines such a corridor would be in a patient after surgery for tetralogy of Fallot who has a vertical ventriculotomy scar on the anterior surface of the right ventricle. The top of the scar could define one boundary and the pulmonary valve annulus, the other boundary. Electrical excitation could travel through this corridor either from left to right or from right to left. If, for instance, excitation emerged from the right side of the corridor it could travel down the front of the heart, around the bottom of the ventriculotomy scar, and then back up the front of the heart, coursing along the left side of the scar to re-enter the corridor from the left. Slow conduction through this corridor is necessary to allow recovery of excitability of cardiac tissue in the outer loop of the circuit, which traverses around the ventriculotomy scar. Small near-field electrical potentials may be recorded from these corridors in diastole. Such a simple reentry circuit invites ablation across the corridor between the inert boundaries, thus interrupting reentry through the corridor in either direction. Other possible well-defined inert boundaries for corridors in the heart would include the AV valves, septal patches, or scars. Difficulty in assessing and treating reentry VT arises when there are multiple reentrant circuits set up by multiple inert boundaries caused by irregular or inhomogeneous scars within the myocardium. When contemplating catheter assessment and treatment of VT, a complete review of the underlying cardiac structure and surgical techniques is needed so that one can anticipate all possible corridors of reentry circuits.

Several approaches are available for mapping and ablation of VT. Single point catheter mapping can be effectively accomplished for focal origin tachycardia. The ventricle of origin is first quickly worked out and then using two ablating catheters during a sustained monomorphic tachycardia, earliest activation is located with one catheter; the second catheter then seeks an even earlier site of activation—locally well ahead of the onset of surface QRS activity. When that point is reached, it becomes a new reference and then the first catheter explores for an even earlier site. This process is continued in an iterative manner until the earliest site is located. Using computer-assisted navigation systems or simple sequential point mapping, this technique can be accomplished with a single mapping catheter. Radiofrequency application at the earliest site as defined will usually yield results marked by a brief acceleration of the tachycardia focus as it warms followed up, by cessation of tachycardia as the focus is extinguished. Pace mapping techniques can be used when there is an inducible, hemodynamically unstable tachycardia precluding prolonged activation mapping. A site in the ventricle is sought where pacing has an exact 12-lead ECG match to the induced tachycardia. If the VT mechanism is triggered automaticity at a focal site, then ablation at the best pace map site has a high likelihood of success. If, on the other hand, the tachycardia is due to a reentry circuit, an exact pace map may be found anywhere along the slowly conducting corridor and a focal ablation is less likely to be successful, since complete transection of the corridor is required. Elegant techniques of establishing the critical slow conduction zone by observing VT behavior during and just after pacing with concealed entrainment are helpful in identifying appropriate ablation targets in reentry VT. The advent of electro-anatomical mapping and non-contact mapping has allowed more precise delineation of early excitation sites from focal tachycardia. These techniques provide three dimensional representation of the heart including delineation of the corridors of slow conduction between low voltage inert zones. The non-contact system offers the benefit of outlining the reentry circuit or origin of tachycardia in a single beat so that sustained tachycardia is not required—an especially important benefit in those with hemodynamically unstable VT. The end point after ablation in these cases is demonstration of bidirectional block across the active critical corridor and lack of inducibility of the target arrhythmia (Figure 7).

Effective ablative therapy in the ventricular myocardium is still limited by the ability to create transmural lesions. Any gap in the line of lesions may led to residual areas of slow conduction and future reentry arrhythmias. Cool catheter tip techniques of ablation delivering greater thermal energy to deeper myocardial layers or electrode arrays designed to simultaneously deliver a coalescence of lesions along a line are strategies that address these concerns. The experience with these techniques in children, however, is limited.

Device Therapy

Developments in cardioverter defibrillator devices have been rapid in the last decade and young patients with VT, as well as older patients with congenital heart disease, may definitely benefit from the technology (Chapter 17). Patients without underlying structural heart disease who may need device therapy would be those with long QT interval syndrome or other primary electrical diseases (e.g. Brugada syndrome). Although arrhythmogenic right ventricular dysplasia would be a presentation unusual in childhood, those patients may need device protection. Patients with dilated or hypertrophic cardiomyopathy may present with lethal ventricular tachyarrhythmias and need device therapy. Among patients with congenital heart disease, the largest group needing device placement are older patients many years after repair of tetralogy of Fallot. In addition, a growing number of patients followed many years after physiologic correction (atrial switch) of transposition will likely develop VT (in addition to the well known atrial arrhythmias) and require ICDs. Finally, the increasing number of patients who have completed the surgical staged therapy for single ventricle presents unique challenges.

In addition to the gradual miniaturization of the devices, a major advance in device therapy for patients with ventricular tachyarrhythmias has been the development of the non-thoracotomy approach. Currently available approaches include transvenous, subcutaneous patches/arrays, and active-can technology, allowing for a variety of vectors of current across the heart. Incorporation of brady and anti-tachycardia pacing into ICD units and combining pace/sense function into defibrillating leads has also benefited those

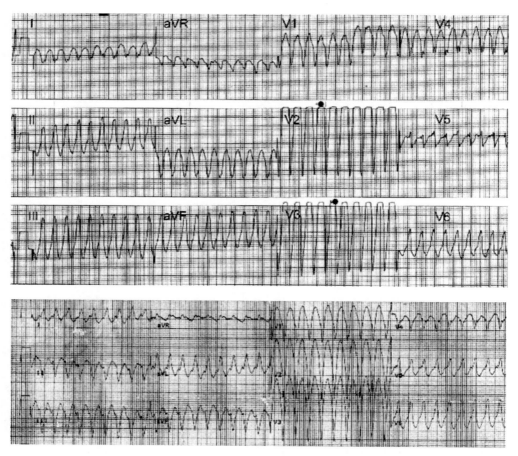

FIGURE 7. Ventricular tachycardia presenting in a 41-year-old male late after repair of tetralogy of Fallot at 9 years of age. Top three tracings: Shows a left bundle branch block and right inferior axis. Bottom panel: From a different clinical presentation in the same patient shows a left bundle branch block and left superior axis. This is probably the same tachycardia circuit being used in opposite directions. At catheterization, a site midway down the medial anterior right ventricular anterior freewall was located, which demonstrated features of concealed entrainment and had mid diastolic potentials characteristic of a slowly conducting zone. Ablation of this site rendered tachycardia of any kind non-inducible.

patients needing ICD therapy. While the availability of transvenous ICD systems is a benefit, the risk of a transvenous system in a young patient includes not only venous thrombosis limiting later access, but also possible future need for lead extraction in the growing patient. Approaches to the heart need to be creative in those small patients with complex congenital heart disease. In patients who have had physiologic correction of transposition, the more posterior orientation of the anatomic left atrium (now the systemic [blue blood] atrium) and left ventricle (the sites of pacing) places the leads in closer proximity to the left phrenic nerve, risking phrenic or diaphragm stimulation. In those young patients who have gone through the 3-stage surgical path (Fontan) to single ventricle palliation, there is no systemic venous route to the ventricle. Important sensing challenges are, therefore, present. In addition, vectors of current for effective defibrillation need to be verified for each patient.

The problem of inappropriate shock in children with VT who require ICDs due either to oversensing or to sinus tachycardia triggering tachycardia detection (i.e. heart rate) has not been fully resolved. Also, the psychological burden of living with an ICD needs to be carefully addressed. The awareness by the child of why the ICD was implanted in the first place can be frightening; but the anticipation that the device may go off or that it might go off and not work in the intended way can also create a chronic psychological stress.

SUMMARY

Ventricular tachycardias in the young are a diverse group of rhythm disorders different than those seen in the adult with ischemic heart disease. There are not only a wide variety of ventricular substrates but also a variety of electrophysiologic mechanisms. The ability of many young patients to tolerate VT allows the clinician to approach the arrhythmia in a deliberate fashion. On the other hand, VT remains an important cause of sudden death in the young and identifying diagnostic possibilities for sudden death in a patient presenting with VT requires a consideration of a variety of diagnostic and treatment strategies. In many young patients, VT, that so often in adults is associated with risk of sudden death, may spontaneously remit. However, ventricular tachycardia after surgery for congenital heart disease calls for a comprehensive evaluation based on a complete understanding of the clinical state of the patient, the natural history of the arrhythmia, the cardiac anatomy, the surgical history, and the electrophysiologic properties of the arrhythmia.

SUGGESTED READING

1. Villain E, Bonnet D, Kachaner J, et al. Incessant idiopathic ventricular tachycardia in infants. Review. In French. *Arch Mal Coeur Vaiss* 1990;83(5):665–671.
2. Glikson M, Constantini N, Grafstein Y, et al. Familial bidirectional ventricular tachycardia. *Eur Heart J* 1991;12:741–745.
3. Gemayel C, Pelliccia A, Thompson P. Arrhythmogenic right ventricular cardiomyopathy. *J Am Coll Cardiol* 2001;38:1773–1781.
4. Bhandari AK, Hong RA, Rahimtoola SH. Triggered activity as a mechanism of recurrent ventricular tachycardia. *Br Heart J* 1988;59(4):501–505.
5. Lerman BB, Stein KM, Markowitz SM. Adenosine-sensitive ventricular tachycardia: a conceptual approach. Review *J Cardiovasc Electrophysiol* 1996;7(6):559–569.
6. DeLacey WA, Nath S, Haines DE, et al. Adenosine and verapamil-sensitive ventricular tachycardia originating from the left ventricle: Radiofrequency catheter ablation. *PACE* 1992;15(12):2240–2244.
7. Towbin JA, Li H, Taggart RT, et al. Evidence of genetic heterogeneity in Romano-Ward long QT syndrome. Analysis of 23 families. *Circulation* 1994;90(6):2635–2644.
8. Brugada P, Brugada J. Right bundle branch block, persistent ST segment elevation and sudden cardiac death: a distinct clinical and electrocardiographic syndrome. *J Am Coll Cardiol* 1992;20:1391–1396.
9. Gaum WE, Biancaniello T, Kaplan S. Accelerated ventricular rhythm in childhood. *Am J Cardiol* 1979;43:162–164.
10. Singh B, Kaul U, Talwar KK, et al. Reversibility of "tachycardia-induced cardiomyopathy" following the cure of idiopathic left ventricular tachycardia using radiofrequency energy. *PACE* 1996;19(9):1391–1392.
11. Carlson MD, White RD, Trohman RG, et al. Right ventricular outflow tract ventricular tachycardia: Detection of previously unrecognized anatomic abnormalities using cine magnetic resonance imagining. *J Am Coll Cardiol* 1994;24(3):720–727.
12. Beaufort-Krol GCM, Bink-Boelkens MTE. Oral propafenone as treatment for incessant supraventricular and ventricular tachycardia in children. *Am J Cardiol* 1993;72:1213–1214.
13. Fenrich AL Jr, Perry JC, Friedman RA. Flecainide and amiodarone: combined therapy for refractory tachyarrhythmias in infancy. *J Am Coll Cardiol* 1995;25(5):1195–1198.
14. Tipple M, Sandor G. Efficacy and safety of oral sotalol in early infancy. *PACE* 1991;14:2062–2065.
15. Sosa E, Scanavacca M, D'Avila A, Pilleggi F. A new technique to perform epicardial mapping in the electrophysiology laboratory. *J Cardiovasc Electrophysiol* 1996;7:531–536.
16. Darbar D, Olgin J, Miller J, Friedman P. Localization of the origin arrhythmias for ablation: from electrocardiography to advanced endocardial

mapping systems. *J Cardiovasc Electrophysiol* 2001;12:1309–1325.
17. Callans DJ, Ren JF, Michele J, Marchlinski FE, Dillon SM. Electroanatomic left ventricular mapping in the porcine model of healed anterior myocardial infarction: correlation with intracardiac echocardiography and pathological analysis. *Circulation* 1999;100:1744–1750.
18. Schilling RJ, Peters NS, Davies W. Feasibility of a noncontact catheter for endocardial mapping of human ventricular tachycardia. *Circulation* 1999;99:2543–2552.
19. Borggrefe M. German multicenter study: chilled ablation of ventricular tachycardia in patients with structural heart disease. *Circulation* 1999;100:1513A.
20. Sears S Jr, Burns J, Handberg E, Sotile W, Conti J. Young at heart: understanding the unique psychosocial adjustment of young implantable cardioverter defibrillator recipients. *PACE* 2001;24:1113–1117.

13

Sick Sinus Syndrome

William A. Scott

Sick sinus syndrome is a clinical entity that has been associated with a variety of arrhythmias. One of the earliest descriptions by Lown stated: "a defect in the elaboration or conduction of sinus impulses characterized by chaotic atrial activity, changing P-wave contour, bradycardia, interspersed with multiple and recurrent ectopic beats with runs of atrial and nodal tachycardia." While there are numerous etiologies described in the literature, sinus node dysfunction in pediatric patients most often occurs secondary to injury to the node itself, its arterial supply and/or its autonomic innervation in the course of cardiac surgical interventions. This injury is usually not limited to the sinoatrial node, and abnormalities in atrial automaticity and conduction are included in the sick sinus syndrome (Table 1). There is evidence that the abnormal variation in heart rate has a destabilizing effect on the atrial tissue, contributing to the progression of the disease. While not specifically part of sinus node disease, these coexisting atrial and, to a lesser degree, atrioventricular conduction disturbances, support the concept that the pathology affecting the sinus node is diffuse.

Conditions associated with a high prevalence of sick sinus syndrome include atrial repair of d-transposition of the great arteries, complete repair for anomalous pulmonary venous drainage and atrial septal defect, and single ventricle palliation with the Fontan operation, although almost any open heart repair may result in sinus node impairment. Congenital sinus node abnormalities are less frequent. Sinus node malformation is associated with left atrial isomerism and sinus node dysfunction may coexist with congenital complete heart block. Familial clustering has been reported in the absence of structural heart disease. Extrinsic causes of sinus node dysfunction include autonomic imbalance and medications. Heavily conditioned athletes have bradycardia and sinus pauses of greater than 2 seconds due to prominent vagal influence. Recently, sleep apnea has been related to sinus node dysfunction. Many antiarrhythmia medications can further impair the sinus node function, particularly for patients with pre-existing abnormalities.

ECG CHARACTERISTICS

The ECG characteristics of sick sinus syndrome are variable and can change within a single patient. Not all patients will manifest all of the electrocardiographic findings noted below. Potential sources of electrocardiographic data include: standard 12–15 lead electrocardiograms, Holter monitoring, event recorders, and rhythm strips from physiologic monitors.

TABLE 1. Sick Sinus Syndrome

Electrocardiographic Manifestations of Sick Sinus Syndrome
  Sinus Bradycardia
  Extreme Sinus Arrhythmia
  Tachycardia-Bradycardia Syndrome
  Sinoatrial Exit Block
  Sinus Pause/Sinus Arrest
  Sinoatrial Reentrant Tachycardia

Electrocardiographic Findings Associated with Sick Sinus Syndrome
  Intra-atrial Reentrant Tachycardia (IART)
  Atrial Fibrillation
  AV Nodal Dysfunction

Clear documentation is not always present in short-term recordings. Multiple leads should be recorded to detail P-wave morphology.

Long-term recordings with Holter provide an overview of overall heart rate variation (Figure 1) and the prevalence of abnormally slow and fast rhythms. Ideally three orthogonal leads are recorded to optimize P-wave recognition (i.e., Lead I, Lead II or III, and a precordial lead). Event recorders, particularly those with continuous storage capabilities, allow for correlation of symptoms to electrocardiographic changes. Newer devices have programmable parameters for automated recording of tachycardia and bradycardia episodes. An implantable device is available for special circumstances that preclude wearing an external monitor.

Exercise testing is useful to assess the chronotropic response. All children and adolescents should be able to attain a heart rate of 180 bpm. Patients with sinus node dysfunction also may have exaggerated slowing of heart rate or pauses in the recovery period.

Sinus Bradycardia

Bradycardia is the most common feature of sinus node dysfunction. Bradycardia may be sustained or paroxysmal. Escape rhythms (Figure 2) arise from the atrium, atrioventricular (AV) node and ventricles, but these tissues are often dysfunctional and the escape rates are frequently lower than expected for age. Healthy children may have brief periods of sinus bradycardia with atrial or junctional escape rhythm in the absence of any sinus node pathology. The normal heart rate is age dependent. The lowest values observed during Holter recording are less than those from ECG recording.

Recording artifacts and other rhythm abnormalities should be excluded. An abnormally low heart rate may be due to second- or third-degree AV block, as well as atrial and junctional extrasystoles. A sudden loss of signal may appear similar to a sinus pause. There are often clues to this type of artifact including resumption of recording with a T-wave or a different time of drop out on another channel (Figure 3). Older Holter recorders using a tape drive mechanism can suddenly speed or slow, mimicking bradycardia or tachycardia. With this type of artifact, all components of the recording are expanded or compressed (Figure 4).

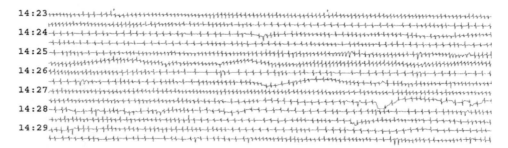

FIGURE 1. Full disclosure printout from a Holter monitor. Individual complexes are too small for detailed evaluation, but abnormal pattern of heart rate variation is apparent.

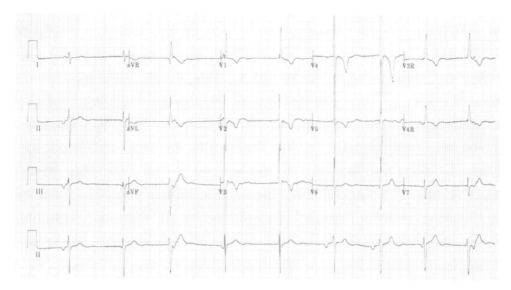

FIGURE 2. Sinus bradycardia with alternating atrial and junctional escape rhythms in a patient 15 years after Mustard repair of d-transposition of the great arteries.

Extreme Sinus Arrhythmia

Sinus arrhythmia is a normal pattern of heart rate acceleration and deceleration present in most children. A significant component of this variation is due to respiratory influences on the autonomic nervous system. Abnormal or extreme sinus arrhythmia is defined as greater than 100% variation in PP intervals. Sinus node exit block may also cause variation of this magnitude.

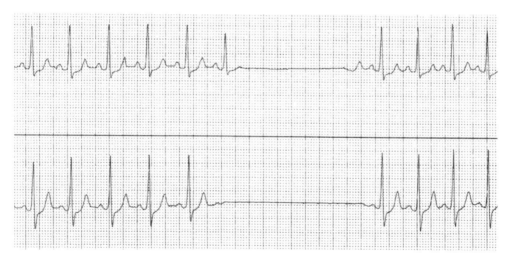

FIGURE 3. Simultaneous lead I (upper panel) and lead III (lower panel) from Holter recording. Apparent sinus pause is actually due to artifact. Signal loss is not simultaneous on both channels and lead I resumes rhythm with a T-wave, which is not possible.

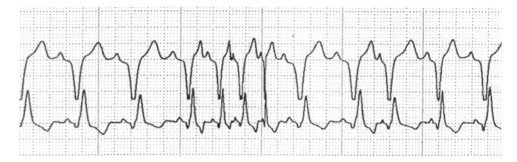

FIGURE 4. Simultaneous lead V1 (upper) and V5 (lower) from Holter recording. Apparent bradycardia and tachycardia due to artifact from changing tape speed. Note that all components of the rhythm (P-wave, QRS complex, and T-wave) are compressed with tachycardia.

Tachycardia–Bradycardia

The tachycardia–bradycardia syndrome is diagnosed in the presence of recurrent prolonged pauses or sustained bradycardia following paroxysms of tachycardia. The tachycardia causes exaggerated overdrive suppression of already impaired automaticity, worsening the bradycardia (Figure 5).

Sinoatrial Exit Block

Sinoatrial exit block is failure of an impulse generated in the sinus node to propagate normally to the atrium. The degree of block is classified just as for the AV node. A direct recording of the sinus node electrogram is necessary to diagnose first degree sinoatrial block. The pattern of PP intervals prior to a pause can be used to infer second-degree block. Mobitz I sinoatrial block (Figure 6) is manifest by a gradual shortening of PP intervals followed by a pause of less than two times the resting cycle length. Mobitz II is presumed to be the mechanism when a sudden pause of two times the resting cycle length is encountered (Figure 7). Both of these are difficult to recognize in the presence of sinus arrhythmia and both patterns may be due to abnormal impulse formation in the sinus node. Neither of these conduction abnormalities can be absolutely proven without direct recording of the sinus node electrogram. Complete or third-degree sinoatrial block is indistinguishable from sinus node arrest on the surface electrocardiogram.

Sinus Pauses and Sinus Arrest

Prolonged pauses are often present due to either sinoatrial node exit block or sinus arrest (Figure 8). Sinus pauses of less than 2 seconds

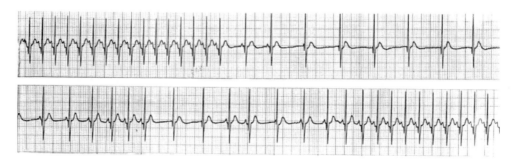

FIGURE 5. Tachycardia–bradycardia with junctional escape in a 2-year-old.

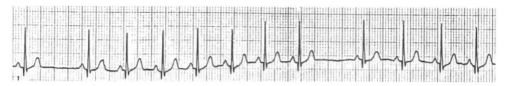

FIGURE 6. Lead I recording suggestive of Mobitz I type sinoatrial exit block. There is a gradual shortening of PP intervals, with a consistent P-wave morphology, followed by a pause of less than two times the resting cycle length. Validation of this phenomenon requires invasive recording of the sinus node electrogram.

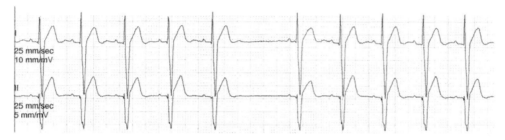

FIGURE 7. Simultaneous lead I and II recording suggestive of Mobitz II type sinoatrial exit block. There is a pause with a PP interval that is twice the resting cycle length. Validation of this phenomenon requires invasive recording of the sinus node electrogram.

are normal in young children and adolescents. A non-sinus rhythm at a rate lower than expected for sinus rhythm is referred to as an escape rhythm and is suggestive of sinus node arrest.

Sinoatrial Node Reentrant Tachycardia

Sinoatrial reentrant tachycardia is uncommon. It results from abnormal conduction within the sinus node or immediate perinodal tissue. The P-waves are identical to those seen with sinus rhythm, but associated with a paroxysmal increase and decrease in rate. This rhythm cannot be distinguished from a focal atrial tachycardia in close proximity to the sinoatrial node.

Atrial Flutter/Fibrillation

Typical atrial flutter is uncommon with sinus node dysfunction. More commonly atrial muscle reentry or intra-atrial reentrant tachycardia (IART) is observed (Figure 9). This arrhythmia most frequently occurs following surgical interventions for congenital

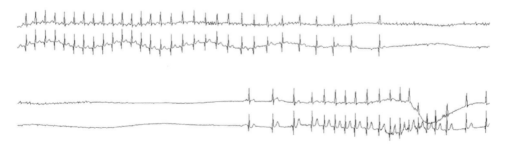

FIGURE 8. Sinus arrest during a breath-holding spell. This represents transient sinus node dysfunction due to a sudden increase in parasympathetic activity.

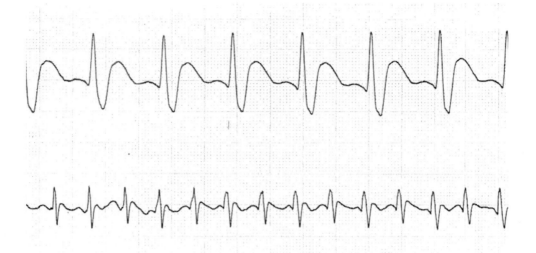

FIGURE 9. Lead I (upper trace) and esophageal electrogram (lower trace) demonstrating intra-atrial reentrant tachycardia with 2:1 conduction. The arrhythmia is not apparent from surface ECG but esophageal recording shows atrial rate twice that of the ventricular rate.

heart disease. The rate is slower than that usually ascribed to typical atrial flutter, usually around 180 to 300 bpm. There is an isoelectric interval between atrial activations, and the typical sawtooth pattern is not present, which confounds recognition of this rhythm. Unlike typical atrial flutter, 1:1 conduction to the ventricles is frequent. Most commonly 2:1 conduction is observed, but again at a rate slower than typical for atrial flutter. Variable conduction ratios may be misinterpreted as extrasystoles. Atrial fibrillation is a late finding with disease progression.

ELECTROPHYSIOLOGIC FEATURES

Electrophysiologic studies have contributed greatly to the overall understanding of sinus node physiology. Typically these studies are conducted invasively with the placement of catheters through the vascular system for endocardial stimulation and recording. Limited studies to assess sinus node function may also be performed with temporary epicardial or transvenous pacing wires (Figure 10), esophageal pacing, and with a previously implanted permanent pacemaker.

Presently, clinical history and noninvasive diagnostic studies primarily guide diagnosis and management. Electrophysiologic study is not routinely performed solely for the diagnosis of sick sinus syndrome because it is invasive, has relatively low sensitivity and specificity, and noninvasive studies provide adequate information for management. Invasive electrophysiologic studies of sinus node function may be performed as an adjunct to hemodynamic study prior to surgical intervention or in association with interventional procedures such as radiofrequency ablation.

The primary measures of sinus node function are the sinoatrial conduction time and the corrected sinus node recovery time. However, atrial refractoriness and intra-atrial conduction times are also prolonged with surgically-induced sick sinus syndrome.

Prolonged Sinoatrial Conduction Time

Sinoatrial conduction time is usually measured by an indirect method (Chapter 3),

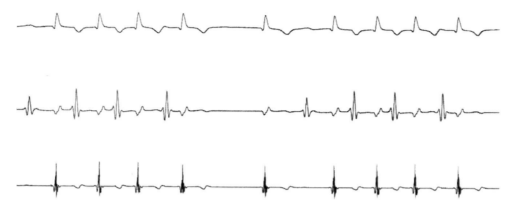

FIGURE 10. Recording of lead I (upper trace), atrial temporary pacing wire (middle trace), and ventricular pacing wire (lower trace). Patient has low amplitude atrial activity with first-degree AV block making rhythm determination difficult. Atrial wire electrogram consistent with sinus arrest or sustained sinoatrial exit block.

Strauss or Narula, which yields a total sinoatrial conduction time (TSACT). The TSACT reflects conduction into and out of the sinus node. These conduction times are not equal; therefore, it is more accurate to report a total sinoatrial conduction time rather than divide the TSACT by 2 and assume that the conduction velocity in both directions is equivalent. Normal values for pediatric patients are <200 msec. With the Strauss method, absence of a zone of reset suggests sinoatrial node entrance block, which is abnormal as well. Sinus arrhythmia invalidates this measurement.

Directly recorded conduction times from the sinus node electrogram reflect only conduction out of the sinus node and are usually shorter than the TSACT. As an independent measure of sinus node function, TSACT is relatively insensitive. Atropine has variable effects on TSACT in sinus node dysfunction. In many patients the measurement normalizes. Further prolongation of the TSACT is thought to be the result of depressed automaticity following enhanced conduction into the sinus node.

Prolonged Sinus Node Recovery Time

The maximum corrected sinus node recovery time (MCSNRT) (Chapter 3) is a measure of recovery of automaticity. Normal values for MCSNRT are <250–270 msec for children, <445 msec for adolescents, and <525 msec for adults. When expressed as a ratio, MCSNRT should be less than 150% of the sinus cycle length. At high rates, shortening of the return cycle may occur due to reflex responses to hypotension, sinus node entrance block, or local release of catecholamines. Abnormal results (Figure 11) may be secondary to excess vagal tone and normalize following autonomic blockade whereas patients with true dysfunction show no improvement following autonomic blockade. The specificity of this test may be enhanced by coadministration of disopyramide or autonomic blockade, but this has not been conclusively demonstrated in children. Other findings suggestive of sinus node dysfunction include secondary pauses, and a prolonged total recovery time, defined as greater than 5–6 beats to return to the resting cycle length. If MSCNRT occurs at long pacing cycle length, this supports the diagnosis of sinus node disease.

TREATMENT

Prospective, randomized clinical trials are not available to guide management in pediatric patients. The presence of sick sinus syndrome does not mandate therapy. Most

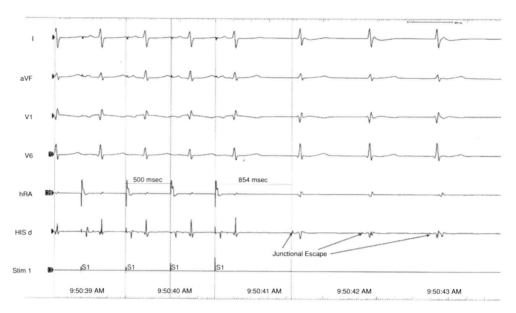

FIGURE 11. Abnormal response to overdrive pacing. The patient had a sinus cycle length of 550 msec. The sinus node should recover inn <275 msec. Here the recovery is prolonged (304 msec) and a junctional focus emerges before the sinus node. The junctional rhythm persists for 10 beats following termination of pacing.

patients with mild to moderate sinus node dysfunction are asymptomatic and there is not a significant risk for sudden death in the asymptomatic individual.

Acute symptomatic bradycardia may be treated with atropine, isoproterenol, or transcutaneous pacing. Temporary transvenous or transesophageal pacing can be established until a permanent pacing system can be implanted.

There is little role for drug therapy for the bradycardia associated with chronic sinus node dysfunction. Symptomatic bradycardia is treated with permanent pacing. Symptomatic bradycardia due to sinus node dysfunction is a clear indication (Class I) for pacemaker placement. Asymptomatic episodes of heart rates under 40 bpm or pauses >3 seconds in the child are less clear indications for intervention (Class IIa), and in the adolescent (Class IIb). Consideration should be given to the patient's size, cardiac anatomy, and AV nodal function, as well as the presence of any tachyarrhythmias. Atrial pacing has significant advantages over ventricular pacing for the treatment of sick sinus syndrome; therefore most patients receive atrial pacemakers. A rate–response algorithm is programmed for those who appear to have significant chronotropic incompetence. The rate response can be optimized with exercise testing and ambulatory Holter monitoring. Patients with any degree of AV nodal dysfunction, or for those where there is concern about the reliability of atrial pacing, receive dual-chamber pacemakers. Initially the AV delay may be programmed to allow for native AV conduction to occur. Later progression of AV node dysfunction can be accommodated with adjustment of the AV interval. Patients who have had Mustard/Senning and Fontan palliations for their congenital heart disease often have scarred atria complicating epicardial lead placement. Transvenous placement, while technically feasible, can significantly increase the risk of thrombosis in the Fontan baffle. Achieving and maintaining adequate sensing and stimulation thresholds may be

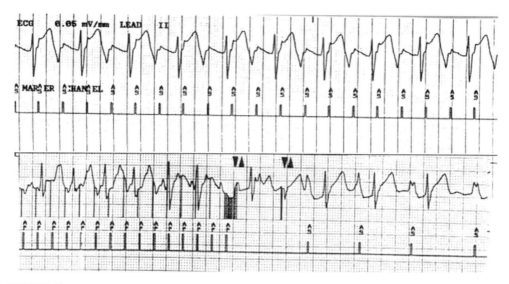

FIGURE 12. Intra-atrial reentrant tachycardia following Fontan repair. The patient has a dual-chamber pacemaker and atrial sensed events (AS) show 2:1 conduction. In the lower panel, sinus rhythm is restored following overdrive stimulation through the pacemaker.

very challenging and rate support with ventricular pacing alone may have to be accepted.

Coexisting tachycardias complicate dual-chamber pacing. Rapid atrial rates that are tracked by the pacemaker may result in hemodynamic compromise from the high rate. Reducing the programmed upper rate limit is one method to avoid inappropriate tracking of tachycardia. Another is mode switching where the pacemaker automatically reverts to ventricular pacing only if the atrial rate exceeds a programmed threshold.

Symptomatic tachycardia is initially managed with pharmacologic therapy. Detailed discussions of the management of IART and atrial tachycardia are found in chapters 4–10 of this text. The choice of medication must include consideration of the underlying heart disease, cardiac function, and the state of the remainder of the conduction system. Digoxin may not significantly worsen bradycardia in all patients and may help to decrease AV conduction of rapid tachycardias. Beta-blocking agents, sotalol, and amiodarone may all inhibit sinus or escape pacemaker automaticity such that permanent pacing is needed to for rate support. Pharmacologic therapy should be initiated with continuous monitoring in the hospital. Catheter ablation has had suboptimal results; many patients require surgical revision concomitant with cryo or radio frequency ablation of arrhythmia substrates. Select patients with infrequent arrhythmias that are reproducibly terminated by overdrive pacing (Figure 12) may be candidates for antitachycardia pacemakers.

PROGNOSIS

Sick sinus syndrome is a progressive disease and deterioration of sinus node function continues into adulthood for survivors of surgery for congenital heart disease. More than 50% of patients with d-transposition who have undergone venous repair develop some degree of sinus node dysfunction, and up to 20% will require pacing for sinus node dysfunction during long-term follow-up. A similar pattern has been observed for the Fontan operation, but recent modifications to the procedure may substantially decrease the prevalence of sick sinus syndrome. Atrial pacing may slow the progression to atrial fibrillation. Sudden death related to bradycardia or prolonged pauses are rare. The development of AV nodal dysfunction is also progressive, but relatively slow in both adult and pediatric studies. Over time the tachyarrhythmias are increasingly difficult to control and may require additional medical, catheter, or surgical interventions. Tachyarrhythmias have been noted as a risk factor for sudden death following surgery for congenital heart disease.

SUGGESTED READING

1. Cohen MI, Rhodes LA. Sinus node dysfunction and atrial tachycardia after the Fontan procedure: The scope of the problem. *Semin Thorac Cardiovasc Surg Pediatr Card Surg Annu* 1998;1:41–52.
2. Crawford MH, Bernstein SJ, Deedwania PC, et al. ACC/AHA Guidelines for Ambulatory Electrocardiography. A report of the American College of Cardiology/American Heart Association Task Force on Practice Guidelines (Committee to Revise the Guidelines for Ambulatory Electrocardiography). Developed in collaboration with the North American Society for Pacing and Electrophysiology. *J Am Coll Cardiol* 1999;34:912–948.
3. Dilawar M, Bradley SM, Saul JP, et al. Sinus node dysfunction after intraatrial lateral tunnel and extracardiac conduit Fontan procedures. *Pediatr Cardiol* 2003; 24(3):284–288.
4. Garrigue S, Bordier P, Jais P, et al. Benefit of atrial pacing in sleep apnea syndrome. *N Engl J Med* 2002;346:404–412.
5. Gelatt M, Hamilton RM, McCrindle BW, et al. Arrhythmia and mortality after the Mustard procedure: A 30-year single-center experience. *J Am Coll Cardiol* 1997;29:194–201.
6. Gregoratos G, Abrams J, Epstein AE, et al. ACC/AHA/NASPE 2002 guideline update for implantation of cardiac pacemakers and antiarrhythmia devices: summary article. A report of the American College of Cardiology/American Heart Association Task Force on Practice Guidelines (ACC/AHA/NASPE Committee to Update the 1998 Pacemaker Guidelines). *J Cardiovasc Electrophysiol* 2002;13:1183–1199.
7. Helbing WA, Hansen B, Ottenkamp J, et al. Long-term results of atrial correction for transposition of the great arteries. Comparison of Mustard and Senning operations. *J Thorac Cardiovasc Surg* 1994;108:363–372.
8. Janousek J, Paul T, Luhmer I, et al. Atrial baffle procedures for complete transposition of the great arteries: natural course of sinus node dysfunction and risk factors for dysrhythmias and sudden death. *Z Kardiol* 1994;83:933–938.
9. Kardelen F, Celiker A, Ozer S, et al. Sinus node dysfunction in children and adolescents: treatment by implantation of a permanent pacemaker in 26 patients. *Turk J Pediatr* 2002;44:312–316.
10. Kavey RE, Gaum WE, Byrum CJ, et al. Loss of sinus rhythm after total cavopulmonary connection. *Circulation* 1995;92:II304–308.
11. Meijboom F, Szatmari A, Deckers JW, et al. Long-term follow-up (10 to 17 years) after Mustard repair for transposition of the great arteries. *J Thorac Cardiovasc Surg* 1996;111:1158–1168.
12. Menon A, Silverman ED, Gow RM, Hamilton RM. Chronotropic competence of the sinus node in congenital complete heart block. *Am J Cardiol* 1998;82:1119–1121, A9.
13. Oberhoffer R, von Bernuth G, Lang D, et al. Sinus node dysfunction in children without heart defect. *Z Kardiol* 1994;83:502–506.
14. Ozer S, Schaffer M. Sinus node reentrant tachycardia in a neonate. *Pacing Clin Electrophysiol* 2001;24:1038–1040.
15. Paul T, Ziemer G, Luhmer L, et al. Early and late atrial dysrhythmias after modified Fontan operation. *Pediatr Med Chir* 1998;20:9–11.

14

First- and Second-Degree Atrioventricular Block

William A. Scott

The atrioventricular (AV) node is a complex structure. Anatomic descriptions of the AV node remain a subject of debate and significant gaps remain in the understanding of AV node physiology. Despite these gaps, there are clinically distinct patterns of abnormal AV nodal conduction that provide insight to pathologic mechanisms. Recognition of these patterns, and the natural history associated with each, is crucial for appropriate patient management.

Electrical conduction through the myocardium (Chapter 2) is dependent on the individual myocyte, cell-to-cell conduction, and conduction through the whole organ. The more rapidly a local region of the cell membrane is able to change its potential (inside relative to the outside), the more rapid is conduction down the length of the myocyte. Therefore, cells that depend on the rapid sodium current for the upstroke of phase 0 of the action potential conduct the signals rapidly, while cells that are dependent on the slower calcium current for phase 0 conduct signals more slowly. Although the various phases of the action potential of cell membranes are generated by more than one ion current, this notion is demonstrated by comparing the conduction velocity in the AV node (dominant calcium dependent action potentials) with atrial and ventricular tissue (dominant sodium channel dependent action potentials). This conduction through the AV node is strongly influenced by autonomic innervation to the node, a number of forms of cardiac abnormalities, as well as the patient's basal state.

Abnormal conduction to the ventricles can result from intrinsic AV nodal or infranodal (the His-Purkinje system) disease including inflammation, infection, and degenerative changes including cardiomyopathy and apoptosis, or from extrinsic causes, including abnormal autonomic tone, electrolyte imbalance, hypothermia, and medication effects. First-degree AV block is almost always due to abnormal conduction in the atrium or AV node. Up to 20% of patients with first-degree AV block and congenital heart disease, most notably AV septal defect and Ebstein's anomaly, have prolonged intra-atrial conduction. Transient prolongation or failure of AV conduction may result from concealed conduction of atrial, junctional, or ventricular extrasystoles (Figures 1 and 2). Second-degree AV block may occur within the AV node,

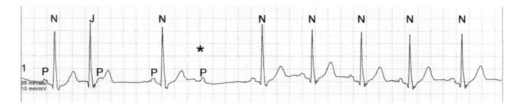

FIGURE 1. Sinus rhythm with a junctional premature beat (J), which results in AV node refractoriness and block of the next sinus beat. Subsequent nonconducted atrial impulse may be due to concealed junctional extrasystole, again making the AV node refractory without activating atrium or ventricle. Proof of this phenomenon requires intracardiac recordings.

the His-Purkinje system, or at the ventricular level.

Similar to the sinus node, the AV node is innervated by the autonomic nervous system. Sympathetic stimulation results in relative enhancement and parasympathetic stimulation, and relative depression of conduction. The sympathetic innervation of the conduction system dominates in infancy but shifts to a balance of sympathetic and parasympathetic innervation by adulthood. Autonomic innervation is less prominent in the His-Purkinje system and has less influence on conduction in these areas. PR interval prolongation in association with bradycardia usually reflects increased vagal tone whereas PR interval prolongation with normal sinus rates is suggestive of AV nodal dysfunction. Functional first- and second-degree AV block can also present during rapid atrial pacing (Figure 3), where, unlike sinus tachycardia, there is no sympathetic enhancement of AV nodal conduction.

There are a group of patients in whom AV conduction is at risk (Table 1). The progression to complete heart block associated with l-transposition of the great arteries is well recognized. At least two separate genetic mutations have been identified with an autosomal dominant form of inheritance. A variety of neuromuscular diseases have also been associated with progressive AV block. Recently a clear association has been established between a mutation of *NKX2.5* and progressive AV block in conjunction with some forms of congenital heart disease (Chapters 15 and 18). In general, block within the His-Purkinje system is more likely to progress to complete heart block.

ECG CHARACTERISTICS

Multiple channel ECG recordings are often indispensable for detection of P-wave morphology and PR intervals. Recording of

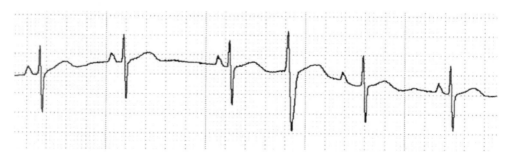

FIGURE 2. Sinus rhythm with interpolated PVC. Concealed conduction into the AV node results in relative refractoriness and first-degree block of subsequent sinus impulse.

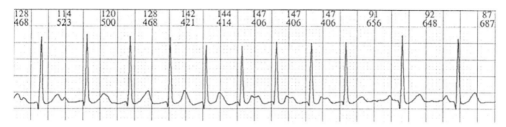

FIGURE 3. Atrial tachycardia that transitions to sinus rhythm. There is first-degree AV block during the tachycardia that resolves with the return of sinus rhythm.

atrial activity from esophageal or temporary epicardial pacing leads after surgery (Figure 4) are also very useful when P-waves are indistinct. Long-term recordings from Holter monitoring often reveal patterns not apparent in the relatively short electrocardiogram.

First-Degree AV Block

First-degree AV block (Figure 5) is defined as a PR interval above the normal range for age, but with persistent 1:1 AV conduction. The normal PR interval also decreases with increasing heart rate. Age appropriate PR intervals are summarized in Table 2. The PR interval can be very long (Figure 6) but usually, in the absence of heart disease, does not progress. The delay is located in the AV node mediated through excessive parasympathetic tone. Exercise, both recreational and during stress testing, induces parasympathetic withdrawal resulting in normalization of AV conduction and the PR interval.

TABLE 1. Conditions Associated with Progression to Complete Heart Block

Familial/Inherited AV Block
Neuromuscular disease
   Emery-Dreifuss
   Myotonic Dystrophy
   Limb-Girdle
   Kearns-Sayre
   Peroneal muscular atrophy
l-Transposition of the great arteries
NKX2.5 deletion

Second-Degree AV Block

Several distinct patterns of second-degree AV block can be recognized. Consistent periodicity (i.e., dropping every 3rd, 4th, or 5th beat, etc.) is frequently present. The ratio of P- to R-waves provides a description of the pattern (i.e., 3:2, 4:3). This pattern of "grouped beats" should always suggest second-degree AV block. Regular non-conducted atrial extrasystoles may also present with this pattern and are distinguished by the irregular PP interval, as well as the different P-wave morphologies (Figure 7).

Mobitz I (Wenckebach)

With typical Mobitz I, or Wenckebach (Figures 8 and 9), there is gradual prolongation of the PR interval prior to a non-conducted beat. The greatest increase in PR interval is between the first and second conducted beats of a series. The lesser increment on subsequent beats leads to a shortening of the RR interval. Following the non-conducted beat, the normal PR interval is restored resulting in an RR interval that is less than twice the sinus rate.

Atypical Wenckebach, which may be more common than the typical form, refers to other patterns of PR prolongation in association with the appearance of a dropped beat (no conduction to the ventricles)—AV block. This pattern often occurs in the presence of sinus arrhythmia and with longer runs of conducted beats between block cycles.

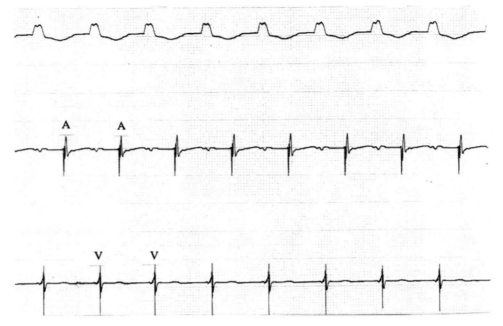

FIGURE 4. Surface ECG (upper signal), atrial wire recording (middle signal), and ventricular wire recording in a patient with first-degree AV block following surgical repair of an AV septal defect. Atrial activity was not apparent from the surface ECG. Atrial wire recordings confirm 1:1 AV relationship with long AV conduction time.

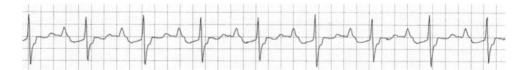

FIGURE 5. Sinus rhythm with first-degree AV block. PR interval 0.36 seconds.

Mobitz II

The PR interval does not vary prior to non-conducted beats with Mobitz II second-degree AV block, and, in the absence of a supraventricular arrhythmia, the RR interval is constant (Figure 9). Since there is no change in PR or RR interval during the conducted beats, the RR interval following a non-conducted beat should be twice that of a conducted RR interval.

2:1

When every other beat is non-conducted in a 2:1 pattern, insufficient information is present to distinguish the pattern of Mobitz I from Mobitz II. Long-term recordings may reveal other ratios of conduction allowing discrimination of these two entities (Figures 9 and 10). Long QT syndrome may

TABLE 2. Maximum Normal PR Interval

Age	PR (sec)
0 to 3 days	0.16
4 to 30 days	0.14
1 to 3 months	0.13
4 to 6 months	0.15
7 to 12 months	0.16
1 to 5 years	0.16
6 to 12 years	0.17
>12 years	0.20

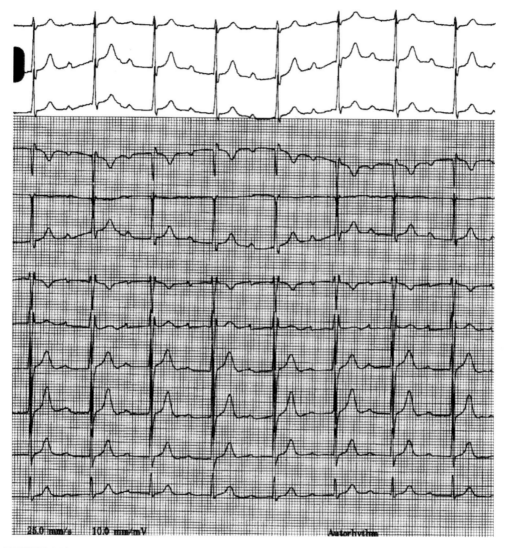

FIGURE 6. Sixteen-year-old boy with first-degree heart block—PR interval 0.36 sec. Asymptomatic. With exercise, the PR interval shortened to normal with 1:1 conduction.

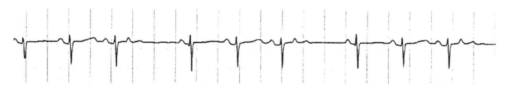

FIGURE 7. Sinus rhythm with frequent nonconducted atrial extrasystoles in a quadrigeminal pattern. The grouped beats mimic second-degree AV block, but the variation in P-wave timing and morphology are distinguishing features.

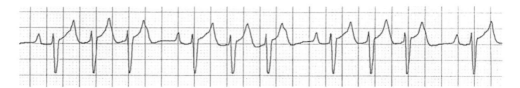

FIGURE 8. Sinus rhythm with first-degree and Mobitz I second-degree AV block with a 4:3 conduction ratio.

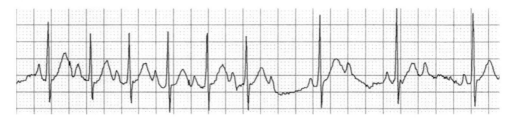

FIGURE 9. Sinus rhythm with Mobitz I second-degree AV block that transitions to 2:1 block.

present with 2:1 conduction as the ventricles, due to the mutated K+ ion channel (LQTS1) delaying ventricular repolarization, are refractory to successive sinus impulses (Figure 10). It is a poor prognostic sign.

Advanced

Advanced AV block is present when two or more impulses are not conducted in the absence of complete heart block (Figure 11). Patients with advanced AV block may progress to complete heart block (Chapter 15).

ELECTROPHYSIOLOGIC FEATURES

Electrophysiologic studies have provided invaluable insights into AV node physiology (Chapter 3). Currently, clinical history and non-invasive diagnostic studies primarily guide diagnosis and management. Electrophysiologic study is not routinely performed solely for the assessment of first- and second-degree AV block because surface electrocardiograms provide adequate information for management. Invasive electrophysiologic studies of AV node function can be performed as an adjunct to hemodynamic study when deemed useful.

Measurements relevant to AV conduction include the intra-atrial conduction time (from the high right atrium near the sinus node to the low septal right atrium near the AV node), the AH interval, a measure of AV nodal conduction, and the HV interval, which reflects conduction through the His-Purkinje system to the ventricles. Intracardiac recordings allow distinction of the level at which block occurs (Figures 12 and 13).

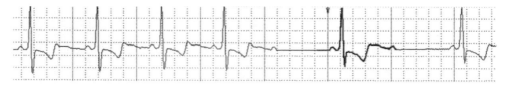

FIGURE 10. Sinus rhythm with stable PR interval that transitions to 2:1 block. In the absence of prior PR interval prolongation, this most likely represents Mobitz II AV block.

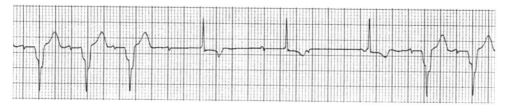

FIGURE 11. Sinus rhythm in a patient with nodoventricular pathway. Advanced AV block associated with narrow complex escape rhythm.

Slow intra-atrial or AV nodal conduction is almost always the mechanism for first-degree AV block and is confirmed by recording a prolonged intra atrial conduction time or AH interval. Similarly, Mobitz I block is almost always within the AV node such that the AH interval prolongs and there is no His bundle deflection with the onset of AV block. In contrast, Mobitz II block is more frequently infranodal. The AH interval typically remains constant and there is a His bundle deflection resulting in a non-conducted impulse and no ventricular activation.

Patients with first-degree or Mobitz I second-degree AV block often have prolonged effective and functional refractory periods for the atrium or the AV node. Normally, the effective refractory period of the His-Purkinje system and ventricular tissue is shorter than the functional refractory period for the AV node. Thus, block distal to the His during programmed stimulation is abnormal though exceeding rare in the young.

PROGNOSIS AND TREATMENT

Most children with first- and second-degree AV block do not experience progression to complete heart block and most do not

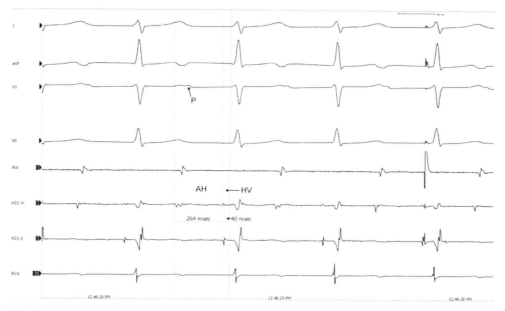

FIGURE 12. Intracardiac study of a patient with first-degree AV block. Note prolonged AH and Normal HV intervals localizing conduction delay to the AV node.

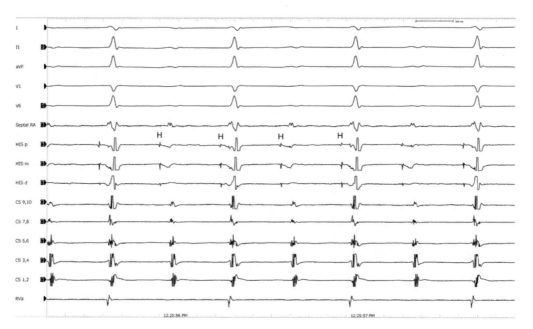

FIGURE 13. Supraventricular tachycardia with 2:1 block. Note His depolarization (H) for conducted and blocked impulses localizing block to distal conduction system.

require treatment. Transient first- and second-degree AV block may be observed during Holter recordings of healthy children and adolescents, particularly those who are athletes and particularly at night with predominantly sympathetic withdrawal. Avoidance of medications known to slow AV conduction is prudent, especially in those patients at risk for AV conduction system disease. For patients who are already receiving these medications, the potential benefit must be weighed against the risk of impaired conduction.

Patients who are acutely symptomatic with second-degree AV block are uncommon, but can be treated with atropine, isoproterenol, and temporary pacing. With some infectious diseases, such as Lyme carditis, the block may resolve entirely. There are no chronic medical management options for patients with significant AV block. Consensus guidelines exist only for advanced AV block, long QT syndrome with 2:1 block or greater, and progressive AV block related to neuromuscular disease (Table 3). Patients with associated structural heart disease, a family history of progressive AV block or sudden death, and those known to carry the mutation of *NKX2.5*, are likely to require permanent pacing as well.

TABLE 3. Conditions Requiring Permanent Pacing

Advanced AV block persisting more than 7 days after surgery
Advanced AV block with symptomatic bradycardia
Long QT with 2:1 block or third-degree AV block
Neuromuscular disease with any AV block

SUGGESTED READING

1. Mymin D, Mathewson FA, Tate RB, Manfreda J. The natural history of primary first-degree atrioventricular heart block. *N Eng J Med* 1986;315:1183–1187.
2. Zipes DP. Second-degree atrioventricular block. *Circulation* 1979;60:465–472.
3. Askanase AD, Friedman DM, Copel J, et al. Spectrum and progression of conduction abnormalities in infants born to mothers with anti-SSA/Ro-SSB/La antibodies. *Lupus* 2002;11:145–151.

4. Benson DW, Silberbach GM, Kavanaugh-McHugh A, et al. Mutations in the cardiac transcription factor NKX2.5 affect diverse cardiac developmental pathways. *J Clin Invest* 1999;104:1567–1573.
5. Chow LT, Chow SS, Anderson RH, Gosling JA. Autonomic innervation of the human cardiac conduction system: changes from infancy to senility—an immunohistochemical and histochemical analysis. *Anat Rec* 2001;264:169–182.
6. Cohen MI, Wieand TS, Rhodes LA, Vetter VL. Electrophysiologic properties of the atrioventricular node in pediatric patients. *J Am Coll Cardiol* 1997;29:403–407.
7. Fatkin D, MacRae C, Sasaki T, et al. Missense mutations in the rod domain of the lamin A/C gene as causes of dilated cardiomyopathy and conduction-system disease. *N Engl J Med* 1999;341:1715–1724.
8. Gregoratos G, Abrams J, Epstein AE, et al. ACC/AHA/NASPE 2002 guideline update for implantation of cardiac pacemakers and antiarrhythmia devices: summary article: a report of the American College of Cardiology/American Heart Association Task Force on Practice Guidelines (ACC/AHA/NASPE Committee to Update the 1998 Pacemaker Guidelines). *Circulation* 2002;106: 2145–2161.
9. James TN, St Martin E, Willis PW, 3rd, Lohr TO. Apoptosis as a possible cause of gradual development of complete heart block and fatal arrhythmias associated with absence of the AV node, sinus node, and internodal pathways. *Circulation* 1996;93:1424–1438.
10. Kafali G, Elsharshari H, Ozer S, et al. Incidence of dysrhythmias in congenitally corrected transposition of the great arteries. *Turk J Pediatr* 2002;44:219–223.
11. Lupoglazoff JM, Cheav T, Baroudi G, et al. Homozygous SCN5A mutation in long-QT syndrome with functional two-to-one atrioventricular block. *Circ Res* 2001;89:E16–21.
12. Mazgalev TN, Ho SY, Anderson RH. Anatomic-electrophysiological correlations concerning the pathways for atrioventricular conduction. *Circulation* 2001;103:2660–2667.
13. Racker DK, Kadish AH. Proximal atrioventricular bundle, atrioventricular node, and distal atrioventricular bundle are distinct anatomic structures with unique histological characteristics and innervation. *Circulation* 2000;101:1049–1059.
14. Wang DW, Viswanathan PC, Balser JR, et al. Clinical, genetic, and biophysical characterization of SCN5A mutations associated with atrioventricular conduction block. *Circulation* 2002;105:341–346.

15

Complete Heart Block—Third-Degree Heart Block

Mohamad Al-Ahdab

Complete atrioventricular (AV) block can be defined as interruption in the transmission of the cardiac impulse from the atria to the ventricles due to an anatomical or functional impairment in the AV conduction system. The conduction disturbance can be transient or permanent.

Congenital complete heart block (CHB), the most common and important form in children, was first described in 1901 by Morquio, who also noted a familial occurrence and an association with Stokes-Adams attacks and death. The presence of fetal bradycardia (40–80 bpm) as a manifestation of CHB was first noted in 1921. The incidence of congenital CHB in the general population varies between 1 in 15,000 to 1 in 22,000 live-born infants.

ETIOLOGY

In the absence of congenital heart disease, neonatal lupus is responsible for 60% to 90% of cases of congenital CHB. Antibodies from a mother with an autoimmune connective tissue disorder, most frequently lupus erythematosus, cross the placenta to the fetus during the first trimester and account for almost all cases presenting in utero or during the neonatal period. Rarely it may explain a few cases occurring later (5% in one report). Other causes include myocarditis and various structural cardiac defects, particularly congenitally corrected transposition of the great arteries, AV discordance, or polysplenia with AV canal defect. Several genetic disorders such as familial atrial septal defect and Kearns-Sayre syndrome (Chapter 18) have been identified (Table 1). In most cases, CHB is characterized pathologically by fibrous tissue that either replaces the AV node and its surrounding tissue or by an interruption between the atrial myocardium and the AV node; other lesions that can occur include congenital absence of the AV node. The net effect is that the block is usually at the level of the AV node. The heart is otherwise structurally normal in these children.

Neonatal Lupus

Complete heart block, hepatobiliary disease, malar rash, thrombocytopenia and, less frequently, myocarditis comprise the neonatal lupus primarily presenting in utero or in the neonate. Frequently the only manifestation of neonatal lupus, and by extension an

TABLE 1. Congenital Complete Heart Block (1 in 20,000 to 25,000 Live Births)

No associated structural heart disease
  Accounts for 67% to 75% of affected infants
  Immune-mediated complete heart block
    Accounts for 80% of congenital complete heart block in the absence of structural defect
    Associated with maternal autoantibodies (anti-SSA/Ro antibodies and anti-SSB/La antibodies) which are the putative etiologic agents
    Because the mentioned antibodies are of the IgG class, transplacental passage and complete heart block do not appear until after 16 to 20 weeks of gestation
    High fetal wastage
    At least 80% of mothers with affected infants will have autoantibodies; most mothers will either have or develop overt symptoms of a systemic connective tissue disorder
    Discontinuity within the cardiac conduction tissue at level of the atrial axis, within the nodal-ventricular conduction tissue, or within the intraventricular conduction tissue*
  Other associations: (applies to complete heart block at birth as well as letter)
    Tumors and neoplasia
    Myocarditis and infections
    Familial, genetic, and metabolic
    Long QT syndrome
  Congestive heart failure occurs but less commonly
  Fair to good postnatal prognosis
  Pacemaker in the newborn period indicated by presence of congestive heart failure or a very slow heart rate (<50 to 55 bpm), or at an older age by symptoms

Associated with complex structural heart disease
  Accounts for 25% to 33% of affected infants
  Most common forms of cardiac malformations:
    Corrected (I-loop) transposition of the great arteries
    Single ventricle
    Defects in atrial and ventricular septation and looping
  Guarded prognosis: high fetal and newborn mortality (even with pacemaker)
  Usually with congestive heart failure
  Early permanent cardiac pacemaker frequently necessary

*See Ho et al., Am J Cardiol 1986, 58:291–294.

autoimmune abnormality in the mother, is CHB in the newborn.

Neonatal lupus is due to transplacental passage of maternal anti-Ro/SSA and/or anti-La/SSB antibodies. Among women with such antibodies, CHB occurs in approximately 2% of pregnancies. Once such a woman has given birth to an infant with CHB, the recurrence rate of CHB in subsequent pregnancies is about 15%; another 6% have an isolated rash consistent with neonatal lupus.

Anti-Ro/SSA and/or anti-La/SSB antibodies bind to fetal cardiac tissue, leading to immune-mediated injury to the AV node and its surrounding tissue. Both Ro/SSA and La/SSB antigens are abundant in fetal heart tissue between 18 and 24 weeks. Apoptosis induces translocation of Ro/SSA and La/SSB to the surface of fetal cardiomyocytes; anti-Ro and anti-La antibodies then bind to the surface of the fetal cardiomyocytes and induce the release of tumor necrosis factor by macrophages, resulting in fibrosis. In addition to inducing tissue damage, anti-Ro/SSA and/or anti-La/SSB antibodies inhibit calcium channel activation of the cardiac L- and T-type calcium channels themselves; L-type channels are crucial to action potential propagation and conduction in the AV node. The sinoatrial (SA) node also may be involved; sinus bradycardia has been described in 3.8% percent of fetuses but is usually not permanent.

CLINICAL MANIFESTATION

The manifestations of CHB vary with the age at presentation. Patients with the neonatal lupus syndrome tend to present earlier than those with CHB not due to neonatal lupus.

Presentation in Utero

Congenital heart block may present with fetal bradycardia between 18 and 28 weeks of gestation. Almost all of these cases (95% in one series) are due to neonatal lupus, demonstrated by the presence of anti-Ro/SSA and/or anti-La/SSB antibodies in the maternal serum. In utero detection is made by echocardiography, which can estimate the fetal PR interval (Chapter 19). Complications in utero include hydrops fetalis, myocarditis, endocardial fibroelastosis, pericardial effusion, and spontaneous intrauterine fetal death. In one report, among 29 cases diagnosed in utero, there was one therapeutic abortion and six intrauterine fetal deaths; a lower rate of intrauterine death (6 of 87) was seen in another study. Second-degree block detected in utero can progress to CHB.

Infants who present with heart block in utero, but who survive until birth, have a high neonatal mortality rate. In one report, 6 of 22 such infants (27%) died within one week of birth. In another series, 10 of 107 (9%) died within the first three months. Infants born before 34 weeks have a higher mortality rate than those born later (52% vs. 9%). Infants with first- or second-degree heart block at birth can progress to CHB.

Presentation in the Neonate

As in the fetus, the cardinal finding in CHB in the neonate is a slow heart rate. In addition to bradycardia, other clinical clues in the neonate include intermittent cannon waves in the neck, a first heart sound that varies in intensity, and intermittent gallops and murmurs. As with cases presenting in utero, almost all presenting in the neonatal period (90% in one series) are due to neonatal lupus. The newborn at greatest risk has a rapid atrial rate, often 150 bpm or faster, and a ventricular rate less than 50 bpm. Similar to acquired CHB, the ECG most commonly shows a narrow QRS complex due to a junctional or AV nodal escape or ectopic rhythm (Figure 1). First- or second-degree heart block found in infants at birth can progress to CHB. The outcome for

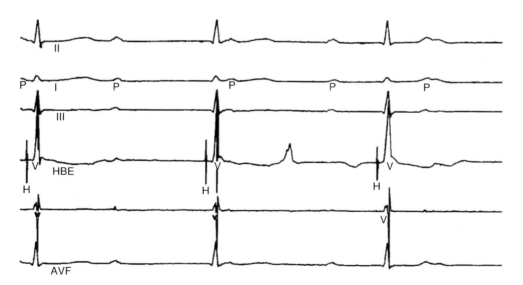

FIGURE 1. Leads II, I, III, and aVF with a His bundle electrogram (HBE) and ventricular (V) electrogram demonstrating CHB with a His bundle (H) escape rhythm. The ventricular rate is 45 bpm. Reprinted with permission from Ho SY et al. The anatomy of congenital heart block. *J Am Cardiol* 1986;58(3):292–294 and Elsevier.

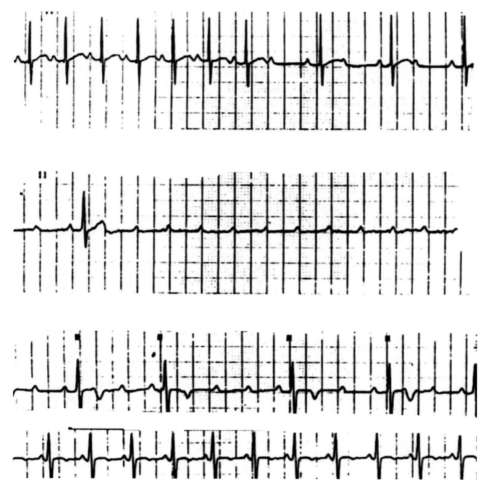

FIGURE 2. Monitor electrocardiographic tracing in a 10-year-old boy presenting with syncope. The tracing shows intermittent CHB with long pauses. A pacemaker was implanted although his AV conduction improved. Myocarditis was presumed but unproven (negative cardiac biopsy). Reprinted with permission from Ho SY et al. The anatomy of congenital heart block. *J Am Cardiol* 1986;58(3):292–294 and Elsevier.

patients diagnosed as neonates is better than for those diagnosed in utero. In the above cited review, 33 patients presented in the perinatal period; five had signs of heart failure, but none had hydrops fetalis. None died within the first six months, but two died at 0.9 and 1.5 years of age. Our experience has been not as grave.

Presentation in Childhood

As many as 40% of cases of congenital heart block do not present until childhood (mean age five to six years) (Figure 2). Few of these patients (5%) have neonatal lupus. The diagnosis is usually made when by presentation such as syncope or by detecting a slow pulse. Heart block is confirmed by ECG or by ambulatory ECG monitoring.

Complete heart block may be intermittent when first detected, but usually becomes persistent in later childhood. It has been suggested that most unexplained CHB diagnosed for the first time beyond infancy is congenital in origin and has escaped notice because of a higher ventricular rate and the absence of symptoms. However, given that prenatal

TABLE 2. Disorders that Can Cause Third-Degree Heart Block

Fibrosis and Sclerosis: Fibrosis and sclerosis of the conduction system accounts for about one-half of cases of AV block.

Lenegre's disease has been traditionally used to describe a progressive, fibrotic, sclerodegenerative affliction of the conduction system in younger individuals associated with slow progression to CHB and may be hereditary.

Lev's disease has referred to "sclerosis of the left side of the cardiac skeleton" in older patients, such as that associated with calcific involvement of the aortic and mitral rings.

Familial Disease: Familial AV conduction block (Chapter 18).
Valvular Disease: Calcification and fibrosis of the aortic or mitral valve rings can extend into the conducting system.
Cardiomyopathies: Includes hypertrophic obstructive cardiomyopathy and infiltrative processes such as amyloidosis and sarcoidosis.
Hyperthyroidism, **myxedema**, and **thyrotoxic periodic paralysis**.
Neuromuscular heredodegenerative disease, **dermatomyositis**, **rheumatoid disease**, and **Paget's disease**.
Infections: Myocarditis due to rheumatic fever, diphtheria, viruses, systemic lupus erythematosus, toxoplasmosis, bacterial endocarditis, syphilis, and Lyme disease.
Malignancies: Such as Hodgkin's disease and other lymphomas; multiple myeloma; and cardiac tumors.
Drugs: A variety of drugs can impair conduction and cause AV block (Chapter 21).
Ischemic heart disease.

---

ultrasound is now well-developed and in wide use, it may be that cases currently missed in fetal life have preserved AV conduction at birth and acquire progressive AV nodal disease thereafter. Consistent with this notion are two observations. First, in a report from one center, the number of childhood case referrals remained constant from 1980 to 1998 despite the introduction of fetal echocardiography and the wide availability of heart rate monitoring during pregnancy and labor. Second, in a series of 102 patients who were asymptomatic through age 15 and were followed for 7 to 30 years thereafter, a slow decline in ventricular rate was noted with increasing age, with a mean heart rate at age 15 of 46 bpm and a mean heart rate after age 40 of 39 bpm.

Some patients present with bradycardia-related symptoms, including reduced exercise tolerance, presyncope, or syncope. Sudden death has also been described. In the above review of 102 patients who were without symptoms through age 15, 27 (26%) had a subsequent syncopal episode, eight of which were fatal. Six of these eight episodes represented a first syncopal episode.

Many other disorders can disrupt the AV conduction system (Table 2). These disorders are rare in the young.

TREATMENT

Management of congenital heart block in utero and in the perinatal period can include steroid therapy if associated with anti-Ro/SSA and anti-La/SSB antibodies, and isoproterenol and/or pacemaker insertion immediately postpartum.

The principal therapeutic decision after the immediate perinatal period involves the need for pacemaker placement. Most patients ultimately have a pacemaker inserted, regardless of the time of onset of the syndrome. In one study, by 20 years of age, only 11% of neonatal and 12% of childhood cases had not required pacemaker implantation. In another report that included 40 patients free of symptoms at age 15, pacemakers were required in 90% by age 60 (Chapter 17).

The Report of the American College of Cardiology/American Heart Association/ North American Society for Pacing and Electrophysiology Task Force on Practice Guidelines (Committee on Pacemaker Implantation) outlines management of third-degree AV block in children, adolescents, and patients with congenital heart disease.

TABLE 3. Class I Indications for Pacemaker Implant in Children

Symptomatic bradycardia (syncope or presyncope)
Moderate to marked exercise intolerance
Heart failure thought related to the bradycardia
Left ventricular dysfunction or low cardiac output
A wide QRS escape rhythm or block below the His bundle
Complex ventricular arrhythmias
In an infant, ventricular rates <50–55 bpm or <70 bpm when associated with congenital heart disease
Sustained pause-dependent VT, with or without prolonged QT, in which the efficacy of pacing is thoroughly documented
Advanced second- or third-degree AV block persisting at least seven days after cardiac surgery

Class I Conditions: Class I conditions are those for which there is evidence and/or general agreement that a permanent pacemaker should be implanted. This class includes patients with advanced second- or third-degree heart block, which is permanent or intermittent (Table 3).

These guidelines are reasonable but should be tailored to the patients needs. Infants with CHB but otherwise a normally structural heart may be followed without a pacemaker even if the heart rate is in the 40s when asleep. Equally important is the variation in the heart rate and the overall status of the child. In contrast, an infant with CHB (congenital or surgical) and structural heart disease should receive a pacemaker. Other useful guidelines are outlined in Table 4.

Class II Conditions: Class II conditions are those for which permanent pacemakers are frequently used, but there is divergence of opinion with respect to the necessity of their insertion (Table 5). Some of these conditions reflect advanced but not complete heart block.

TABLE 4. Other Customized Guidelines

Cardiac enlargement
QT interval prolongation, which may represent a substrate for ventricular arrhythmias (Class II by the ACC/AHA/NASPE guidelines)
Ventricular arrhythmias related to a slow rate or that can be abolished by a more rapid heart rate
Ectopic rhythms or other medical conditions are present that require drugs that suppress the automaticity of escape pacemakers and result in symptomatic bradycardia

LONG-TERM PROGNOSIS

Complete heart block, as noted, presenting in utero or the neonatal period due to neonatal lupus, is associated with a significant early mortality. Of 175 cases described in two reports, 29 (17%) died either in utero or within the first three months of life. Infants and young children with CHB who are asymptomatic usually remain so until later childhood, adolescence, or adulthood. Children with a mean heart rate below 50 bpm and evidence of an unstable junctional escape rhythm may benefit from early pacemaker implant. One particular risk related to the AV block and the ventricular bradycardia is the development of torsades de pointes (Figure 3).

Those patients who do not experience symptoms or syncopal attacks may nonetheless experience physiologic consequences of bradycardia. The ventricular rate tends to fall slowly with age. To compensate for the slow heart rate, the heart enlarges to produce a higher stroke volume; in some cases, this can lead to voltage criteria for left ventricular enlargement and nonspecific ST-T wave changes as well as to heart failure.

In general, the prognosis for the majority of young patients following pacemaker implantation for isolated congenital CHB is excellent. However, one report evaluated 16 patients (10 with neonatal lupus) in whom a pacemaker was implanted within the first two weeks of life; 12 developed heart failure before age 24. The major findings on myocardial biopsy were hypertrophy and interstitial fibrosis. During follow-up, four patients died from progressive heart failure and seven required transplantation. In another study of 149 patients followed for 10 years, 6% developed a dilated cardiomyopathy by

TABLE 5. Class II Indications

Second- or third-degree AV block within the bundle of His in an asymptomatic patient (consider as Class I)
Prolonged subsidiary pacemaker recovery time with a pause greater than three seconds
Transient surgical second- or third-degree AV block that reverts to bifascicular block
Asymptomatic second- or third-degree AV block and a ventricular rate below 50 bpm when awake beyond the first year of life
Complete AV block with double or triple rest cycle length pauses or minimal heart rate variability
Asymptomatic neonate with congenital CHB and bradycardia in relation to age
Long QT syndrome (especially with ventricular arrhythmias)
Congenital heart disease and impaired hemodynamics due to sinus bradycardia or loss of AV synchrony
Neuromuscular disease with any degree of AV block (including first-degree AV block), with or without symptoms, because there may be unpredictable progression of AV conduction disease

6.5 years of age; risk factors included anti-Ro/SSA or anti-La/SSB antibodies, increased heart size at initial evaluation, and the absence of improvement with a pacemaker.

While the development of heart failure in such patients may be a consequence of myocardial fibrosis associated with CHB, another factor may be the long-term consequences of right ventricular pacing with consequent ventricular asynchrony. Recently, a report compared 23 patients with congenital CHB, each of whom had a pacemaker, to 30 matched healthy control subjects. Echocardiography was performed before pacemaker implantation and after at least five years of right ventricular pacing in the CHB patients. The CHB patients with pacemakers, not surprisingly when compared to perfectly normal controls, developed asynchronous left ventricular contraction, an increase in left ventricular end-diastolic diameter, a decrease in cardiac output, and a decrease in exercise

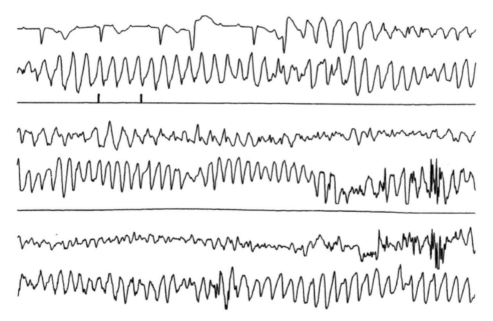

FIGURE 3. Eight-year-old boy with congenital CHB developed ventricular extrasystoles and torsades de pointes. He spontaneously converted and a dual-chamber pacemaker was implanted. Reprinted with permission from Ho SY et al. The anatomy of congenital heart block. *J Am Cardiol* 1986;58(3):292–294 and Elsevier.

performance. When pacemaker therapy is initially placed in children, there are often limited choices. Biventricular pacing, when the patient is of sufficient size, may, in part, address this observation (Chapter 17).

SUGGESTED READING

1. Morquio L. Sur une maladie infantile et familiale characterisée par des modifications permanentes du pouls, des attaques syncopales et epileptiforme et la mort subite. *Arch Méd d'Enfants* 1901;4:467.
2. White P, Eustis R. Congenital heart block. *Am J Dis Child* 1921;22:299.
3. Michaelsson M, Engle MA. Congenital complete heart block: An international study of the natural history. *Cardiovasc Clin* 1972;4:85–101.
4. Johansen AS, Herlin T. Neonatal lupus syndrome. Association with complete congenital atrioventricular block. *Ugeskr Laeger* 1998;160:2521–2525.
5. Ross BA. Congenital complete atrioventricular block. *Pediatr Clin North Am* 1990;37:69–78.
6. Jaeggi ET, Hamilton RM, Silverman ED, et al. Outcome of children with fetal, neonatal or childhood diagnosis of isolated congenital atrioventricular block. *J Am Coll Cardiol* 2002;39:130–137.
7. Ho SY, Esscler E, Anderson RH, et al. Anatomy of congenital complete heart block and relation to maternal anti-Ro antibodies. *Am J Cardiol* 1986;58:291–294.
8. Anderson RH, Wenick ACG, Losekoot TG, et al. Congenitally complete heart block: Developmental aspects. *Circulation* 1977;56:90–101.
9. Rosen KM, Mehta A, Rahimtoola SH, et al. Sites of congenital and surgical heart block as defined by His bundle electrocardiography. *Circulation* 1971;44:833–841.
10. Lev M, Silverman J, Fitzmaurice FM, et al. Lack of connection between the atria and the more peripheral conduction system in congenital atrioventricular block. *Am J Cardiol* 1971;27:481–490.
11. Lev M, Cuadros H, Paul MH. Interruption of the atrioventricular bundle with congenital atrioventricular block. *Circulation* 1971;43:703–710.
12. James TN, Spencer MS, Kloepfler JC. De subitaneis mortibus, XXI. Adult onset syncope with comments on the nature of congenital heart block and the morphogenesis of the human atrioventricular septal junction. *Circulation* 1976;54:1001–1009.
13. Reid JM, Coleman EN, Doig W. Complete congenital heart block. Report of 35 cases. *Br Heart J* 1982;48:236–239.
14. Nasrallah AT, Gillette PC, Mullins CE. Congenital and surgical atrioventricular block within the His bundle. *Am J Cardiol* 1975;36:914–920.
15. Buyon JP, Hiebert R, Copel J, et al. Autoimmune-associated congenital heart block: Demographics, mortality, morbidity and recurrence rates obtained from a National Neonatal Lupus Registry. *J Am Coll Cardiol* 1998;31:1658–1666.
16. Sholler GF, Walsh EP. Congenital complete heart block in patients without anatomic cardiac defects. *Am Heart J* 1989;118:1193–1198.
17. Brucato A, Frass, M, Franceschini F, et al. Risk of congenital complete heart block in newborns of mothers with anti-Ro/SSA antibodies detected by counterimmunoelectrophoresis: a prospective study of 100 women. *Arthritis Rheum* 2001;44:1832–1835.
18. Buyon, JP, Kim, MY, Copel, JA, Friedman, DM. Anti-Ro/SSA antibodies and congenital heart block: necessary but not sufficient. *Arthritis Rheum* 2001;44:1723–1727.
19. Alexander E, Buyon JP, Provost TT, et al. Anti-Ro/SS-A antibodies in the pathophysiology of congenital heart block in neonatal lupus syndrome: An experimental model. *Arthritis Rheum* 1992;35:176–189.
20. Miranda-Carus ME, Askanase AD, Clancy RM, et al. Anti-SSA/Ro and anti-SSB/La autoantibodies bind the surface of apoptotic fetal cardiocytes and promote TNF-alpha secretion by macrophages. *J Immunol* 2000;165:5345–5351.
21. Garcia S, Nascimento JH, Bonfa E, et al. Cellular mechanism of the conduction abnormalities induced by serum from anti-Ro/SSA positive patients in rabbit heart. *J Clin Invest* 1994;93:718–724.
22. Xiao GQ, Hu K, Boutjdir M. Direct inhibition of expressed cardiac L- and T-type calcium channels by IgG from mothers whose children have congenital heart block. *Circulation* 2001;103:1599–1604.
23. Askanase AD, Friedman DM, Copel J, et al. Spectrum and progression of conduction abnormalities in infants born to mothers with anti-SSA/Ro-SSB/La antibodies. *Lupus* 2002;11:145–151.
24. Cruz RB, Viana VS, Nishioka SA, et al. Is isolated congenital heart block associated to neonatal lupus requiring pacemaker a distinct cardiac syndrome?. *Pacing Clin Electrophysiol* 2004;27:615–620.
25. Nield LE, Silverman ED, Taylor GP, et al. Maternal anti-Ro and anti-La antibody-associated endocardial fibroelastosis. *Circulation* 2002;105:843–848.
26. Glickstein JS, Buyon J, Friedman D. Pulsed Doppler echocardiographic assessment of the fetal PR interval. *Am J Cardiol* 2000; 86:236–239.

27. Waltuck J, Buyon JP. Autoantibody-associated congenital heart block: outcome in mothers and children. *Ann Intern Med* 1994;120:544–551.
28. Dewey RC, Capeless MA, Levy AM. Use of ambulatory electrocardiographic monitoring to identify high-risk patients with congenital complete heart block. *N Engl J Med* 1987;316:835–839.
29. Michaelsson M, Jonzon A, Riesenfeld T. Isolated congenital complete atrioventricular block in adult life. A prospective study. *Circulation* 1995;92:442–449.
30. Pinsky WW, Gillette PC, Garson A Jr, McNamara DG. Diagnosis, management, and long-term results of patients with congenital complete atrioventricular block. *Pediatrics* 1982;69:728–733.
31. Karpawich PP, Gillette PC, Garson AJ, et al. Congenital complete atrioventricular block: Clinical and electrophysiologic predictors of need for pacemaker insertion. *Am J Cardiol* 1981;48:1098–1102.
32. Gregoratos G, Cheitlin MD, Conill A, et al. ACC/AHA Guidelines for Implantation of Cardiac Pacemakers and Antiarrhythmia Devices: Executive Summary. A report of the American College of Cardiology/American Heart Association Task Force on Practice Guidelines (Committee on Pacemaker Implantation). *Circulation* 1998;97: 1325–1335.
33. Gregoratos G, Abrams J, Epstein AE, et al. ACC/AHA/NASPE 2002 Guideline update for implantation of cardiac pacemakers and antiarrhythmia devices: summary article. A report of the American College of Cardiology/American Heart Association task force on practice guidelines (ACC/AHA/NASPE committee to update the 1998 pacemaker guidelines). *Circulation* 2002;106: 2145–2161.
34. Winkler RB, Freed MD, Nadas AS. Exercise-induced ventricular ectopy in children and young adults with complete heart block. *Am Heart J* 1980;99:87–92.
35. McHenry MM. Factors influencing longevity in adults with congenital complete heart block. *Am J Cardiol* 1972;29:416–421.
36. Reybrouck T, Vanden Eynde BB, Dumoulin M, et al. Cardiorespiratory response to exercise in congenital complete atrioventricular block. *Am J Cardiol* 1989;64:896–899.
37. Kertesz NJ, Friedman RA, Colan SD, et al. Left ventricular mechanics and geometry in patients with congenital complete heart block. *Circulation* 1997;96:3430–3435.
38. Moak JP, Barron KS, Hougen TJ, et al. Congenital heart block: Development of late-onset cardiomyopathy, a previously underappreciated sequela. *J Am Coll Cardiol* 2001;37:238–242.
39. Udink ten Cate FE, Breur JM, Cohen MI, et al. Dilated cardiomyopathy in isolated congenital complete atrioventricular block: early and long-term risk in children. *J Am Coll Cardiol* 2001;37:1129–1134.
40. Pordon CM, Moodie DS. Adults with congenital complete heart block: 25-year follow-up. *Cleve Clin J Med* 1992;59:587–590.
41. Groves AM, Allan LD, Rosenthal E. Outcome of isolated congenital complete heart block diagnosed in utero. *Heart* 1996;75:190–194.
42. Thambo JB, Bordachar P, Garrigue S, et al. Detrimental ventricular remodeling in patients with congenital complete heart block and chronic right ventricular apical pacing. *Circulation* 2004;110: 3766–3772.
43. Karpawich PP. Chronic right ventricular pacing and cardiac performance: the pediatric perspective. *Pacing Clin Electrophysiol* 2004;27(6 Pt 2):844–849.

16

Syncope

Margaret Strieper, Robert M. Campbell, William A. Scott

Syncope, defined as the temporary loss of consciousness and postural tone resulting from an abrupt, transient decrease in cerebral blood flow, has emerged over the last decade as one of the most common reasons for a pediatric cardiology referral. Although many individuals will experience syncope at least once during their lifetime, it is usually self-limited and benign. Rarely, it may be the first warning sign of a serious condition including arrhythmias, structural heart disease, or non-cardiac disease (Table 1). Patients with recurrent syncopal episodes, syncope during exercise, emotion/stress-induced syncope, syncope resulting in injury, syncope in the driving-age pediatric patient, or syncope in patients with congenital heart disease require investigation. Recurrent syncope may cause a major impact on lifestyle, interfering with school and/or sports. This chapter presents a differential diagnosis (Table 1) of syncope in children, outlines in detail neurocardiogenic syncope (NCS), and reviews different evaluation and treatment strategies.

DIAGNOSTIC EVALUATION

Given the many possible causes of syncope, the diagnostic evaluation can be quite involved and expensive and the specific etiology may never be determined. Therefore, a carefully planned approach rather than a "shotgun" diagnostic strategy is important. The patient history, family history, physical examination, and an electrocardiogram are fundamental and direct the remainder of the evaluation. Table 2 details the components of a comprehensive syncope evaluation. The patient history is the cornerstone on which the syncope evaluation is constructed and the diagnosis dependent; it is often, along with the physical examination and an ECG, all that is necessary. Important historical details from the patient include: the time of day of the event (early morning is typical), the state of hydration and nutrition at the time of the event (when last had fluid or food intake), the environmental conditions (i.e., ambient temperature), the patient's activity immediately prior to the syncopal episode, the frequency and duration of the episodes, and any aura, prodrome, or specific symptoms and signs prior to the episode. Witnesses, if available, should provide details regarding the patient's condition prior to the syncope, duration of loss of consciousness, any injuries or seizure-like movements, heart rate during episode (rarely available), and duration and nature of recovery (often patients are sleepy after neurocardiogenic syncope). Medications (prescriptions and/or over-the-counter) used by the

TABLE 1. Differential Diagnosis

Neurocardiogenic Syncope

Arrhythmias
  Channelopathies
  Complete Heart Block
  Sick Sinus Syndrome
  Tachyarrhythmias–Supraventricular
    and Ventricular Tachycardia,

Cardiac–Structural
  Cardiomyopathy–Hypertrophic/Dilated
  Coronary Artery Anomalies
  Tumor
  Left Ventricular Outflow Obstruction
  Primary Pulmonary Arterial Hypertension
  Eisenmenger Syndrome
  Mitral Valve Prolapse

Medications
  Recreational (illegal)
  Antiarrhythmic
  Diuretics
  Vasodilators
  Producing QT prolongation

Neurologic
  Seizure
  Vertigo
  Migraine
  Tumor

Psychiatric
  Conversion Reaction
  Panic Attack
  Hysteria
  Hyperventilation

Metabolic
  Hypoxia
  Hypoglycemia

Reprinted with permission from Strieper MJ, Distinguishing benign syncope from life threatening cardiac causes of syncope, Semin Pediatr Neurol 2005;12:32–38 and Elsevier.

patient are critical historical points particularly regarding proarrhythmic agents such as QT prolonging medications. Information regarding prior diagnostic reports and/or consultations can prevent duplicate testing.

For some patients whose evaluation lacks internal consistency, other studies may be advisable, such as echocardiography to examine for cardiomyopathy, myocarditis, anomalous coronary arteries, pulmonary arterial hypertension, or arrhythmogenic right ventricular dysplasia. A rare patient may warrant cardiac catheterization, including hemodynamic, angiographic, and electrophysiologic evaluation, along with right ventricular endomyocardial biopsy, to exclude potential structural, functional, and arrhythmic abnormalities, particularly before clearance to resume activities.

Family history is vital in the evaluation of syncope. It is not uncommon to find a history of multiple family members who experienced syncope during adolescence. Many of the older family members may also report a history of low blood pressure, and many families limit salt due a hypertensive family member who is on a salt-restricted diet. However, if the family history is positive for recurrent syncope, it is also important to consider other familial disorders by specific questioning about the presence of hypertrophic or dilated cardiomyopathy, long QT syndrome (and other ion channelopathies), primary pulmonary hypertension, or arrhythmogenic right ventricular dysplasia. Families should be queried regarding sudden unexplained death in children or young adults (i.e., drownings, sudden cardiac death, sudden infant death syndrome, and car accidents), seizures, or familial congenital deafness (Table 2). Noting the person or source providing the family history and an estimate of its reliability can be of future use; it may be helpful to request further details from additional family members, particularly if a genetic disorder is suspected.

During the patient examination, orthostatic vital signs (magnitude of decrease in blood pressure relative to change from supine to erect position) should be obtained. Many patients will manifest a mild (up to a 30 mmHg decrease in blood pressure) but asymptomatic orthostatic change with upright positioning. An ECG should be obtained on every patient who experiences syncope, particularly if it is recurrent, occurs with exercise, and is not associated with the characteristic symptoms of neurocardiogenic syncope. The ECG should be evaluated for heart rate, corrected QT interval (gender-related), T-wave morphologic change, T-wave alternans, or any ventricular arrhythmia. The ECG should also be evaluated for preexcitation syndromes, AV

TABLE 2. Syncope Evaluation

Patient History*
　Age at onset
　Time of day
　Frequency (increasing or decreasing over time)
　Prodrome (see text: patient may be amnestic)
　Situation (activity, place, body position)
　Patient and witness description of event
　Duration of loss of consciousness
　Symptoms upon recovery (sleepy)
　Medications (prescription, OTC)
　Concomitant disease
　Prior evaluation and results
Family History*
　Sudden death at young age
　SIDS
　Syncope
　Seizures
　Accidental death (e.g., drowning, automobile accidents)
　Pacemaker/Defibrillator
　Congenital deafness
　Cardiomyopathy
Patient Exam*
　General condition (habitus, phenotype, hydration, nutritional state, thyroid)
　Cardiac exam
　Blood pressure–sitting, lying, standing (for orthostatic response)
　Pulse–strength, rate, UE/LE difference
　Heart murmurs suggesting anatomic disease
　Musculoskeletal exam
　Inherited connective tissue disorder phenotype
　Neurologic exam

ECG*
　Rate and rhythm
　AV Conduction (heart block)
　Intraventricular conduction (Brugada syndrome, Arrhythmogenic Right Ventricular Dysplasia)
　QTc Interval (male ≤ 0.44 sec; female ≤ 0.46 sec)
　T-wave morphology
Exercise Testing**
　Especially rule out LQTS, exercise-induced syncope
Noninvasive Imaging**
　Echo Doppler
　　Anatomic/structural assessment
　　Tumors
　　Pulmonary artery pressure estimate
　　Outflow track gradients
　　Abnormal origin left coronary artery
　　Coronary aneurysms
　MRI
　　Arrhythmogenic right ventricular dysplasia
　　Coronary arteries
　　Tumors
Transtelephonic ECG/Holter**
　Implantable recorder
Head-Up Tilt Testing (HUT)**
Catheterization**
　Hemodynamic
　Angiographic
　Electrophysiologic study
　Biopsy
　Right ventricular endomyocardial biopsy

Note: LE = Lower extremity
UE = Upper extremity
SIDS = Sudden Infant Death Syndrome
*All patients
**As clinically indicated

---

conduction, or features of Brugada syndrome (see Chapters 4, 13–15, 18).

NEUROCARDIOGENIC SYNCOPE—SIMPLE FAINTING, VASOVAGAL SYNCOPE

Pathophysiology

The pathophysiologic mechanisms underlying neurocardiogenic syncope (NCS) are not completely understood. However, there is a general consensus that a cardiac-central nervous system reflex is involved (Figure 1). The most common initiating event is prolonged (or the abrupt assumption of) upright position (sitting or standing), which subjects the patient to gravitationally mediated venous pooling in the lower extremities and pelvis. This causes an abrupt central hypovolemia (compared to the immediate preexisting state), leading to a decrease in venous return and stroke volume. In addition, an emotional or physical stress (e.g., pain or fright) or a reflex mechanism related to hair grooming, glutition (swallowing), or micturation may initiate this sequence by stimulating a reflex

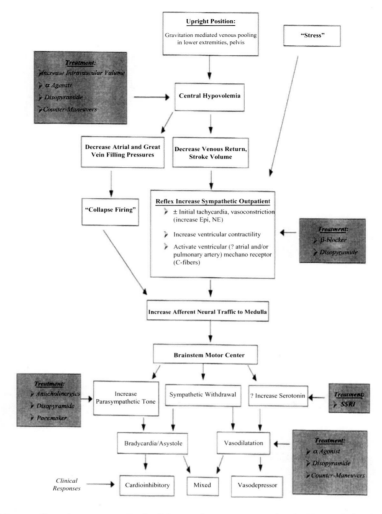

FIGURE 1. Neurocardiogenic syncope pathophysiology and treatments. Reprinted with permission from Strieper MJ, Distinguishing benign syncope from life threatening cardiac causes of syncope, Semin Pediatr Neurol 2005;12:32–38 and Elsevier.

increase in sympathetic output manifested as tachycardia and vasoconstriction along with an increase in ventricular contractility. Activation of C-fiber mechanoreceptors increases afferent neural traffic to the central nervous system (medulla), stimulating the brain-stem motor center and causing several and often combined possible responses, such as the following: (1) an increase in parasympathetic activity, causing profound bradycardia or asystole; (2) a sympathetic withdrawal resulting in peripheral vasodilatation (venous and arterial), a decrease in systemic blood pressure, and decrease in heart rate; and (3) an increase in serotonin concentration, also resulting in peripheral vasodilatation and marked decrease in the systemic blood pressure. As a result of the loss of consciousness, postural tone and the upright state, the patient falls to a supine position restoring venous return and the central circulating blood volume (i.e., heart and lungs) followed by rapid normalization of blood pressure and heart rate. The loss of consciousness is usually short (generally ≤1–2 minutes). Excretory incontinence is uncommon. Seizures rarely occur as a result of the sudden prolonged decrease in cerebral perfusion. During recovery, return to sentience is rapid but post event fatigue is common.

Clinical Presentation—NCS and other Considerations

A prodrome lasting from several seconds to 1–2 minutes and consisting of nausea, epigastric discomfort, a clammy and cold sweat, pallor, dizziness, lightheadedness, tunnel vision, headache, and weakness is highly characteristic and strongly suggests simple fainting, vasovagal or neurocardiogenic syncope. If the prodrome is of sufficient duration, patients may learn to recognize their symptoms and lie down to relieve the symptoms and prevent syncope. Some patients with profound bradycardia or asystole may have little or no prodrome, causing a sudden loss of consciousness that may result in injury. Absence of a significant prodrome also raises the possibility of structural, functional, or arrhythmic causes for syncope. On the other hand, palpitations, chest discomfort, and a sudden loss of consciousness as well as a prompt recovery are more compatible with an isolated cardiac event. Other symptoms such as atypical precordial chest pain or tightness in the chest, breathlessness, acrocyanosis, tingling in the hands or feet, and a sense of alarm or anxiety are compatible with the hyperventilation syndrome.

If "seizures" (tonic-clonic movements) occur as a result of cerebral hypoperfusion and anoxia, the patient can be confused as having a primary neurologic abnormality. Formal neurologic consultation and/or neurologic testing should be considered if seizures are part of the presentation. Interestingly, longstanding complaints of intermittent abdominal pain and nausea, most likely due to a profound increase in vagal tone, constitute an unusual and rarely suspected presentation of NCS. This symptom complex, as a manifestation of NCS, may be associated with a positive head-up tilt (HUT) evaluation, and may respond favorably to NCS treatment. In addition, there may be a link between NCS and chronic fatigue syndrome, which also has findings of hypotension, headaches, and postexertional fatigue. However, it is unlikely that an otherwise healthy pre-adolescent child would exhibit the chronic fatigue syndrome.

Patients who present with syncope during exercise are also a challenge. Gradual loss of consciousness, associated with a presyncopal prodrome and occurring after exercise suggests the possibility of exercise-induced NCS, provoked by pre-existing unrecognized dehydration, exercise-induced catecholamine-enhanced ventricular contractility, and peripheral (from the shift in blood flow from the abdominal circulation to the working skeletal muscles) and cerebral (induced by the respiratory alkalosis of the exercise related hyperventilation) vasoconstriction. On the other hand, sudden syncope during exercise and without prodrome raises the suspicion for a more serious underlying cardiac structural or functional cause, including arrhythmias, and warrants further investigation.

Postural orthostatic tachycardia syndrome (POTS) has been described as a specific variation of NCS in adults; it may be relatively common in children as well. Most of these patients present with symptoms of rapid palpitations suggesting a primary tachycardia, but on further questioning, also have associated symptoms suggesting low blood pressure (dizziness, lightheadedness). Hyperthyroidism can present in this manner. When this clinical situation presents, ECG and BP monitoring should be performed to evaluate for primary tachycardia. Sinus tachycardia (130–160 bpm) during a symptomatic hypotensive phase strongly supports the diagnosis of POTS and helps to exclude primary tachycardia. Treatment is directed towards prevention of venous pooling and/or intravascular volume depletion.

Breath-holding spells most likely represent a variation of NCS in the small child. These spells generally occur in young children

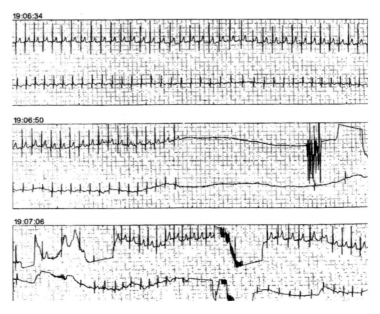

FIGURE 2. Breath-holding spell in 11-month boy. Note tachycardia followed by abrupt asystole.

between the ages of 9 months and 4 years of age. As a result of painful stimuli or emotional upset, these episodes usually begin with crying. Breath holding generally occurs during expiration, which can cause hypoxemia (inducing hypoxemic syncope) and a decrease in venous return (provoking the NCS reflex). Evaluation with continuous ECG monitoring generally reveals significant bradycardia or prolonged asystole (10–15 seconds) (Figure 2). With loss of consciousness, spontaneous respirations resume and the patient generally shows a rapid return to full consciousness with normalization of heart rate. Although there is no proven treatment for breath-holding spells, which generally resolve spontaneously with age, a trial of anticholinergic agents (belladonna) along with parental counseling may reduce their recurrence. Another clinical entity, reflex anoxic seizures, is caused by sudden asystole without breath holding. Detection is difficult because of the abrupt onset and immediate recovery; the event may be limited to a sudden "face plant" followed by prompt recovery. An implanted loop recorder can be useful. These patients may respond to permanent pacing or vagolytic therapy.

Other clinical scenarios highly consistent with NCS are syncope during hair combing, micturation, deglutition, Valsalva maneuvers, hot morning showers, or blood drawing or donation.

A "stressful" milieu is often in the background of 20%–25% patients with recurrent syncope. It is often prudent to introduce the notion of stress-related factors early in the patient evaluation, thereby introducing the possibility that psychiatric issues may be involved. Several findings may suggest the presence of a psychiatric role in the genesis of "syncope." First, patients who experience the onset of syncope in the supine position [excluding a strong physical provocative stress (phlebotomy) or an arrhythmia] may have a psychiatric cause. Second, patients do not often rapidly recover from their loss of consciousness even after resuming the supine position. Third, loss of consciousness is frequently prolonged, by witnessed accounts, often up to several hours. Finally, there is a marked indifference to syncope and its ramifications. These patients often experience a conversion reaction, "lose consciousness" but maintain blood pressure and heart rate during the HUT table test. This HUT finding is important, since the patient-specific diagnosis directs the appropriate patient recommendations.

Head-Up Tilt Table Test—Indications

Current indications for a head-up tilt (HUT) table test include: (1) recurrent unexplained syncope with no evidence by exam, or ECG, for a structural or arrhythmia abnormality; (2) syncope resulting in injury; (3) exercise-induced syncope; (4) syncope in the driving age patient; (5) recurrent troublesome pre-syncope causing severe patient incapacitation that interferes with the events of daily living; or (6) failed empiric treatment for NCS. Syncope that is characterized by a typical history, is infrequent, and is not associated with injury or a potentially adverse event (driving a car, exercise) does not require confirmation by a HUT test.

HUT Protocol

A number of HUT protocols have been advanced. In one format (Figure 3), the child is NPO for at least two hours prior to the procedure. Following EMLA resistance to demand and preparation of the IV site, a peripheral IV is placed and intravenous fluids infused at 40 mL/hour. A continuous and non-invasive digital arterial pulse waveform is displayed and recorded allowing for non-invasive beat-to-beat representation of the arterial pressure during HUT testing (Figure 3a). The use of automated Dinamap or manual blood pressure recordings is not optimal during the presence of HUT-induced hypotension. After 20 minutes supine stabilization (Figure 3b), the patient is tilted to the 80° upright position (Figure 3c). This HUT test continues for up to 20 minutes, unless a positive response (clinical symptoms reproduced and

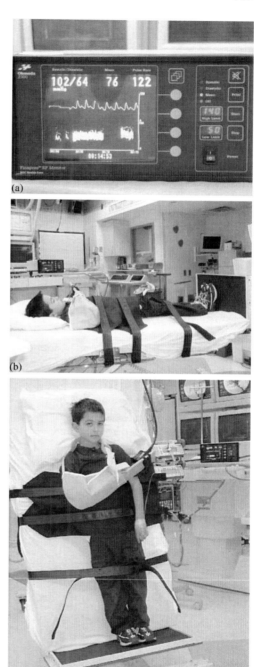

FIGURE 3. a. Finapres monitor for continuous heart rate and non-invasive blood pressure measurement. b. The patient positioned supine on the tilt table with IV, EKG, and blood pressure monitor. c. Tilt to 80° upright position, with "seat belt" restraint.

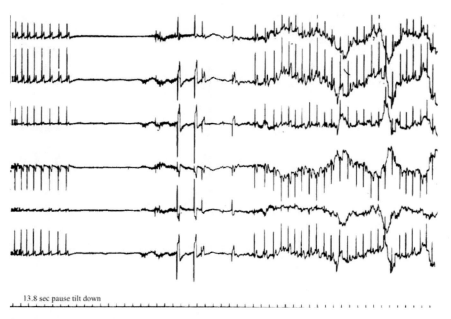

FIGURE 4. Cardioinhibitory response in 15-year-old girl after 4 minutes of HUT—13 seconds asystole. Note the slowing of the heart rate before asystole (paper speed 12.5 mm/sec).

hypotension-bradycardia) occurs. The patient is then returned to the supine position.

If this baseline test is negative (no symptoms, no hypotension or bradycardia), isoproterenol is infused at a rate sufficient to increase the supine heart rate to about 125% of baseline. Once isoproterenol administration is stabilized for approximately 10 minutes, the patient is again tilted to an 80° upright position and maintained for 20 minutes or until a positive HUT response occurs.

An alternative protocol calls for initial supine stabilization for 5 minutes followed by 80° tilt for 40 minutes. If the response is negative, the patient is returned to the supine position for 5 minutes and nitroglycerin (0.4 mg sublingually for patients ≥ 40 kg; for patients ≤ 40 kg, 0.2 mg) is administered. The patient is tilted to 80° for 20 minutes. Hypotension, bradycardia or both, sufficient to induce the clinical symptom complex, constitute a positive test. If either the baseline or provocative agent HUT is positive, the test is terminated. If both the baseline and provocative HUTs are negative, the HUT test response is negative.

There are three positive HUT NCS clinical responses. The first clinical response is the *vasodepressor* response, defined as a $\geq 50\%$ decrease in mean arterial blood pressure, often with preservation of the heart rate or only a mild increase in heart rate. This represents approximately 40% of positive tilt table responses. Second, patients may have a *mixed* hypotensive and bradycardic response, defined as a $\geq 50\%$ decrease in mean arterial pressure and a $\geq 50\%$ decrease from the maximum heart rate during HUT testing. This represents approximately 50% of the positive HUT responses. Finally, a *cardioinhibitory* response is defined as sudden severe bradycardia or asystole, which occurs in approximately 5%–10% of HUT positive patients (Figure 4).

These clinical responses are not necessarily patient-specific nor are they necessarily reproducible; there could be considerable variation day-to-day and HUT-to-HUT, with different responses expressed at different times.

Although the HUT test is regarded in some quarters as the "gold standard" for

NCS diagnosis, it should be noted that it is only a supportive test, confirming the clinical history. HUT engages a complex physiologic response with a highly provocative and exaggerated physiologic stress in individuals who may or may not be particularly vulnerable to this stress. HUT is easiest to interpret when clinical symptoms and significant hypotension and/or bradycardia are produced during the episode. Only then can this test be interpreted as truly positive. False positive HUT responses can occur, when patients experience hypotension or bradycardia but the clinical symptoms are not exactly reproduced. Negative HUT responses may represent either a true negative (patient does not suffer from NCS) or a false negative (patient is NCS positive, but HUT negative). It is not possible at this time to distinguish clinically false negative results from true negative HUT results. There is no widely accepted standard HUT protocol. Considerable variation in tilt angle, duration of tilt, and monitoring exists between centers. This lack of a standard protocol confounds the interpretation of HUT responses between centers. Sensitivity and specificity for HUT test parameters is by report variable and limited data only in pediatric patients are available.

NCS TREATMENT

The purpose of treatment for NCS is to prevent recurrent syncope. Additionally, pre-syncopal symptoms may also be eliminated. However, efforts to eliminate all pre-syncopal symptoms are often associated with drug side effects and the use of multiple medications. There are only limited randomized treatment trials for pediatric patients. The opinion that any treatment for pediatric syncope patients is better than no treatment has not been established. Additionally, NCS almost always resolves spontaneously within several months to years after onset. Apparent therapeutic drug efficacy may simply represent spontaneous resolution of NCS. Occurrences of syncopal episodes are unpredictable and some patients may be event free for several months or longer. Therefore, duration of treatment and assessment of drug effect is further confounded. A daily diary to record syncope and pre-syncopal episodes may help when evaluating symptoms both before and after treatment.

Volume Expansion

Maintaining adequate intravascular volume is the foundation for treatment for NCS. Patients are instructed to avoid caffeinated beverages (due to the renal diuretic effect), but any and all other fluids are encouraged. Specially, liquid such as "sport drinks" with additional sodium are preferable for maintaining hydration. Up to 60 to 90 ounces a day are recommended for adolescents. Salt tablets may be prescribed, but they are relatively large, difficult to swallow, and will frequently cause nausea. Alternatively, patients are encouraged to liberally use salt with meals and eat non-fatty salted snacks (i.e., popcorn without butter).

Although fluid administration alone helps to alleviate symptoms, it may not alone be adequate to prevent most patient symptoms or recurrent syncope. However, drug therapy alone, in the absence of adequate of intravascular volume, is rarely as effective as when combined with good hydration.

Counter Maneuvers

Some simple patient physical maneuvers may help to ameliorate or abort pre-syncopal symptoms/syncope. These maneuvers generally involve exercising the muscles of the leg (leg pumping and tensioning, leg crossing, squatting, elevation of the legs above the level of the heart); a recent study demonstrated a therapeutic effect of isometric handgrip to abort impending NCS by increasing systemic blood pressure. These simple maneuvers can be taught to even young children. However, it is important that these patients have an adequate prodrome to allow them time to perform the maneuvers prior to the onset of true syncope. The possible beneficial effects of conditioning by using prolonged exposure to the upright position in pediatric NCS patients are often stymied by lack of patient compliance due to the time required for appropriate conditioning.

Fludrocortisone

Florinef (fludrocortisone) has several effects. It increases intravascular volume and sodium, often at the expense of an increased urinary loss of potassium. It also may augment the peripheral effects of catedrolamines, leading to both venous and/or arterial vasoconstriction. Together with adequate intravascular volume through the use of oral hydration, significant decrease in venous return may not occur and NCS may be prevented. Florinef has been demonstrated to produce a positive therapeutic response in a non-blinded pediatric population, and fludrocortisone acetate and atenolol treatment (for pediatric NCS) have been found to be equally effective. Florinef is generally well tolerated with a low incidence of side effects. Monitoring for hypokalemia, as a result of increased urinary loss, is important. When receiving Florinef and alpha agonist medications, patients must be monitored for the onset of hypertension, due to the combined vasoconstrictive effects.

Alpha Agonist Agents

Alpha agonist agents work through their vasoconstrictor effects; venoconstriction helps to maintain ventricular volume, while arterial constriction offsets the hypotensive response. Acute intravenous phenylephrine hydrochloride is effective in pediatric patients for blocking NCS responses on follow-up HUT, after an initial positive HUT. However, pure oral alpha agonist agents are not readily available. Another possible alpha agonist is generic pseudoephedrine hydrochloride. Although pseudoephedrine hydrochloride is also known to have mild beta agonist effects, this agent is generally well tolerated in pediatric patients. Pseudoephedrine hydrochloride, together with adequate hydration strategies, is often an effective mitrial treatment for NCS. Up to 75% to 80% of HUT positive patients may respond to this treatment, with complete elimination of recurrent syncope and either improvement in or elimination of presyncopal symptoms. Patients who fail with this treatment are treated concomitantly with Florinef. Together these two medications may provide up to 85% to 90% control for tilt positive pediatric patients. Pseudoephedrine side effects, in less than 10% of patients, include fatigue, most certainly a by-product of drug-induced insomnia and restless sleeping. A few patients complain of tremulousness, appetite suppression, or irritability. Alternatively, midodrine (another alpha agonist agent) has recently been an advocated as a treatment option at some centers.

Disopyramide

Disopyramide, an anti-arrhythmic agent, has been employed as a second line NCS treatment. Data suggest several mechanistic actions, including peripheral anticholinergic effect (to combat bradycardia), negative inotropic properties (preventing activation of the C-fibers), and smooth muscle vasoconstriction, preventing both venous and arterial vasodilatation. Despite the concern for possible drug side effects, specifically proarrhythmia, this drug is generally tolerated quite well by pediatric patients and often will prove effective when first line agents fail.

Beta-Blockers

Beta-blockers were one of the first drug treatment options described for NCS. These medications operate through several mechanisms, including block of sympathetically mediated increase in ventricular contractility and activation of the C-fibers, and also by prevention of sympathetically mediated arterial vasodilatation. Some patients may manifest side effects, which may mask the therapeutic benefits or aggravate the pre-syncopal symptoms. Care should be taken when using this agent with patients who suffer from POTS, since block of the compensatory sinus tachycardia may actually worsen the patient's clinical status, causing syncope.

Anticholinergic Agents

This drug group may be most effective for patients who manifest profound parasympathetic-driven bradycardia, asystole or a parasympathetic component to peripheral arterial vasodilatation leading to hypotension. Even when effective, pronounced drug side effects including dry mouth, urinary retention, constipation, and visual blurring often make long-term treatment with these drugs intolerable. A recent study has demonstrated a possible therapeutic benefit with the use of glycopyrrolate for breath-holding spells in young infants.

Serotonin Reuptake Inhibitors

Because serotonin may lead to vasodilatation, promoting a vasodepressor or mixed NCS clinical response, serotonin reuptake inhibitors (SSRIs) have been advocated by some for the treatment of drug refractory NCS. The mechanism of action is still uncertain, although it has been speculated that these agents may desensitize central serotonin receptors or decrease serotonin release to the peripheral circulation. These drugs may also be applicable for patient subsets when anxiety or emotional upset triggers the NCS episode or in whom the emotional response to recurrent syncope is also problematic clinically.

Pacemaker Therapy

While initially thought to be an ideal therapy for patients with predominant bradycardia/asystole, many centers have learned that pacing alone is often ineffective in preventing all episodes of NCS. Many patients, although adequately paced, will still be symptomatic as a result of vasodepression and hypotension. Patients with asystolic HUT responses can often be managed effectively with medications, obviating the need for pacemaker therapy. Since medical therapy generally will prevent recurrent syncope, and NCS is often self-limited, the decision to implant a permanent pacemaker becomes a difficult decision, especially in a young person. Rarely does one need to implant a pacemaker; our experience is confined to 5 patients over the past 12 years. Several of these patients were ≤ 2 years old with profound asystole (probably representing reflex and anoxic seizures); these patients responded well to pacing, with the elimination of syncopal events.

PROGNOSIS

Most patients show spontaneous resolution of their NCS syncope and pre-syncope in 6–12 months following onset of episodes. Based on our experience, 5% to 10% of NCS patients will show symptoms over an extended period of time, often 3 to 5 years. The reasons for these differing clinical courses are unknown.

A special concern with NCS is the driving-age patient. Even following diagnosis of NCS as the definite cause of the patient's syncope, and the institution of effective therapy, these patients should be restricted from driving for at least 2 to 4 months. Different states have guidelines about restrictions from driving for patients with different medical issues, including seizure disorders, life-threatening arrhythmias, and syncope. It is important to be certain that a definite diagnosis has been made and that treatment has proven effective before the patient can safely return to driving. Physicians involved in care of these patients should give written instructions regarding driving restrictions, and document these restrictions in detail very specifically in the patient's chart.

SUMMARY

Syncope is a common clinical event. An organized, detailed approach to diagnosis, with particularly close attention paid to the details of the history, is most likely to result in a time and cost effective evaluation. To optimally define treatment, restrictions, and prognosis, patient-specific diagnosis is important. Although NCS is the most common cause of pediatric syncope, potentially life-threatening etiologies should not be overlooked.

SUGGESTED READING

1. Calkins H, Byrne M, El-Atassi R, et al. The economic burden of unrecognized vasodepressor syncope. *Am J Med* 1993;95:473–479.
2. Sheldon R, Rose S, Ritchie D, et al. Historical criteria that distinguish syncope from seizures. *J Am Coll Cardiol* 2002;40:142–148.
3. Strieper MJ, Auld D, Hulse JE, Campbell RM. Evaluation of recurrent pediatric syncope: role of tilt table testing. *Pediatrics* 1994;93:660–662.
4. Kosinski D, Grubb BP, Temesy-Armos P. Pathophysiological aspects of neurocardiogenic eurocardiogenic syncope. *Pacing Clin Electrophysiol* 1995;18:716–721.
5. Dickinson CJ. Fainting precipitated by collapse firing of venous baroreceptors. *Lancet* 1993;342:970–972.
6. Bou-Holaigh I, Rowe P, Kan J, Calkins H. The relationship between neurally-mediated hypotension and the chronic fatigue syndrome. *JAMA* 1995;274:961–967.
7. Grubb BP, Kosinski DJ. Boehm K, Kip K. The postural orthostatic tachycardia syndrome: a neurocardiogenic variant identified during head-up tilt table testing. *Pacing Clin Electrophysiol* 1997;20:2205–2212.
8. Lombroso CT, Lerman P. Breath holding spells (cyanotic and pallid infantile syncope). *Pediatrics* 1967;39:563–567.
9. Berkowitz JB, Auld D, Hulse JE, Campbell RM. Tilt table evaluation for control pediatric patients: comparison to symptomatic patients. *Clin Cardiol* 1995;18:521–525.
10. Ross B, Hughes S, Kolm P. Efficacy of fludrocortisone and salt for treatment of neurally-mediated syncope in children and adolescents. *Pacing Clin Electrophysiol* 1992;15:506.
11. Scott WA, Pongiglione G, Bromberg BI, et al. Randomized comparison of atenolol and fludrocortisone acetate in the treatment of pediatric neurally-mediated syncope. *Am J Cardiol* 1995;76:400–402.
12. Strieper MJ, Campbell RM. Efficacy of alpha-adrenergic agonist therapy for prevention of pediatric neurocardiogenic syncope. *J Am Coll Cardiol* 1993;22:594–597.
13. Hulse JE, Strieper MJ, Auld D, Campbell RM. Pseudoephedrine therapy for pediatric neurocardiogenic syncope. *Ped Cardiol* 1994;15:259.
14. Jankovic J, Golden JL, Hiner BC, et al. Neurogenic orthostatic hypotension: a double blind, placebo-controlled study with midodrine. *Am J Med* 1993;95:38–48.
15. Milstein S, Buetikofer J, Durmigran A, et al. Usefulness of disopyramide for prevention of upright tilt-induced hypotension bradycardia. *Am J Cardiol* 1990;65:1334–1344.
16. Cox MM, Perlman B, Mayor MR, et al. Acute and long-term β-adrenergic blockade for patients with neurocardiogenic syncope. *J Am Coll Cardiol* 1995;26:1293–1298.
17. Grubb BP, Kosinski D. Serotonin and syncope: an emerging connection? *Eur J Cardiac Pacing Electrophysiol* 1996;5;306–314.
18. Brignole M, Croci Francesco, Menozzi C, et al. Isometric arm counter-pressure maneuvers to abort impending vasovagal syncope. *J Am Coll Cardiol* 2002;40:2053–2059.
19. Deal B, Strieper M, Scagliotti D, Hulse E, Auld D, Campbell RM, Strasburger J, Benson W. The medical therapy of cardioinhibitory syncope in pediatric patients. *PACE* 1997;20:1759–1761.
20. Linzer M, Felder A, Hacket A, et al. Psychiatric syncope. *Psychosomatics* 1990;31:181–188.
21. Kapoor WN, Fortunato M, Hanusa BH, Schulberg HC. Psychiatric illnesses in patients with syncope. *Am J Med* 1995;99:505–512.
22. *Syncope: Mechanisms and Management.* Edited by BP Grubb and B Olshansky. Futura Publishing Company, Inc. Armonk, NY, 1998.

17

Cardiac Pacemakers and Implantable Cardioverter-Defibrillators

Gerald A. Serwer and Ian H. Law

Until recently the need for cardiac pacing in the young patient has been confined to the treatment of bradyarrhythmias. While this continues to be the main reason for permanent cardiac pacing, implantable devices to manage tachyarrhythmias have begun to be employed. In addition, the use of biventricular pacing to treat poor myocardial function and congestive heart failure has been investigated in children but remains in its infancy. This chapter reviews the current indications for cardiac pacing, particularly in reference to the treatment of tachyarrhythmias, the current pacemaker technology, the variety of available devices and electrodes, implantation techniques, and chronic device and electrode testing and follow-up, particularly with relevance to optimal long-term management of the child who requires device implantation.

INDICATIONS

Indications for device placement can be divided into three categories: bradyarrhythmias, tachyarrhythmias, and heart failure.

Bradyarrhythmias

Recommended guidelines have been published, but the need for a pacemaker must be individualized for each patient. Indications in our practices include complete heart block, sick sinus syndrome, drug-induced bradycardias, and control of tachyarrhythmias (Figure 1). The presence of second- or third-degree atrioventricular (AV) block with associated symptomatology suggestive of chronic or intermittent low cardiac output such as syncopal or presyncopal events or chronic exercise intolerance, is an indication for pacemaker placement. In addition, in patients with complete heart block, the presence of an increasing cardiac size by chest x-ray or echocardiogram, the presence of decreased ventricular function, a prolonged QT interval, and the appearance of wide QRS tachycardias intermixed with a narrow QRS rhythm would lead to the recommendation for device implantation. In our experience, the use of solely a heart rate below a specific level has not been used as an exclusive indicator for device placement. While some have recommended a rate of 50 bpm as the lowest acceptable rate,

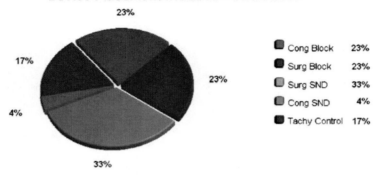

FIGURE 1. Percentage of patients with each indication for device implantation. All are new implants since 2000. Cong Block = non-surgically related heart block; Surg Block = surgically related heart block; Surg SND = surgically related sinus node dysfunction; Cong SND = non-surgically related sinus node dysfunction; Tachy Control = device placed for control of tachyarrhythmia.

many children do well without symptoms for many years with such heart rates.

Pacemaker implantation is mandatory for all patients with surgically-induced and permanent (≥ 10 days) complete heart block. Heart rates tend not to be stable with a high incidence of acute asystole. No patient is discharged following cardiac surgery without pacemaker implantation unless there is 100% intact atrioventricular conduction. While later recovery of conduction can occur, it is unlikely and pacemaker placement is usually performed at 10 to 14 days. For patients with return of AV conduction, they must have documented 100% AV conduction by 24-hour Holter monitoring to avert pacemaker implantation.

For patients with late onset complete heart block following cardiac surgery, pacemaker placement is also recommended without requiring associated symptomatology. For these patients the heart block is usually not persistent, but intermittent. The risk of sudden syncopal episodes is felt to be elevated and thus pacemaker implantation is recommended.

The largest and the most rapidly expanding group of patients that require device implantation are those with sinus node dysfunction. While most patients have undergone previous cardiac surgery, some have no other cardiac disease. Documented low atrial rates with or without junctional escape rhythms along with symptoms of intermittent or chronic low cardiac output warrant pacemaker implantation as the symptoms tend to worsen with time. The younger patients with sinus node dysfunction following cardiac surgery may not require pacing, whereas patients in their late teenage and early adult years may now require device implantation. Patients who have undergone extensive atrial surgery such as the Fontan procedure or the atrial switch repair of transposition of the great arteries are not only at high risk for developing sinus node disease but also for atrial tachycardias that may in part be unmasked by low atrial rates. Control of atrial tachycardias can be greatly aided by maintenance of a minimal atrial rate of 70–80 bpm.

Tachyarrhythmias

Children who require treatment with antiarrhythmic agents for control of tachyarrhythmias that result in resting bradycardia will often benefit from pacemaker implantation. However, these children are often better suited by implantation of an antitachycardia device with bradycardia pacing support.

Children with hypertrophic obstructive cardiomyopathy and significant left ventricular outflow tract gradients have been shown in some studies to benefit from dual chamber pacing. However, this is still a very controversial recommendation and has been rarely employed in our practice. These children may require a combination of bradycardia pacing and an implantable defibrillator following surgical relief of the obstruction for chest pain.

Indications for placement of an antitachycardia device in children are also somewhat imprecise. Antitachycardia devices can be divided into two groups. The first group is capable of delivering antitachycardia pacing and defibrillator shocks and the second can perform only antitachycardia pacing. A patient who has been resuscitated from a sudden cardiac death event or has been presyncopal with documented ventricular tachycardia requires placement of an implantable cardioverter defibrillator (ICD)—i.e., secondary prevention. However, primary prevention for those children that are felt to be at risk for having this type of syncopal episode is beginning to be explored. Children who have had cardiac conditions known to have a high incidence of ventricular fibrillation such as tetralogy of Fallot or cardiomyopathy and who have had documented significant ventricular ectopy are increasingly undergoing ICD placement. A long QRS duration in a patient following tetralogy of Fallot repair has also been suggested as an indication for ICD placement. However, this is very controversial and is currently not used in our practice as a sole criterion for ICD placement.

Atrial flutter is also being treated in specific situations with cardiac pacing. By keeping the base atrial rate above a predefined number, usually in the 70 to 80 range, the burden of atrial flutter may be decreased. In addition, when this is coupled with antitachycardia pacing, control of atrial flutter may be significantly improved. The recent introduction of an antitachycardia dual-chamber rate responsive pacemaker may lead to an increase in this therapy for atrial tachyarrhythmias in children with sinus node dysfunction who are not amenable to ablation techniques. Pacing algorithms for prevention of atrial flutter have been proposed but to date have not been shown to be effective.

Heart Failure

For patients with heart failure, cardiac resynchronization rests on the idea that the difference in activation between the right and left ventricles due to intraventricular conduction disturbances, such as left and right bundle branch block, leads to decreased ventricular performance. Pacing both ventricles in a coupled manner may lead to more normal biventricular excitation (i.e., resynchronization) and thus improve ventricular function. This technology is an established technique to improve cardiac function and quality of life in adults with significant congestive heart failure, and ventricular dyschrony manifest by a decreased left ventricular ejection fraction and a wide QRS complex. The treatment of children with biventricular pacing for improvement in ventricular performance and increased cardiac output is in its early years. There are short-term results from a multicenter study indicating that cardiac resynchronization is feasible in children and young adults with congenital heart disease, but the indications for biventricular pacing and a candidate population of young patients has not been fully defined.

IMPLANTATION CHOICES

When it has been decided that a child will benefit from cardiac pacing, several decisions must be reached to provide the most optimal therapy for these children. The first decision is whether or not epicardial or endocardial electrode implantation is needed and the second is whether dual-chamber or single-chamber pacing is in the child's best interest. Once these decisions have been reached, attention can be turned towards the choice of appropriate electrodes and, finally, the implantable device itself. One must consider not only what is in the child's immediate best interest, but how this interest may change over time. Such decisions must take into consideration the potential for future adverse events that might be avoided or better treated with more appropriate initial choices of implantation route and devices.

Epicardial versus Endocardial Placement

The first decision to be made following the recommendation for pacemaker implantation is whether the electrodes are to be implanted on the epicardial or endocardial surface. This decision is clearly dependent by the availability of venous access to the atrium and the pulmonary ventricle, as well as patient size and the presence or absence of any other conditions considered contraindications for endocardial electrode placement. Patients in whom there is no venous route to the ventricle and atrium must undergo an epicardial implant. Often superior vena caval obstruction, if narrow, can be opened by stent placement sufficiently wide to permit immediate passage of an endocardial electrode. While there have been isolated reports of inferior vena caval passage with introduction of the electrodes in the femoral region and generator implantation in the abdomen, this is not considered optimal due to the problems encountered with growth and the fact that such implantation may prohibit any further use of the femoral vessels for central venous access.

Patients in whom there is no pulmonary ventricle (e.g., patients with single ventricles following the Fontan procedure) are not candidates for transvenous pacing. Again, there have been reports of pacemaker electrode placement within the systemic ventricle, but this approach can be associated with embolization and resultant CNS injury. Even patients who have been maintained on Coumadin at adequate levels may still experience microemboli from the pacing lead. Thus, in our practice, electrodes in the systemic ventricle are not implanted.

Controversy exists as to whether or not one should implant the transvenous electrode in the Fontan circuit. With the low flow state and the potential for clot formation along with the possibility of right to left shunting through either a fenestration in the Fontan baffle or baffle leaks, any thrombi have the potential to cause significant problems. This is not only problematic for systemic embolization but also for pulmonary embolization with a subsequent increase in pulmonary vascular resistance and decrease in pulmonary flow. In our practice such an approach is not used.

Patient size has long been debated as to its effects on the choice of electrode route. While reports of infants undergoing transvenous implantation abound, the long-term sequelae have not been well studied. With the recently developed smaller (5 French) transvenous leads, great vein obstruction becomes less of an issue. However, were there to be significant SVC or innominate vein obstruction in a young child, future electrode placement would be more difficult. Also the effects on tricuspid valve function in the presence of multiple electrode leads are also a concern. While electrode extraction is a possible solution, it does carry a significant risk. At the current time in our practice the transvenous approach is not used for children less than 20 kg in weight for a single lead and less than 30 kg for those requiring a dual lead system. An obvious advantage of the epicardial approach is the increase in the potential sites for pacing in an individual who will require life-long pacing and multiple revisions as pacing leads fail. As electrodes become smaller and electrode extraction becomes safer, this consideration may change.

Other contraindications to endocardial electrode placement are the presence of right to left intracardiac shunting, a hypercoagulability state, and pulmonary vascular obstructive disease. In these patients in whom any small emboli to the lungs would further

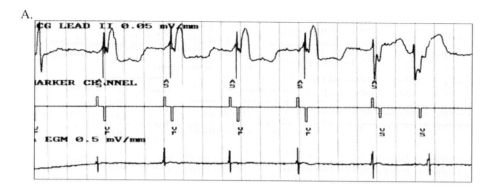

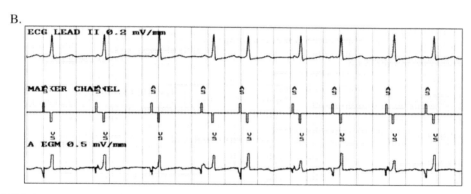

FIGURE 2. Atrial electrograms from a patient with a bipolar electrode, A, versus a unipolar electrode, B. Top tracing is the body surface EKG; middle tracing is the pacemaker marker channel; third tracing is the atrial electrogram. Note the lack of a far field R-wave in Panel A even with ventricular pacing compared to the large far field R-wave in panel B that exceeds the atrial amplitude.

increase pulmonary resistance, transvenous leads are avoided.

Bipolar versus Unipolar Electrode Usage

The next issue that must be decided is whether one will utilize bipolar or unipolar electrodes. Initially, epicardial implantation always required unipolar electrodes, as this was all that was generally available. The introduction of several types of bipolar electrodes has eliminated this issue. Bipolar implantation, in both transvenous and epicardial settings, has the advantage of creating less extracardiac stimulation and less difficulty with myo-potential extracardiac sensing. When bipolar atrial electrodes are used, there is less difficulty with far field R-wave sensing, which can be a major issue if one is using an antitachycardia device (Figure 2). In addition, bipolar electrodes tend to have a higher impedance than do unipolar electrodes, thus decreasing current drain for similar output settings with subsequent increased generator longevity. This is a consequence of a smaller surface area, as the second electrode in a unipolar system is the device casing ("can") with a large surface area.

Two types of epicardial electrodes are available (Figure 3). The first, intramyocardial electrodes, are those that penetrate the epicardial surface (Figure 3A–C). These electrodes are either barb tipped, which lodge within the myocardium, or a helical corkscrew electrode that burrows into the myocardium. Both of these work adequately, but cautions must be observed. Both types of electrodes must be fixed to the epicardial surface with sutures.

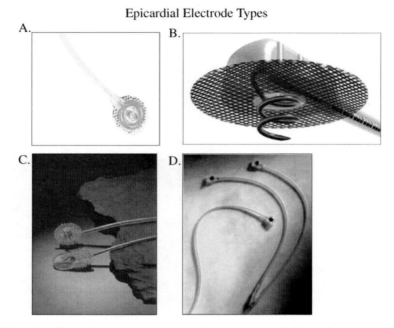

FIGURE 3. Examples of both bipolar and unipolar epicardial electrodes. A. Guidant helical bipolar intramyocardial electrode. B. Medtronic 5069 unipolar helical intramyocardial electrode. C. Medtronic 4951 intramyocardial barbed electrode (lower electrode) compared with the 5069 electrode (upper electrode). D. Medtronic 4965 steroid eluting epimyocardial unipolar electrode (left) and the 4968 bipolar steroid eluting epimyocardial electrode (right). A courtesy of Guidant, Inc. B-D courtesy of Medtronic, Inc.

Failure to suture the electrode to the epicardial surface results in excessive motion of the electrode that can result in fracture of the electrode or increased fibrosis around the electrode due to excessive irritation. Both events result in loss of pacing.

The intramyocardial bipolar electrode has a central helix surrounded by a circular plate that sits on the epicardial surface. This electrode has the advantage of being able to penetrate surface scarring or fat but the ground portion of the electrode that sits on the surface is still prone to being affected by it. As it is intramyocardial, it is not steroid eluting. Care must also be taken with the corkscrew electrode. Only the distal portion of the electrode of either type is active. The electrode should not be placed completely transmural with the tip of the electrode sitting within the blood pool of the ventricular cavity. This results in a low impedance circuit through the blood and prevents myocardial stimulation. The proximal portion has an insulating coating to decrease the surface area and increase the electrode impedance but limits the electrode surface area that is in contact with the myocardium. For this reason, helical electrodes cannot be used in the atrium.

The second type of epicardial electrode is a flat epimyocardial electrode plate that sits on the epicardial surface. The epimyocardial bipolar electrode has two separate heads, each of which must be sutured to the epicardium individually, which increases implant time and requires two separate areas of viable myocardium (Figure 3D). However, its sensing vector can be optimally chosen to avoid far field R-wave sensing in the atrium and maximizing R-wave amplitude in the ventricle. It is also steroid eluting. These electrodes do not penetrate into the myocardium and therefore do not cause as much irritation and subsequent fibrosis as the intramyocardial type of electrodes. In addition, they are steroid eluting, further decreasing the fibrosis around the tip. However, if there is significant scar tissue or epicardial fat present, it can be difficult to find a section of heart muscle that can be stimulated. Thus, implant times may be increased. Yet the improved thresholds and higher impedances make this electrode the epicardial implant of choice in our practice for both the atrium and ventricle.

Endocardial Electrodes

Endocardial electrodes can be either unipolar or bipolar. Early unipolar electrodes were smaller than their bipolar equivalents and thus were preferred in children. With advances in electrode technology, this size difference has disappeared, leaving almost no advantages to unipolar transvenous electrodes. The transvenous bipolar electrode has a relatively long rigid tip compared to the unipolar equivalent. In small atrial chambers such as those following the atrial switch repair of transposition of the great arteries, the unipolar electrode may be advantageous. However, only bipolar electrodes can be used with the antitachycardia devices.

The major decision to be made for transvenous implantation is between the methods of electrode fixation to the endocardial surface. As for the epicardial route, electrodes can either sit on the endocardial surface held in place by tines that lodge in the trabecular recesses (passive fixation) or can penetrate the endocardial surface via a helix that serves not only as the active electrode but also to hold the electrode in place (active fixation). The cross-sectional areas of the two electrode types are comparable and the time needed to implant them is similar. For atrial electrodes and ventricular electrodes implanted in the morphologic left ventricle in patients with L-transposition of the great arteries or in patients with D-transposition of the great arteries following the atrial switch operation, active fixation electrodes are much more widely used. When right ventricular pacing at sites other that the apex is needed, active fixation electrodes are required. Active fixation electrodes may be easier to remove.

In our practices, bipolar electrodes are used for both epicardial and endocardial approaches. The choice between passive and active fixation is more fluid and both continue to be used. Because extracardiac and far field R-wave sensing is less and both the acute chronic thresholds are as good as or better than the intramyocardial electrodes, for epicardial implants bipolar epimyocardial electrodes are used. There are times, however, when this electrode type cannot be used due to excessive myocardial scar formation from prior cardiac surgical procedures; intramyocardial electrodes are then required.

Dual versus Single-Chamber Pacemakers

The next major decision that must be made is whether a single- or dual-chamber pacemaker is to be used. Pacemakers are classified according to the nomenclature schema devised by the North American Society for Pacing and Electrophysiology (NASPE) and the British Pacing and Electrophysiology Group (BPEG), (Figure 4A). This scheme has a five character descriptor for each device. The first column refers to the chamber or chambers being paced, the second to the chambers(s) being sensed, the third to the action taken upon a spontaneous beat occurring, the fourth to the presence or absence of activity sensing, and the fifth to the use of multisite pacing. A similar schema has also been proposed for implantable cardioverters-defibrillators (Figure 4B). For the purposes of this discussion, single-chamber pacemakers are defined as those that only pace and sense in the ventricle (or atrium). For patients that have intact atrioventricular conduction, but require pacing due to sick sinus syndrome, atrial pacing, while technically single-chamber pacing, produces the same advantages as dual-chamber pacing. It is acceptable to utilize single-chamber atrial pacing in those patients who have documented reliable

HRS/BPEG Pacemaker Classification

Chamber Paced	Chamber Sensed	Response to Sensing	Rate Modulation	Multisite Pacing
V	V	T	R	A
A	A	I	O	V
D	D	D	-----	D
O	O	O	-----	O

A

HRS/BPEG Defibrillator Code

Chamber Shocked	Antitach Pacing Chamber	Tachy Detection	Antibrady Pacing Chamber
A	A	E	A
V	V	H	V
D	D	-----	D
O	O	-----	O

B

FIGURE 4. A. HRS/BPEG schema for pacemaker classification (Bernstein et al., *PACE* 2002;25:260). B. HRS/BPEG schema for defibrillator classification (Bernstein et al., *PACE* 1993;16:1776). See text for details.

atrioventricular conduction and in whom ventricular pacing is unlikely to be needed in the future. The only exception to this is for patients who require antitachycardia devices. These pacemakers, even when programmed in the single chamber for atrial brady pacing, require a ventricular electrode for accurate detection of atrial tachyarrhythmias.

Single-chamber ventricular pacing can often be used in the young, small patient with congenital complete heart block and an otherwise structurally normal heart. These patients tend to have good myocardial function and placement of only a ventricular epicardial electrode requires a much smaller surgical procedure. However, whenever there is a question of the adequacy of myocardial function, dual chamber pacing with preservation of atrioventricular synchrony is preferred. Numerous studies have shown the superiority of dual-chamber versus single-chamber pacing for maintenance of cardiac output in the patient with structural cardiac disease. In addition, heart rate variability is improved when one utilizes the patient's sinus node to set the heart rate rather than an activity sensor. While activity sensors do provide some heart rate variability in older ambulatory patients, they do not provide as physiologic a heart rate response as does the patient's sinus node. In the non-ambulatory neonate, the sensor is incapable of detecting the type of activity needed to vary the heart rate because activity sensors are not reliable and are of no benefit.

Dual-chamber pacemakers are slightly larger due to the need for the pacemaker to accept two electrodes, but usually the difference is not clinically significant. Size differences are usually more a function of battery size rather than connector size. Standard single- and dual-chamber pacemakers can be implanted in the smallest neonate. Device longevity is not significantly affected by the added burden of atrial sensing. If atrial pacing is needed much of the time, then a dual-chamber device is a better choice.

Dual-chamber is preferred in most cases. In our experience, dual-chamber or atrial single-chamber devices are used in approximately 75% of the cases (Figure 5).

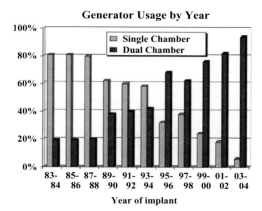

FIGURE 5. Percentage of patients undergoing dual- versus single-chamber pacemaker implants at the University of Michigan Congenital Heart Center versus the year of implant. Overall about 75% of implants are dual-chamber devices.

Single-chamber ventricular devices are used in some neonates with congenital complete heart block and normal cardiac structure and function, in patients who rarely require pacing but have a device to prevent an acute bradycardia secondary to drug therapy or intrinsic electrical system disease, and in the rare situation when an atrial electrode cannot be implanted.

ICDs

The application of an implantable cardioverter defibrillator (ICD) to children is a recent, but an increasingly employed therapy for those with life-threatening arrhythmias. The appropriate selection is similar to that for pacemakers involving a choice between a dual-chamber or single-chamber device, as well as a transvenous versus epicardial approach. Similar to pacemakers, a dual-chamber ICD requires an additional electrode, but the size difference is not significant. If the patient's status calls for dual-chamber pacing, then a dual-chamber ICD is necessary. For patients that do not require pacing except potentially following a defibrillatory episode, then single-chamber devices may be acceptable; however, a single-chamber device does not have the ability to discriminate sinus tachycardia from slow ventricular tachycardia nor to treat atrial tachyarrhythmias, should one develop. If there have been documented atrial tachyarrhythmias in addition to ventricular tachyarrhythmias, a dual-chamber device is required.

For patients that do not have venous access to the ventricle or are already undergoing a median sternotomy for cardiac surgery, an epicardial approach can be used. Epicardial placement of the defibrillatory patches (Figure 6) can be performed with adequate defibrillatory thresholds obtained; however, bipolar epicardial sensing electrodes are required. In the rare small patient, a combination of transvenous sensing electrode leads and epicardial defibrillatory electrode leads (rather than patches) can be utilized.

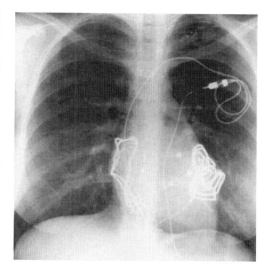

FIGURE 6. Chest x-ray of a patient with epicardial defibrillatory electrodes and transvenous sensing and pacing electrodes. Note the use of an electrode extender to tunnel the transvenous electrode to the abdominally placed ICD (not shown). Used with permission from Serwer, et al., Pediatric pacing and defibrillator usage, in *Clinical Cardiac Pacing and Defibrillation*, 2nd edition, Ellenbogen KA, Kay GN, Willkoff BL, eds., W.B. Saunders, Philadelphia, 2000 and Elsevier.

Cardiac Resynchronization Devices

Cardiac resynchronization can be performed in patients with heart failure/impaired ventricular function undergoing placement of a dual-chamber pacemaker or implantable cardioverter defibrillator. As this patient population by definition has impaired ventricular function, tracking the atrial activity can enhance ventricular filling and enable programming options, such as varying the atrioventricular delay to optimize cardiac output. In smaller children and patients with congenital heart disease whose anatomy limits access to the coronary sinus, an epicardial lead may be required to pace the left ventricle. However, advances in transvenous lead technology including smaller lead shafts along with variety of lead placement and fixation techniques have expanded the number of young candidates for transvenous cardiac resynchronization systems.

IMPLANTATION TECHNIQUES

It is recognized that there are many ways to implant both epicardial and transvenous or endocardial electrodes. The general methodology utilized in our practices and deviations from the standard techniques that have been found useful in selected settings are outlined. While there may be different techniques that work equally well, those described have proved successful in our hands.

Epicardial Implant Techniques

For epicardial unipolar pacing, a suitable site on the epicardial surface, free of scar tissue, is chosen. The location of this site is often governed more by the suitability of the tissue than by consideration for location. For intramyocardial electrodes, a site is chosen on the left ventricle as the myocardium is thick enough to accommodate the intramyocardial electrode. For epimyocardial electrodes, a site either on the left or right ventricle is adequate depending on the ease to which it can be approached by the surgeon and the potential for viability of the underlying myocardium. The electrode should also be sutured to the myocardial surface even though the intramyocardial portion of the electrode itself will provide fixation. Without suture fixation, the electrode will be subject to much more stress with each cardiac contraction, potentially resulting in premature electrode fracture and/or increased fibrosis around the electrode resulting in threshold elevation.

For atrial epicardial electrodes, a site should be chosen on the right atrium rather than the left atrium. If a left atrial pacing site is used, there is a longer delay between atrial activation and ultimate atrioventricular conduction with various effects on atrioventricular synchrony. Because the conduction time from the left atrium to the right atrium must occur before the impulse reaches a normally conducting atrioventricular conduction system, programming the optimal atrioventricular interval can be extremely difficult. Even when there is no atrioventricular conduction, left atrial pacing should be avoided (Figure 7).

For epicardial bipolar pacing, the same considerations apply. However, the placement of the second electrode or anode becomes more critical. The second electrode needs to be positioned approximately two centimeters from the cathodal electrode. Furthermore, the

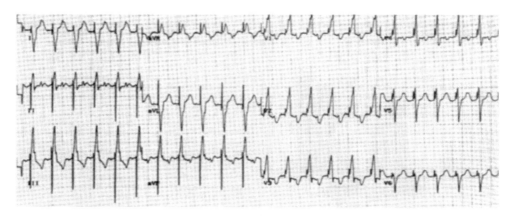

FIGURE 7. Twelve lead EKG from a patient with left atrial electrode placement showing atrial sensing with ventricular pacing. Note the P to ventricular pace time is 220 msec even though the programmed AV interval is 150 msec. This is due to the time necessary for the impulse to propagate from the right atrium to the left atrial electrode. Thus, to maintain an appropriate AV interval, the programmed value must be shorter than usual. Used with permission from Serwer, et al., Pediatric pacing and defibrillator usage, in *Clinical Cardiac Pacing and Defibrillation*, 2nd edition, Ellenbogen KA, Kay GN, Willkoff BL, eds., W.B. Saunders, Philadelphia, 2000 and Elsevier.

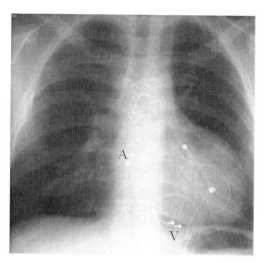

FIGURE 8. Chest x-ray showing appropriate placement of atrial (A) and ventricular (V) bipolar epicardial electrodes. Older unipolar electrodes are also seen. A line connecting the two poles of the atrial electrodes (A) would be perpendicular to the long axis of the heart while such a line for the ventricular electrodes (V) would be parallel to it.

line drawn between the two ventricular electrodes should be parallel to the long axis of the ventricle; in contrast, the line drawn between the atrial electrodes should be perpendicular to the long axis of the ventricle. This configuration tends to minimize far field R-wave sensing, which is particularly important for atrial tachycardia sensing when antitachycardia devices are used (Figure 8).

When applying epimyocardial electrodes, care must be taken to wet the electrode prior to its application to the epimyocardial surface. These electrodes tend to have a small surface area and any intervening material that is adherent to the surface, including air, tends to insulate the surface causing high acute thresholds. If acute high thresholds are found, it is often worth waiting for a brief period to see if the thresholds will decrease as the electrode settles into its location. Care must also be taken when suturing the epimyocardial electrode to the surface so that the underlying myocardium is not buckled, which separates the active part of the electrode itself from the underlying myocardium. Often initial thresholds in excess of 2 V at 0.5 msec pulse width duration will decrease to less than 1 volt 24 hours after implantation. Thus, if it is felt that the underlying myocardium appears quite viable and the electrode appears to be in good contact with the underlying myocardial surface, thresholds of up to 2.5 to 3.0 V can easily be accepted with the expectation that these will improve quickly and provide reliable and stable long-term pacing.

Endocardial Electrode Placement

Endocardial or transvenous electrode placement requires vascular access. If there is any question about the potential vascular accessibility, a venous angiogram is performed to evaluate the anatomy, position, and size of the innominate vein. The innominate vein is then accessed percutaneously using the Seldinger technique and a guidewire advanced into the right atrium. To avoid entrapment of the electrode between the clavicle and first rib, which can lead to electrode crush and subsequent failure, the subclavian vein should be entered as far distally as possible. Once vascular access is achieved, the incision can be made utilizing the entry point of the wire as the most proximal end of the incision. Alternatively, entry into the vein can be performed after the incision for the pocket is made, accepting the small risk of no access and thus an unusable pocket. Dissection should be carried deep enough to allow adequate tissue to cover the pacemaker. A pocket that is created too shallow will result in the potential for skin breakdown or a pocket that is cosmetically unappealing. This can be of major concern to the patient, especially children. Due to the lack of subcutaneous tissue in many children, many pacemaker pockets are created under the pectoral muscle. A more cosmetic incision, particularly important in girls and young women, is one placed laterally and more vertically in the anterior axillary line. With increased use of bipolar pacing, extracardiac stimulation of the pectoral muscle is no longer an issue. In addition, with the use of high impedance electrodes and the elimination of the insulated coating around the pacemaker can, current flows are low and diffuse enough at the level of the pectoral muscle to often avoid pectoral muscle stimulation even with unipolar electrodes.

Once the pocket has been created, the previously placed wire is dissected so that it is now free within the pacemaker pocket. The tissue surrounding the wire is gently dissected to open a passage for the placement of the electrode introducer. To avoid blood loss even when inserting the pacing lead, the peal-away introducer (sheath) should have a hemostatic valve. It is then placed over the wire and positioned in the superior vena cava. The final position of the end of the introducer must be carefully assessed. The end must either lie within the innominate vein prior to it joining the superior vena cava or must have already turned inferiorly and have entered the superior vena cava. Positioning the tip of the introducer at the innominate vein/superior vena caval junction increases the difficulty in introducing the electrode into the right atrium and the risk of superior vena caval perforation by the pacing electrode. In addition, if the electrode proves unsatisfactory, it can be easily replaced without again percutaneously entering the subclavian vein.

When placing a dual-chamber system, the ventricular electrode is placed first. While apical right ventricular pacing has been widely used, the intraventricular septum appears to afford improved right ventricular long-term function. However, the first priority must be to find a position within the right ventricle with low pacing thresholds, adequate spontaneous R-wave amplitude, and a highly stable position. Once in position, the electrode is thoroughly tested.

To introduce the atrial lead, a guidewire is introduced through the sheath next to the ventricular lead and the introducer is removed. A second introducer is then advanced over the retained guidewire, but only after the ventricular lead has been positioned. This sequence does not result in the ventricular lead's dislodgement. Following placement of the second sheath, the atrial electrode is introduced into the atrium and its electrode tip is positioned on either the atrial septum or in the right atrial appendage. The J-shaped stylets are often useful in larger children, but the size of the curve is often too large for the smaller child. For smaller children, a J-shaped stylet with a smaller curve is fashioned from a straight stylet. Steerable stylets have recently been introduced that are useful for positioning of the atrial electrode, particularly in the child with abnormal atrial anatomy.

Following electrode testing, the introducer is removed and the electrode positions of both atrial and ventricular electrodes are verified with fluoroscopy. Adequate lead lengths must be left to be sure the lead will not be dislodged when a deep breath is taken and the heart moves downward or when the heart is displaced inferiorly with standing. Placement of a large loop within the atrium to accommodate growth has been suggested, but has been met with variable success. Too often the loop adheres to the endocardial surface and the advantage of having a loop within the atrium is negated. If a loop that is too large remains, it may move within the atrium causing atrial arrhythmias, and, in some cases, prolapsing across the tricuspid valve resulting in significant tricuspid regurgitation. A generous curve is all that is usually necessary.

Coronary Sinus Lead Placement

When cardiac resynchronization is prescribed, a transvenous lead is placed in the coronary sinus to pace the left ventricle. At first, coronary sinus leads were stylet driven, limiting access to distal coronary veins. The introduction of smaller leads that can be advanced over angiographic wires has greatly simplified access to smaller, distal, more tortuous veins.

Prior to placing a coronary sinus lead, the coronary sinus anatomy must be determined. Congenital heart disease with associated coronary venous anomalies as well as smaller patient size may preclude transvenous cardiac resynchronization. Current coronary sinus lead placement systems use a long 8 French sheath. A variety of curves are available to suit the right atrial size and coronary sinus location. Advancing the 8 French sheath into the coronary sinus can be aided by using a steerable catheter or curved catheters and guidewires. Once the 8 French sheath has been advanced into the proximal coronary sinus, a venous angiogram is obtained using a balloon with an end hole catheter. The distal veins are best visualized if the initial contrast injection is performed with the balloon inflated but not overinflated. To identify branching veins that are on the posterolateral aspect of the left ventricle, images are taken in both the AP and LAO projections. In general, suitable venous sites for left ventricular electrode lead placement are approximately halfway out the coronary sinus, and half the distance between the atrioventricular groove and the apex of the left ventricle (Figure 9). Ideal sites are not always available and alternate sites more distal in the coronary sinus are often adequate.

Once a suitable vein has been identified, the lead is placed. Because the venogram serves as a roadmap for lead placement, maintaining the same fluoroscopic camera angle is extremely important. The lead may be advanced into the coronary vein by either first advancing the lead through the 8 French sheath into the coronary sinus, then back loading the angiographic wire, or conversely, guiding the angiographic wire into the desired position, then advancing the coronary sinus lead over the wire. Techniques may depend on lead type and manufacturer guidelines. Angiographic wires of variable sizes and stiffness are available to aid in negotiating the venous anatomy. The angiographic wire should be advanced well into the desired vein to insure stability of position while advancing the lead over the wire. Lead position is maintained by passive fixation of soft tines at the tip of the lead, or a fixed curve on the lead. If the final position of the lead is in a large vein, the probability of subsequent dislodgement is much greater.

Prior to sheath removal, sensing and pacing threshold testing is performed to assure

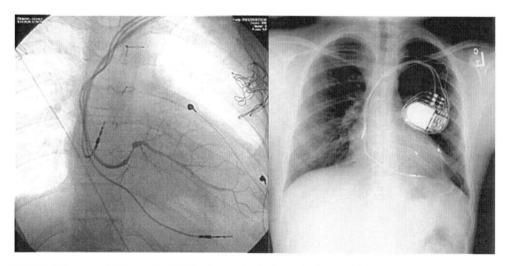

FIGURE 9. Ten-year-old trisomy 21 girl following repair of atrioventricular septal defect and biventricular pacemaker implant for complete heart block and cardiac resynchronization—Left panel: Coronary sinus angiogram showing large branching vein approximately one-third of the distance along the coronary sinus with no significant continuation. Right panel: PA chest x-ray demonstrating final position of the left ventricular lead (arrow) one-half the distance to the apex.

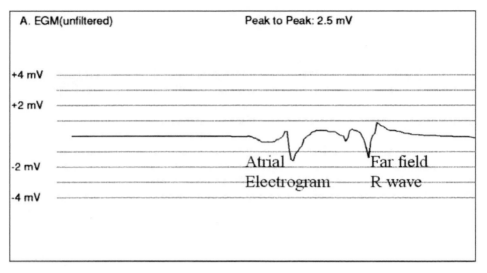

FIGURE 10. Atrial electrogram from an atrial electrode obtained at the time of electrode placement showing the electrogram amplitude and the amplitude of the far field R wave.

appropriate lead function. The likelihood of phrenic nerve capture is much greater than that seen in right atrial and right ventricular leads so high output pacing to test for phrenic nerve capture and diaphragmatic pacing is essential. The left ventricle pacing threshold can be established in a unipolar mode using an indifferent electrode within the dissected pocket or the cathode of the right ventricular lead. The original coronary sinus leads were unipolar, limiting pacing options. Bipolar leads are now available and newer generators can be programmed to pace the LV either unipolar or bipolar.

Once adequate electrode lead placement has been assured, the lead pins are inserted into the generator. The leads are then secured to the floor of the pacemaker pocket using the collar on the lead. The generator is then placed in the pocket with the electrodes coiled beneath the pacemaker generator. Placement of the electrodes under the generator facilitates future generator replacement and minimizes the risk of electrode damage when generator replacement is necessary.

Acute Electrode Testing

At the time of initial placement, all electrodes must be tested to assure adequate performance. Thresholds are determined as the minimum voltage necessary to achieve 100% capture at a given pulse width. Typically, in our laboratories voltage thresholds are determined at pulse widths of 1.0 msec, 0.5 msec, 0.3 msec, and 0.1 msec. Electrode impedance is also measured at a pulse amplitude of 5 V and a pulse width of 0.5 msec. When spontaneous activity is present, the maximal R/P amplitude and the slew rate are directly measured from the recorded spontaneous electrogram. In addition, the atrial electrogram also provides information concerning the degree of far field R-wave sensing (Figure 10).

Appropriate thresholds (i.e., the minimal voltage at 0.5 msec pulse duration that achieves 100% capture) are individualized. Ideally, one should see voltages less than 1 V at 0.5 msec. However, there are some patients in whom higher thresholds must be accepted either because of a limited number of places to position the electrode or the presence of intrinsic myocardial disease that creates threshold elevation. In all instances it must be determined that the threshold will reliably pace the heart given the maximal output of the device chosen. When a biventricular pacing system is implanted, both right and left ventricular pacing thresholds need to be established separately, as well as a biventricular pacing threshold. Criteria for ideal left ventricular lead placement in the acute setting have not been established. Proposed predictors of long-term success have included narrowing the QRS, improvement in cardiac output by Fick or thermodilution measurement, or improvement in regional ventricular wall motion by echocardiography.

Acceptable values for electrode impedance are determined by the design of the electrode and are outlined by the manufacturer.

ICD Placement

Similar to brady-devices, ICDs can be placed either epicardially or endocardially and placement of ICDs shares many similarities to placement of bradycardia devices. When placing an epicardial ICD, the difference involves placement of the epicardial patch electrodes. The optimal placement for these electrodes varies from patient to patient, but typically they are placed in an anterior/posterior position and are sutured to the pericardial sac rather than the myocardial surface. However, significant scarring can result making subsequent removal of the patches extremely difficult. Two patches are not always required if the device can is used as the second electrode. Placement of the one patch must then take into account the ultimate position of the device to obtain a vector that maximizes current flow through the heart.

Endocardial placement requires that the ventricular coil is positioned wholly within the right ventricle. If it extends through the tricuspid valve into the right atrium, the defibrillatory thresholds tend to be elevated and the defibrillatory efficiency is poor. In smaller children this problem can be addressed either by attaching the electrode tip to the ventricular septum and looping the coil at the right ventricular apex or by placing the electrode tip in the right ventricular apex and then pushing the lead to position the coil along the right ventricular septal surface. The key to obtaining low defibrillatory thresholds is to provide a large area of contact between the coil and the ventricular myocardium.

The next choice that must be made is whether to use a dual-coil electrode with the second coil positioned in the superior vena cava or a single-coil electrode with the can of the ICD being the second electrode. For smaller children, the inter-coil distance is often too long and results in placement of the superior vena cava coil within the innominate vein and, in some cases, outside the vascular space, placing the coil at risk of being crushed between the clavicle and the first rib (Figure 11). Thus, in small children, a single coil electrode lead is often used. However, as dual-coil electrode lead sizes have decreased, similar to the single-coil electrode leads, and

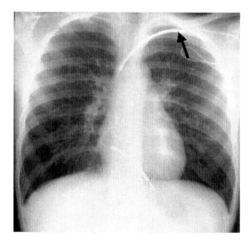

FIGURE 11. Chest x-ray showing fracture (arrow) of the SVC coil of the defibrillatory electrode due to being crushed between the clavicle and first rib.

as smaller inter-coil distances have become available, dual-coils have found increased use in small children. In addition, because of their versatility when choosing a defibrillatory vector between the right ventricular coil and either the superior vena cava coil, ICD can, or both, dual-coil leads provide additional ways to lower the defibrillatory thresholds.

Once the electrode leads have been positioned, defibrillatory thresholds are determined. Because there is risk in inducing ventricular fibrillation, our practice has been to test the defibrillatory capability of the device utilizing a 10-J output. If the 10-J shock is successful in converting the patient from ventricular defibrillation on two separate trials (at least 5 minutes apart), testing at lower energy settings is not performed. Given the output of the current devices, acute defibrillatory thresholds of less than 15 J are considered adequate and do not require replacement or repositioning of the electrodes. Defibrillatory thresholds between 15 and 20 J are considered marginal and the decision as to electrode repositioning is based on the difficulty in achieving the original position. Thresholds greater than 20 J are considered inadequate and potentially place the patient at risk for the failure of the device to convert ventricular fibrillation. If the initial defibrillatory thresholds are elevated, the polarity of the shocking pulse is first reversed in an effort to lower the thresholds or to alter the shocking vector. If that is unsuccessful, the leads are then repositioned or a subcutaneous array is considered. In our experience it is rarely necessary in children to utilize a subcutaneous array. In older patients who have had multiple cardiac procedures, the need for subcutaneous array, while still low, is increased.

Our practice is not to repeat defibrillatory threshold determination following initial placement and initial determination of thresholds unless there has been some clinical change in the patient. DFT testing is rarely significantly changed from that determined at the initial placement unless there had been some change in measured defibrillatory electrode impedance, electrode position on chest x-ray, or a failed defibrillatory attempt.

DEVICE FOLLOW-UP

Proper follow-up testing of the pacemaker and electrodes is critical to ensure continued appropriate function. Of all follow-up problems, electrode lead malfunction is most common; generator malfunction, and, in time, battery depletion can occur, and all must be equally and thoroughly assessed at the time of each follow-up visit. Our schedule is to see the child two weeks following implantation to make certain that the incisions are healing well and there has been no acute electrode dislodgement. A repeat examination is scheduled at six weeks and at that time chronic, long-term settings are programmed. Subsequent visits are scheduled at three months, six months, and one year. Patients with congenital complete heart block and a structurally normal heart are seen approximately once per year until the pacemaker is five years post implantation, at which time the follow-up period is decreased to six months. Patients with structural cardiac disease are seen at least every six months regardless of pacemaker age. Between clinic visits transtelephonic pacemaker evaluation is performed every other month during the first five years and once per month thereafter. The development of manufacturer web-based home pacemaker interrogation will undoubtedly influence this follow-up schedule. The nature of the follow-up evaluation in both the clinic and by transtelephonic evaluation is uniform, but allowances must be made on an individual basis.

Clinic Evaluation

When a patient returns for a clinic visit, a thorough history and physical examination is obtained, followed by an evaluation of both the pacemaker and all electrode leads. Further testing is dependent on the patient's underlying cardiac disease.

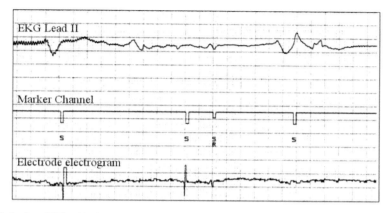

FIGURE 12. Surface ECG with simultaneous Marker Channel and electrogram showing ventricular sensing to be occurring with no ventricular activity evident on the surface ECG indicative of electrode fracture. Elevated electrode impedance was present on device interrogation.

Device interrogation is performed specifically noting battery impedance and battery voltage. Telemetry measurement of battery voltage can be somewhat variable from reading to reading, and it can be difficult to ascertain true changes in battery voltage from one interrogation to the next. Battery impedance, however, is more reproducible and gives one a better indication of battery depletion. As the battery is depleted, the impedance rises to values usually in excess of 3,000 Ω; however, this depends on the properties of each device.

Electrode lead evaluation consists of noting the electrode impedances to ascertain that they have not significantly increased, which suggests fibrosis around the electrode tip or lead filar fracture is suspected when impedance decreases. Lead insulation fracture. An increased impedance can be caused by fracture of some of the filars within the lead yet capture can be maintained. These fractures can also result in oversensing and inappropriate pacing inhibition (Figure 12). Variation in impedance can be noted from visit to visit, not to exceed 200–300 Ω. Some pacemakers monitor electrode impedance variability on a day-to-day basis, which provides useful information when there is concern about electrode tip fibrosis. Large changes in electrode impedance is insufficient as the sole indicator for electrode replacement, but this information together with changes in electrode thresholds is useful in detecting pending electrode lead problems.

Electrode threshold testing—the minimal voltage at 0.5 msec pulse duration that achieves 100% capture—is next performed. Threshold values are typically determined for at least 2 and preferably 4 different pulse amplitudes to better estimate the strength duration curve (Figure 13). Some pacemakers permit determination of the minimum voltage needed to pace at a given pulse width. Either method is acceptable as it allows one to compare changes in thresholds from visit to visit.

The diagnostic data from the pacemaker is reviewed, specifically evaluating the percentage of beats in each chamber that are spontaneous versus paced, and the heart rate variability that is present. This is particularly important when one relies on the activity sensor for heart rate variability. The lack of variability would indicate inappropriate sensor programming. In addition, a large percentage of time spent at higher heart rates indicates likely inappropriate sensor programming.

The presence of high rate episodes may indicate the development of tachyarrhythmias in either chamber. This is particularly relevant

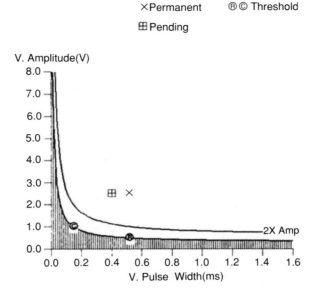

FIGURE 13. Example of a strength–duration curve obtained at clinic follow-up, as well the appropriate output settings selected based upon it.

to patients who have undergone cardiac surgical procedures known to predispose one to tachyarrhythmias. The presence of high rate episodes or very high rates being detected in either chamber would be an indication for further investigation.

Finally, based on the data collected, the programming parameters are reviewed. Some changes are generally required within the first three months of device implantation as the device is "fine-tuned." However, beyond this initial adjustment, changes are often not needed. The most common change is alterations in generator output in order to maintain an adequate safety margin to guarantee 100% capture, but, at the same time, minimize battery current drain. It is our practice to set voltage and pulse width settings high for the first six to eight weeks as the electrodes undergo acute changes associated with initial implantation. After six weeks the settings can often be lowered, which reflect the threshold values determined at the time of the clinic visit. Our general guidelines are to set the output voltage at least 1.7 to 2 times the threshold value for the chosen pulse duration while making certain that the chosen pulse duration is at least twice the threshold value for the chosen pulse amplitude. To make certain both of these criteria are met, a strength duration curve is constructed, as a single threshold setting at given pulse amplitude does not allow one to evaluate the adequacy of the chosen output settings (Figure 13).

CRT Follow-Up

Follow-up of the patient with a cardiac resynchronization device requires additional testing of the left ventricular threshold and may include programming changes to optimize cardiac output. Current devices not only allow varying the atrioventricular delay, but also the interventricular delay. Techniques to optimize the atrioventricular delay include shortening the atrioventricular delay to minimize the QRS duration, and monitoring the "a" and "v" waves in the echocardiogram while adjusting the atrioventricular interval to obtain a non-fused "a-v" complex with optimal "a:v" amplitude ratio. Techniques to optimize the inter-ventricular delay include altering the delay to narrow the QRS duration, or monitoring ventricular wall motion by echocardiogram while adjusting the RV-LV delay to achieve maximal synchronization of left ventricular contraction. The ultimate utility of current methods as well as the development of other noninvasive ways to monitor and optimize the effectiveness of cardiac resynchronization is not yet available.

Transtelephonic Monitoring

Between visits to the pacemaker clinic, patients are monitored by transtelephonic monitoring. In addition, they are encouraged to send transmissions whenever there is a concern about improper pacemaker function. A two-minute electrogram strip followed by a two-minute recording with the magnet placed over the pacemaker is the standard recording. The response of the pacemaker to the magnet is quite variable among manufacturers and one must be familiar with the magnet response of the device to appropriately interpret the tracings. Some devices decrease the generator output for one beat during the magnet application to check threshold adequacy while others do not have this function. The rate at which the pacemaker paces in response to magnet application also varies from device to device, and the pacemaker response to reaching elective replacement time is also variable. In clinics where a multitude of different pacemakers are followed, appropriate reference material delineating magnet activity must be available. Electronic versions as well as printed versions of these materials suitable for both laptop and handheld computers are available on the Internet.

Recording of such electrograms can be difficult in uncooperative children. Often several electrograms must be sent to fully evaluate the pacemaker. A voice mail system to record transmissions sent after normal business hours is often the time most convenient for the patient and family and often receives the best recording. If immediate viewing of the transmission is needed, it can be quickly retrieved and viewed.

Newer computer-based transtelephonic systems that perform actual interrogation and storage (on a server) of pacemaker recorded data similar to that obtained at a clinic visit are now available. These systems allow a more thorough evaluation of the pacemaker and electrode function over the telephone than has been previously available, thus providing assurance to the patient concerning the adequacy of the pacemaker function, as well as potentially decreasing the number of clinic visits.

While electrode malfunction cannot necessarily be anticipated, battery depletion can. Since scheduled clinic visits are routine, semi-elective, rather than emergent, pacemaker replacement can be scheduled, avoiding many patients presenting with unexpected pacemaker malfunction.

ICD Follow-Up

In many instances, ICD follow-up is similar to pacemaker follow-up particularly for the brady pacing functions of the device. All new episodes of device discharge should be examined to be certain that the discharge was appropriate and successful in terminating the tachyarrhythmia. In addition to pacing electrode impedances, impedance of the high voltage circuit is also available in the newer devices, thus assuring the intactness of the defibrillatory electrode leads. For patients in whom high voltage impedances cannot be obtained, periodic chest x-rays are used to look for intactness of the high voltage electrodes. Periodic determination of defibrillatory thresholds are not routinely performed as, in our experience, such thresholds do not change unless accompanied by other indicators such as a failed defibrillatory attempt, changes in pacing and/or high voltage electrode impedances, or changes in electrode position on chest x-ray.

Electrograms recorded from both pacing and high voltage electrodes should routinely be performed to look for changes comparing them to those obtained at prior visits. Particularly no inappropriate sensing of either far field R-wave by atrial electrodes or T-waves by the ventricular electrodes that would lead to inappropriate tachy-detection and inappropriate therapy delivery should be present. Careful examination of all tachyarrhythmia episodes should also be performed to make certain that the tachyarrhythmia waveforms were appropriately sensed as well. If there is any question concerning the adequacy of device sensing or the ability to terminate a tachyarrhythmia, then defibrillatory threshold testing may be needed.

SUMMARY

Patients requiring device implantation need lifelong care. However, with appropriate selection of the device and appropriate follow-up, lifestyle changes attributed to the device can be minimized. Exercise limitations, beyond those needed to protect the device and the overlying skin, are often not necessary beyond those due to the underlying cardiac status. As new technologies are developed that minimize battery drain and increase the time between invasive procedures, device longevity is now reaching 7 to 10 years. Devices for which longevity data are available are obsolete. Electrode lead longevity, however, can more accurately be estimated. In our experience, electrode lead longevity is between 10 to 12 years for transvenous leads and 5 to 8 years for epicardial ones.

The goal of all pacemaker management is to improve cardiac performance and to permit the patient to lead an active and normal lifestyle. As newer indications for device use emerge, the number of patients requiring device implantation will increase making attainment of this goal increasingly important.

SUGGESTED READING

1. Serwer GA, Dorostkar PC, LeRoy SS. Pediatric pacing and defibrillator usage. In *Clinical Cardiac Pacing and Defibrillation*. Ellenbogen KA, Kay GN, Wilkoff BL, eds. W.B. Saunders: Philadelphia, PA 2000.
2. Gregoratos G, Abrams J, Epstein AE, et al. Guideline update for implantation of pacemakers and antiarrhythmic devices: summary article. *Circulation* 2002;106:2145–2161.
3. Weindling SN, Saul JP, Gamble WJ, et al. Duration of complete atrioventricular block after congenital heart disease surgery. *Am J Cardiol* 1998;82:525–527.
4. Rishi F, Hulse JE, Auld DO, et al. Effects of dual-chamber pacing for pediatric patients with hypertrophic obstructive cardiomyopathy. *J Am Coll Cardiol* 1997;29:734–740.
5. Stephanson EA, Casavant D, Tuzi J, et al. on behalf of the ATTEST investigators. Efficacy of atrial antitachycardia pacing using the Medtronic AT500 pacemaker in patients with congenital heart disease. *Am J Cardiol* 2003;92:871–876.
6. Kristensen L, Nielsen JC, Pedersen AK, et al. AV block and changes in pacing mode during long-term follow-up of 399 consecutive patients with sick sinus syndrome treated with an AAI/AAIR pacemaker. *PACE* 2001;24:358–365.
7. Cohen MI, Bush DM, Vetter VL, et al. Permanent epicardial pacing in pediatric patients. *Circulation* 2001;103: 2585–2590.
8. Bauersfeld U, Schonbeck M, Candinas R, et al. Initial experiences with a new steroid-eluting bipolar epicardial pacing lead. *Eur J Card Pacing Electro* 1996;6:222.
9. Serwer GA, Mericle JM, Armstrong BE. Epicardial ventricular pacemaker electrode longevity in children. *Am J Cardiol* 1988;61:104–106.
10. Serwer GA, Uzark K, Dick M II. Endocardial pacing electrode longevity in children. *J Am Coll Cardiol* 1990;15:212A.
11. Cohen MI, Buck K, Tanel RE, et al. Capture management efficacy in children and young adults with endocardial and unipolar epicardial systems. *Europace* 2004;6:248–255.
12. Johns JA, Fuh FA, Burger JD, Hammon JW Jr. Steroid-eluting epicardial pacing leads in pediatric patients: Encouraging early results. *J Am Coll Cardiol* 1992;20:395–401.
13. Karpawich PP, Horenstein MS, Webster P. Site specific right ventricular implant pacing to optimize paced left ventricular function in the young without and without congenital heart disease. *PACE* 2002;25:566.
14. Stefanelli CB, Bradley DJ, Leroy S, et al. Implantable cardioverter defibrillator therapy for

life-threatening arrhythmias in young patients. *J Interven Card Electro* 2002;6:235–244.
15. Fischbach PS, Law IH, Dick M II, et al. Use of a single coil transvenous electrode with an abdominally placed implantable cardioverter defibrillator in children. *PACE* 2000;23:884.
16. Strieper M, Karpawich P, Frias P, et al. Initial experience with cardiac resynchronization therapy for ventricular dysfunction in young patients with surgically operated congenital heart disease. *Am J Cardiol* 2004;94:1352–1354.
17. Lee JC, Shannon K, Boyle NG, et al. Evaluation of safety and efficacy of pacemaker and defibrillator implantation by axillary incision in pediatric patients. *Pacing Clin Electrophysiol* 2004;27(3):304–307.
18. Blom NA, et al. Transvenous biventricular pacing in a child after congenital heart surgery as an alternative therapy for congestive heart failure. *J Cardiovascul Electrophysiol* 2003;14(10):1110–1112.
19. Janousek J, et al. Resynchronization pacing is a useful adjunct to the management of acute heart failure after surgery for congenital heart defects. *Am J Cardiol* 2001;88(2):145–152.
20. Zimmerman FJ, et al. Acute hemodynamic benefit of multisite ventricular pacing after congenital heart surgery. *Ann Thorac Surg* 2003;75(6):1775–1780.
21. Walker F, et al. Long-term outcomes of cardiac pacing in adults with congenital heart disease. *J Am Coll Cardiol* 2004;43(10):1894–1901.

18

Genetic Disorders of the Cardiac Impulse

Mark W.W. Russell and Stephanie Wechsler

Heritable cardiac arrhythmias and cardiac conduction disorders are relatively rare yet often fatal disorders that may involve any phase of the cardiac impulse including impulse generation, propagation, or electrochemical recovery. These disorders include primary abnormalities of cardiac ion channels (e.g., long QT syndrome and Brugada syndrome), primary abnormalities of cardiac function with secondary arrhythmic complications (e.g., hypertrophic and dilated cardiomyopathies) and defects of impulse propagation (e.g., familial complete heart block). Molecular genetic characterization of families with these disorders has led to important advances in the understanding of the generation of arrhythmias, and thus, improved prospects for diagnosis and treatment. Except for the cardiomyopathies, an important and striking commonality among these disorders is that the heart is structurally and functionally normal.

These disorders can be grouped into several main categories: ion channel disorders or channelopathies, cardiomyopathies, and conduction system abnormalities. The ion channelopathies can be further subdivided into those that (1) prolong the action potential duration by delaying repolarization and/or predisposing to early after-depolarization-induced arrhythmia and (2) those that lead to enhanced automaticity and triggered activity (Chapter 2).

ION CHANNELOPATHIES

Ion Channelopathies that Prolong the Action Potential

Long QT Syndrome

Heritable prolongation of the QT interval as a cause for torsades de pointes and sudden death was first reported in the late 1950s and early 1960s. Patients with long QT syndrome (LQTS) were characterized by abnormal prolongation of the QT interval on their 12-lead electrocardiograms and by a predisposition to ventricular arrhythmias and sudden death (Figure 1).

The inheritance pattern has been described both as autosomal *dominant* and as autosomal *recessive* (Figure 2). Detailed molecular genetic studies in families with LQTS show that at least seven different genes, when altered by genetic mutation, are responsible for causing this disorder (Table 1). Six of these genes have been identified and at least one more exists due to identification of families that do not have mutations in any of the six

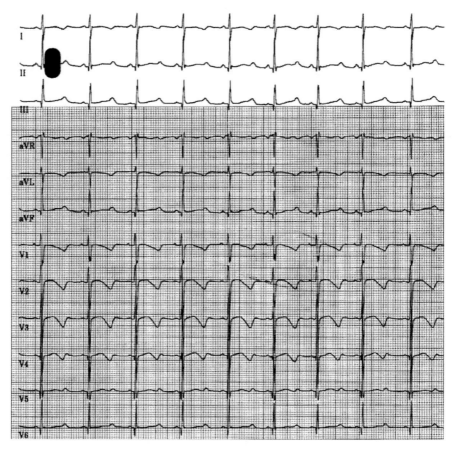

FIGURE 1. 12-Lead ECG in an 8-year old boy with sudden loss of consciousness for 10 seconds. Note the QTc of 0.48 seconds in lead 2 and V5.

known genes. Of the six identified genes, four encode potassium channel subunits, one encodes the cardiac sodium channel *SCN5A*, and one leads to a change in ankyrin-B, a membrane protein that interacts with ion channels.

Electrophysiologic and genetic characterization of LQTS patients has demonstrated that mutations in these different genes can cause distinct manifestations of LQTS, each with unique risks and different responses to therapy. For example, it initially appeared that all patients with LQTS were at risk for cardiac arrhythmias at times of anger and emotion. Thus, beta-blockade was the treatment of choice to prevent arrhythmias and sudden death. Now, it appears that arrhythmia associated with emotion is most characteristic of the most common group of LQTS patients, those with defects of *KCNQ1*, the alpha subunit of the potassium channel that appears to be primarily responsible for the slow potassium (I_{Ks}) current.

I_{Ks} Defects. Defects of *KCNQ1* and *KCNE1*, the alpha and beta subunits of the potassium channel responsible for the I_{Ks} current, can cause both the autosomal dominant and autosomal recessive forms of LQTS. Heterozygosity for mutation of a single copy of either of these genes causes autosomal dominant LQTS, initially known as Romano-Ward syndrome. Mutation of both

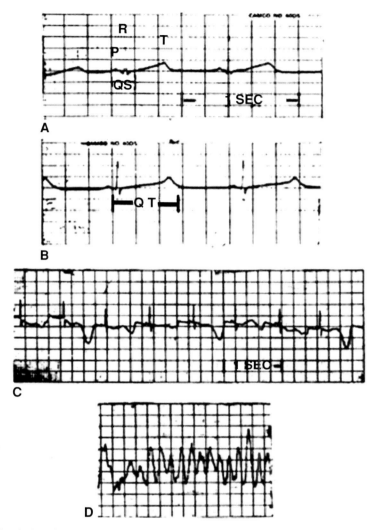

FIGURE 2. Panels A and B: Demonstrating long QTc 0.54 seconds in an 18-month-old boy with sensorineural deafness. Panel C: Marked T-wave alternans. Panel D: Spontaneous ventricular fibrillation, which self-terminated. Reprinted with permission from Wahl, R et al. Congenital deafness with cardiac arrhythmias: The Jervell and Lange-Nielsen syndrome. American Annals of the Deaf 1980;155(1):34–37.

copies of one of these genes (homozygosity) results in severely symptomatic LQTS that is characterized by sensorineural deafness, initially known as Jervell and Lange-Nielsen syndrome.

The I_{Ks} current is responsible for the slowly activating outward potassium current that allows the myocyte to return to its resting potential following a depolarization. Patients with I_{Ks} defects are at greatest risk of arrhythmias and sudden death during adrenergic surges that occur during fright or anger, the "flight or fight response." Beta-blockade treatment has resulted in a significant improvement in survival in LQTS patients with I_{Ks}-type defects. Interestingly, studies have demonstrated that patients with *KCNQ1* defects, and presumably with *KCNE1* defects as well, are particularly susceptible to arrhythmias while swimming. This observation has

TABLE 1. Channelopathies

Diagnosis	Gene locus	Gene symbol	Gene product	Manifestations
Long QT Syndrome	11p15.5	*KCNQ1*	I_{Ks}, α subunit	TdP/Sudden death with stress/exercise; can cause JLN
	7q35-36	*KCNH2*	I_{Kr}, α subunit	TdP/Sudden death with stress/exercise
	3p21-23	*SCN5A*	I_{Na}, α subunit	TdP/Sudden death with rest/sleep
	4q25-27	*ANK2*	ankyrin-B	Sudden death/sinus bradycardia
	21p22.1	*KCNE1*	I_{Ks}, β subunit	TdP/Sudden death with stress/exercise; can cause JLN
	21p22.1	*KCNE2*	I_{Kr}, β subunit	TdP/Sudden death with stress/exercise
Brugada Syndrome	3p21-23	*SCN5A*	I_{Na}, α subunit	Polymorphic VT/VF with rest
Andersen Syndrome	17q23-24	*KCNJ2*	Kir2.2 (inward rectifier)	Polymorphic VT/VF; periodic paralysis
Familial Ventricular Fibrillation	3p21-23	*SCN5A*	I_{Na}, α subunit	Polymorphic VT/VF with rest
Catecholaminergic	1q42	*RYR2*	SR Ca^{2+} releasing channel	Polymorphic VT with stress/exercise
Polymorphic VT	1p11-13	*CASQ2*	SR Ca^{2+} storage protein	Polymorphic VT with stress/exercise

Abbreviations: JLN = Jervell-Lange-Nielson; SR = sarcoplasmic reticulum; TdP = torsades de pointes; VF = ventricular fibrillation; VT = ventricular tachycardia.

led to the recommendation that all patients with the LQTS should not swim alone or without flotation gear. Additional studies may indicate that only those patients with I_{Ks} defects need to avoid swimming.

LQTS-causing mutations may occur anywhere in the genes that encode these two subunits of the potassium channel I_{Ks}, but the more severe defects appear to be those that result in the production of an ion channel subunit with abnormal potassium conduction properties. These abnormal subunits can complex with normal subunits so that the vast majority of I_{Ks} channels within the myocyte are either non-functional or function abnormally. In contrast, there are some mutations that do not produce an abnormal protein product, but rather result in approximately half the number of fully functional I_{Ks} channels, causing a less severe manifestation of the LQTS. In some cases the defect may be mild enough that no repolarization abnormality is noted even with provocative testing such as epinephrine infusion. With some of the mutations, no abnormalities are noted when a patient inherits only one defective copy of the gene and the LQTS only becomes apparent when two altered copies of the gene are inherited as is noted in the Jervell and Lange-Nielsen syndrome.

I_{Kr} Defects. Defects of KCNH2 and KCNE2, the alpha and beta subunits of the potassium channel responsible for the rapid potassium (I_{Kr}) current, can cause the autosomal dominant form of LQTS. I_{Kr} is the current responsible for the initiation of cardiac repolarization following excitation-contraction. Patients with KCNH2 defects have been demonstrated to be at increased risk for arrhythmia in response to auditory stimuli like alarm clocks and bedside phones.

I_{Kr} has some unique properties that affect the clinical manifestations of the LQTS in these patients. The potassium conductivity of the I_{Kr} channel is very sensitive to extracellular concentrations of potassium. In large part, the transmembrane electrochemical gradient is maintained by the difference between the intracellular and extracellular potassium concentrations. The capacity of the channel to conduct an outward potassium current (i.e., I_{Kr}) varies directly with the extracellular potassium concentration. As the extracellular potassium rises, the outward potassium current through the channel increases. Conversely, a low extracellular potassium concentration inhibits the channel's function. Therefore, patients that have *KCNH2* or *KCNE2* mutations that inhibit the function of the I_{Kr} channel are particularly sensitive to low extracellular potassium concentrations that further reduce the conductance of the channel. Recent studies have examined treating I_{Kr}-type LQTS patients with potassium supplements, potassium channel openers such as nicorandil, and potassium sparing agents such as spironolactone. These therapies would more directly address the ion channel dysfunction unique to I_{Kr}-type LQTS patients.

SCN5A Defects. Patients with defects of the cardiac sodium channel, SCN5A, account for less than 10% of the reported LQTS cases. Inheritance is autosomal dominant. These patients are more susceptible to arrhythmias and sudden death during times of bradycardia and may die in their sleep. Compared to patients with potassium channel defects, the frequency of arrhythmic events appears to be less common in patients with sodium channel defects. Yet, their arrhythmias are more likely to be fatal since they are often not observed.

The LQTS-causing *SCN5A* defects described to date impair the ability of the channel to completely inactivate after depolarization, allowing persistent inward leak of Na ions, with repeated opening of the channels during sustained depolarization generating a small, sustained inward depolarizing current. Thus, the *SCN5A* mutations lead to a gain of function for the channel rather than a loss of function (as seen with the potassium channel mutations). This sustained depolarization leads to prolongation of the plateau phase of cardiac depolarization, thereby delaying the onset of

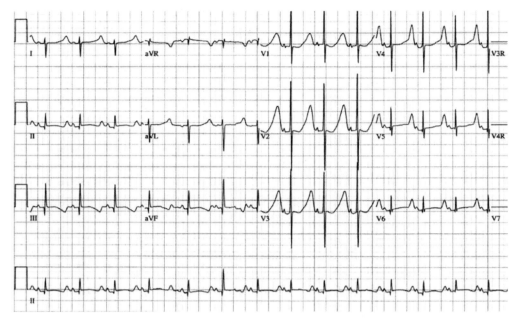

FIGURE 3. 12-Lead electrocardiogram in 2-day-old infant. Note the long flat ST segment and relatively symmetrical T-wave typical of LQTS3 (not yet confirmed by genetic analysis).

repolarization and resulting in excess calcium loading of the myocyte. The increased intracellular calcium may allow the development of early after-depolarizations and the generation of ventricular arrhythmias, particularly at slower heart rates. Therefore, therapies such as atrial pacing to prevent bradycardia and treatment with the sodium channel blocker, mexiletine, have been proposed as more specific and potentially more effective therapies for patients with *SCN5A*-type LQTS (LQT3 patients, Figure 3).

Interestingly, only a subset of *SCN5A* mutations, which is associated with this gain of function of the channel, is capable of causing LQTS. Mutations that produce defects in other parts of the sodium channel that inhibit the channel's ability to conduct inward sodium current or which result in a non-functional channel (loss of function defects) are associated with very different phenotypes such as occurs in Brugada syndrome, idiopathic ventricular fibrillation, and cardiac conduction defects.

Drug-Induced LQTS. It has long been known that some patients with normal QT intervals could have marked QT lengthening and torsades de pointes in response to certain medications (Table 2). These apparently "idiosyncratic" reactions have now been attributed in many cases to inherited, sub-clinical LQTS mutations that only become clinically apparent when there is another compromise to cardiac repolarization as occurs with certain medications that block sodium or potassium channels. The affect of these medicines on the cardiac repolarization may be particularly dramatic when they are combined with other agents that delay their metabolic degradation by inhibiting the P450 system or when they occur in a patient with hypokalemia, an inhibitor of the I_{Kr} current.

Diagnosis. The diagnosis of the LQTS is based on the measurement of the QT interval on the 12-lead electrocardiogram and corrected for heart rate using Bazett's formula

TABLE 2. Drugs to be Avoided in Patients with LQTS

Generic name	Use	Causes TdP	May cause TdP
Albuterol	Asthma		
Amantidine	Anti-viral/Parkinson's		X
Amiodarone	Antiarrhythmic	X*	
Arsenic trioxide	Anti-cancer	X	
Azithromycin	Antibiotic		X
Bepridil	Anti-anginal	X*	
Chloral hydrate	Sedative		X
Chlorpromazine	Anti-psychotic; anti-emetic	X	
Cisapride	GI stimulant	X*	
Clarithromycin	Antibiotic	X	
Cocaine	Local anesthetic		
Disopyramide	Antiarrhythmic	X*	
Dobutamine	Heart failure		
Dofetilide	Antiarrhythmic	X	
Dolasetron	Anti-nausea		X
Domperidone	Anti-nausea	X	
Dopamine	Heart failure		
Droperidol	Sedative; anti-nausea	X	
Ephedrine	Decongestant		
Epinephrine	Anaphylaxis/Allergic reactions		
Erythromycin	Antibiotic	X*	
Felbamate	Anticonvulsant		X
Fenfluramine	Appetite suppressant		
Flecainide	Antiarrhythmic		X
Foscarnet	Anti-viral		X
Fosphenytoin	Anti-convulsant		X
Gatifloxacin	Antibiotic		X
Granisetron	Anti-nausea		X
Halofantrine	Anti-malarial	X*	
Haloperidol	Anti-psychotic	X	
Ibutilide	Antiarrhythmic	X*	
Indapamide	Diuretic		X
Isoproterenol	Allergic reactions		
Isradipine	Anti-hypertensive		X
Levalbuterol	Asthma		
Levofloxacin	Antibiotic		X
Levomethadyl	Pain control	X	
Lithium	Bipolar disorder		X
Mesoridazine	Anti-psychotic	X	
Metaproterenol	Asthma		
Methadone	Pain control	X*	
Midodrine	Syncope		
Moexipril/HCTZ	Anti-hypertensive		X
Moxifloxacin	Antibiotic		X
Naratriptan	Migraines		X
Nicardipine	Anti-hypertensive		X
Norepinephrine	Shock/hypotension		
Octreotide	Acromegaly; GI bleeding		X
Odansetron	Anti-emetic		X

(cont.)

TABLE 2. Drugs to be Avoided in Patients with LQTS (*continued*)

Generic name	Use	Causes TdP	May cause TdP
Pentamidine	Anti-infective (pneumocystis)	X*	
Phentermine	Appetite suppressant		
Phenylephrine	Decongestant		
Phenylpropanolamine	Decongestant; Weight loss		
Pimozide	Anti-psychotic	X*	
Procainamide	Antiarrhythmic	X	
Pseudoephedrine	Decongestant		
Quetiapine	Anti-psychotic		X
Quinidine	Antiarrhythmic	X*	
Risperidone	Anti-psychotic		X
Ritodrine	Inhibit premature labor		
Salmeterol	Asthma		X
Sibutramine	Appetite suppressant		
Sotalol	Antiarrhythmic	X	
Sparfloxacin	Antibiotic	X*	
Sumatriptan	Migraines		X
Tacrolimus	Immunosuppressant		X
Tamoxifen	Anti-cancer		X
Telithromycin	Antibiotic		X
Terbutaline	Asthma		
Thioridazine	Anti-psychotic	X	
Tizanidine	Muscle relaxant		X
Venlafaxine	Antidepressant		X
Voriconazole	Anti-fungal		X
Ziprasidone	Anti-psychotic		X
Zolmitriptan	Migraines		X

Notes: Table derived from data available at the University of Arizona Health Sciences Center, ArizonaCert website (http://www.qtdrugs.org/medical-pros/drug-lists/drug-lists.htm#).
* Females are at a greater risk of TdP than males (usually >2-fold).

$[QT/(R-R)^{1/2}]$. Soon after LQTS was recognized as a clinical entity, the corrected QT intervals of 0.47 in males and 0.48 in females were considered the threshold values for identifying patients with LQTS. As more patients with LQTS were diagnosed and characterized, it became apparent that many patients with LQTS, thus at increased risk for sudden death, had corrected QT intervals that were within the normal range. For example, research studies show *KCNQ1* genetic defects that cause symptomatic LQTS in one family member, can be associated with corrected QT measurements that range from 0.41 to greater than 0.6 in other family members, with a mean QTc in unaffected individuals of 0.41. This overlap of the range of the QTc measurement between normal and abnormal highlights the need for improved diagnostic criteria in identifying patients with LQTS.

Electrophysiologic testing has not proven to be helpful in correctly identifying affected patients. Holter monitoring, provocative testing with epinephrine infusion or exercise ECG testing are more useful in detecting patients with repolarization abnormalities that are not manifest on a brief resting electrocardiogram. In some cases, these tests may be helpful in confirming the LQTS diagnosis, particularly in those subjects with potassium channel defects, yet these tests have not been consistently useful in determining which patients do or do not have LQTS. Because LQTS is a heterogeneous disorder resulting from abnormalities in a number of different ion channels specifying different ionic currents, it is not surprisingly that patients with different underlying defects will respond differently to provocative testing. Since information regarding the specific type of LQTS and the exact mutation is often lacking, it remains a clinical challenge to interpret these provocative tests when caring for an individual patient.

T-wave morphology has been used to estimate which category of genetic defect a LQTS family may have. In general, patients with sodium channel defects have a very late onset but normal appearing T-wave. Patients with I_{Kr} defects often have low, broad, often bifid T-waves. These generalizations are most helpful when the T-wave morphology can be tracked through a large family and patterns can be more easily recognized.

A scoring system has been devised to improve the sensitivity and specificity of the diagnostic criteria for the LQTS. The scoring system categorizes each patient as high or low risk for having the LQTS based on their symptoms, their family history, their measured corrected QT interval, and their T-wave morphology (Table 3). While this appears to be a significant improvement in establishing the LQTS diagnosis, genetic testing has been eagerly anticipated by clinicians and patients as a means to clarify the diagnosis and has the important added benefit of helping to direct gene-specific therapy.

Genetic testing for LQTS has been available on a clinical basis since 2004 in most states in the United States. While there are important benefits of genetic testing to confirm a clinical diagnosis of LQTS, it is important to recognize the limitations of these tests as they currently exist. The commercially available sources of LQTS genetic testing currently perform sequence analysis of the coding exons and intron/exon boundaries for the five most common gene mutations that cause LQTS. Included in the current test are *KCNQ1* and *KCNE1* (the genes that encode the subunits of the I_{Ks} channel) *KCNH2* and *KCNE2* (the

TABLE 3. Diagnostic Criteria in LQTS

Diagnostic Criteria	Points
ECG	
QTc	
0.48 sec	3
0.46–0.47 sec	2
0.45 sec (male)	1
Torsades de pointes	2
T-wave alternans	1
Notched T-wave in 3 leads	1
Low heart rate for age	0.5
Clinical History	
Syncope	
With stress	2
Without stress	1
Congenital deafness	0.5
Family History	
Family members with definite LQTS	1
Unexplained sudden death below age 30 among immediate family members	0.5

Notes: Scoring: 0–1 point = low probability of LQTS; 2–3 points = intermediate probability of LQTS; 4 points or more = high probability of LQTS. Adapted from Schwartz et al. (1993).

genes that encode the subunits of the I_{Kr} channel), and *SCN5A* (the gene that encodes the major subunit of the Ina channel). By current estimates of the frequencies of these specific mutations, this testing is estimated to identify 50%–75% of mutations present in patients with a clinical history strongly suggestive of LQTS. It is important to remember that genetic testing in its current form can identify a particular causative mutation in some patients with LQTS but it cannot "rule out" LQTS since it does not identify all mutations. A genetic test that does not identify a mutation is simply uninformative; it does not rule out the diagnosis.

Management. It is apparent that not all patients with LQTS have the same risks or responses to treatment. The category of LQTS defect, I_{Ks}, I_{Kr}, or sodium channel-type, and its specific mutation can affect the risk of morbidity and mortality associated with the syndrome. In addition, there appear to be other inherited factors that can modify an individual patient's risk, making them either more or less susceptible to a fatal arrhythmia than another patient or family member with the identical genetic defect. When positive for a particular mutation, currently available genetic testing of patients with LQTS can be beneficial in characterizing the specific type of ion channel defect and perhaps determining the best course of treatment.

Since genetic testing for LQTS is relatively new and its clinical yield is only 50%–75% for a specific mutation, the category of genetic defect is often not known to the clinician when deciding on treatment for a patient with LQTS. Therefore, treatment is usually aimed at preventing adrenergic surges, avoiding medications and conditions known to prolong the QT interval, and limiting risk through lifestyle modifications such as avoidance of competitive athletics, close supervision while swimming, and removal of alarm clocks and bedside phones. Hypokalemia should be avoided and dietary intake of potassium-rich foods would theoretically be of benefit. Beta-blockade is usually the first line of treatment and has resulted in a significant decrease in LQTS mortality. Patients who have had a significant cardiac arrest, have persistent symptoms despite beta-blockade therapy, or who have a strong family history of sudden death are candidates for implantation of a transvenous defibrillator (ICD).

Genetic testing has presented a treatment dilemma for clinicians when caring for the genotype positive/phenotype negative patient. It is clear from research studies that LQTS mutations are incompletely penetrant. In other words, the presence of a disease-associated LQTS mutation does not absolutely indicate that the mutation carrier will have clinical symptoms of LQTS. In fact, approximately one-third of LQTS mutation carriers will be clinically asymptomatic with normal QT interval by ECG and provocative testing throughout their lifetimes; however, there is no currently available method to determine which mutation carriers will remain asymptomatic. Since the first clinical symptom of LQTS can be a potentially lethal arrhythmia, it can be difficult to decide whether to initiate treatment for an asymptomatic LQTS carrier. At a minimum, asymptomatic LQTS mutation carriers and all first-degree relatives of patients with LQTS who have not had genetic testing should avoid medications on the list of those known to prolong the QT interval (http://www.sads.org/).

Brugada Syndrome

Brugada syndrome (BrS) is characterized by ST-segment elevation in the right precordial leads (V1-V3) unrelated to other factors such as ischemia, electrolyte abnormalities, or structural cardiopulmonary disease (Figure 4).

Patients with BrS are predisposed to sudden cardiac death due to polymorphic ventricular tachycardia that degenerates to ventricular fibrillation. It is inherited as an autosomal dominant disorder with an incidence of 5 to 66 per 10,000 individuals. There is a much higher incidence in areas of Southeast Asia, as well as a male predominance (8:1). The symptoms, typically syncope or cardiac arrest, usually do not appear until the third to fourth decade of life.

Genetic studies have determined that more than one gene is capable of causing BrS. To date, mutations in only one gene, encoding the *SCN5A* cardiac sodium channel, have been identified. Unlike the *SCN5A* gain of function mutations that cause LQTS, BrS-causing *SCN5A* mutations may occur anywhere in the gene and are all loss of function mutations. These mutations all reduce the sodium channel current, either through a reduction in the number of channels and/or diminished function of nearly half of the channels. The diminished inward sodium current is thought to allow a prominent transient outward potassium current (I_{to}) in the right ventricular epicardium, leading to an epicardial to endocardial voltage gradient. This produces the ST elevation that is characteristic of the disorder and may lead to re-excitation (phase 2 re-entry) in the ventricular epicardium, triggering ventricular arrhythmias.

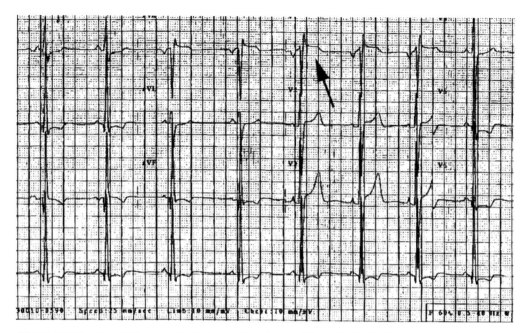

FIGURE 4. Elevation of ST segment in V1 (arrow) in a 15-year-old boy with family history of Brugada syndrome.

Diagnosis. Like LQTS, the diagnosis of BrS relies upon the characteristics of the 12-lead electrocardiogram, but the ECG is also less than optimally sensitive and specific. Three distinct right precordial ST-wave morphologies have been described. In Type I BrS, there is prominent "coved" J-point peak with an ST elevation of >0.2 mV followed by a negative T-wave with essentially no intervening isoelectric phase. In Type II BrS, there is a prominent J-point peak (>0.2 mV) followed by a positive or biphasic T-wave resulting in a saddleback appearance. In Type III BrS, the J-point elevation is less pronounced (>0.1 mV) and may be of either the "cove" or "saddleback" type. The ECG abnormalities may mimic RBBB, especially in patients with Type I BrS. The BrS pattern can be differentiated from RBBB by the absence of a wide S-wave in lead I and in the left lateral precordial leads.

The typical ECG appearance may not always be present but may be unmasked by intravenous infusion of a sodium channel blockers such as ajmaline (1 mg/kg at 10 mg/min), flecainide [2 mg/kg (maximum 150 mg) over 10 minutes], or procainamide (10 mg/kg; 100 mg/min, the only medication FDA approved in the U.S.). The sensitivity and specificity of the provocative testing has not been established and, in a patient that does have BrS, may precipitate significant ventricular arrhythmias including ventricular fibrillation. Furthermore, there is some evidence to suggest that asymptomatic patients, who only demonstrate ECG changes with provocative testing, may have a benign course.

Genetic testing for BrS is now clinically available; however, it is of limited use since only one of the BrS genes have been identified to date. Current genetic testing involves sequencing of the coding exons and intron/exon boundaries in all the exons of the *SCN5A* gene. This test is estimated to only detect a causative mutation in approximately 15%–20% of BrS cases; thus, as is the case for LQTS, an uninformative test (one that fails to detect a mutation) does not eliminate the possible diagnosis of BrS. In the future, as more BrS genes are identified and their manifestations are characterized, genetic testing will likely be of more use to help make the diagnosis and guide the course of therapy.

Management. Treatment is required for all symptomatic BrS patients. Unfortunately, pharmacotherapy has not proven to be beneficial. There is theoretical justification for the use of quinidine to treat this disorder since quinidine, at low doses, is a potent inhibitor of the I_{to} potassium current. To date, quinidine has not been prospectively studied in BrS patients. There is also data to suggest that mexiletine, a sodium channel blocker, may improve the functioning of the mutant BrS protein, allowing it to be transported correctly to the cell membrane where its function can help improve sodium conductance into the myocyte. Nonetheless, the only treatment currently demonstrated to be effective in reducing mortality in BrS patients is placement of an ICD. While ICD placement is clearly indicated in all BrS patients with symptoms of syncope, there is controversy concerning their use in asymptomatic BrS patients, particularly children and young adults. When an individual is identified during the screening of a family with symptomatic BrS, it could be argued that ICD placement should be considered.

Approximately 30% of asymptomatic individuals will have a ventricular arrhythmia within three years of evaluation, the same percentage as symptomatic individuals that will experience a recurrent ventricular arrhythmia within three years. No clinical or electrophysiologic criteria have been demonstrated to predict which BrS patients will experience a ventricular arrhythmia. Currently, placement of an ICD is recommended in asymptomatic individuals if sustained ventricular arrhythmias can be induced on electrophysiologic testing.

Andersen Syndrome

Andersen syndrome is characterized by potassium-sensitive periodic paralysis, ventricular ectopy, and dysmorphic features. The phenotype can be highly variable but may include short stature, mandibular hypoplasia, low-set ears, and clinodactyly. The paralysis may occur at low, normal, or high serum potassium levels and is usually responsive to potassium treatment. There can be prolongation of the QT interval, but that is a minor criterion for the diagnosis. The ventricular arrhythmias, characteristically bidirectional ventricular tachycardia (Figure 5) may lead

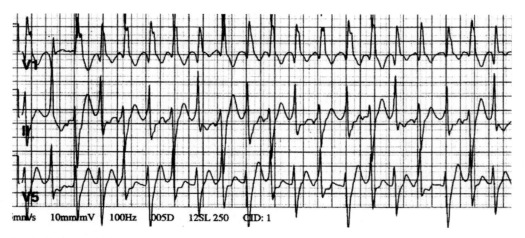

FIGURE 5. Rhythm strip (V1, II, V5) in a 16-year-old girl with clinodactyly, short stature, small mandible, and a mutation in the *KCNJ2* gene. Note the bidirectional pattern (alternating axes) of the ventricular arrhythmia. She received an ICD but has had no discharges in two years.

to syncope and sudden death. In at least some cases, the syndrome is due to a mutation of the gene encoding for the *KCNJ2* potassium channel. Disease-causing *KCNJ2* mutations are thought to decrease the strength of the channel interactions with phosphatidylinositol 4, 5-bisphosphate. Recently, treatment with amiodarone and acetazolamide was demonstrated to relieve the cardiac and skeletal muscle manifestations of Andersen syndrome in a patient with a *KCNJ2* mutation.

Ion Channelopathies that Enhance Automaticity

Catecholaminergic Polymorphic Ventricular Tachycardia/Familial Ventricular Tachycardia/Arrhythmogenic Right Ventricular Dysplasia Type 2

Catecholaminergic polymorphic ventricular tachycardia (CPVT), (Figure 6) familial ventricular tachycardia (FVT), and arrhythmogenic right ventricular dysplasia type 2 (ARVD2) have recently been demonstrated to be due to abnormal calcium release from the sarcoplasmic reticulum. All of these disorders can be due to mutations in the cardiac ryanodine receptor type 2, the channel that releases calcium from sarcoplasmic reticulum stores. Thus, it appears that these disorders are best considered as variations of a single clinical entity. All are relatively rare heritable disorders and are characterized by adrenergic-induced ventricular tachyarrhythmias and sudden death. In about one-third of cases, CVPT has been determined to result from an autosomal dominant mutation in the cardiac ryanodine receptor type 2 (*RYR2*), the channel that releases calcium from sarcoplasmic reticulum (SR) stores. Other cases of CVPT have autosomal recessive inheritance of homozygous mutations in calsequestrin, a molecule that helps to sequester calcium in the SR. ARVD2 and FVT both result from autosomal dominant mutations of *RYR2*. FVT is a loosely-defined disorder characterized by an inherited predisposition to ventricular arrhythmias and this term has been applied to patients with BrS, ARVD, and CPVT.

Unlike some LQTS and cardiomyopathy patients who also demonstrate adrenergic-induced ventricular tachycardias, the QT interval and the heart are normal in patients with CPVT, FVT, or ARVD2. Some patients with ARVD2 have been reported to have some mild thinning and fibrous-fatty replacement of the right ventricular myocardium but to a far less extent than other forms of ARVD.

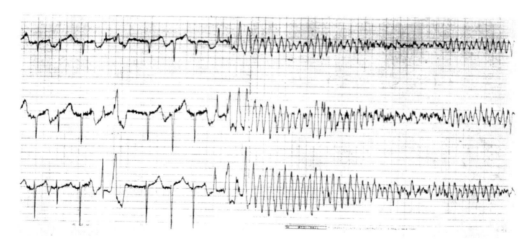

FIGURE 6. ECG tracing in a 12-year-old girl with a normal QTc and recurrent polymorphic VT or torsades de pointes.

Patients with CPVT develop symptoms of syncope and sudden death early in childhood, usually around the age of 8, with more severely affected individuals demonstrating an earlier onset of symptoms. Symptoms usually occur with stress or exercise and there is a 30%–50% incidence of sudden death by 20–30 years of age if it is left untreated. Syncope and sudden death are caused by exercise or stress-related ventricular tachycardias. In patients with CPVT, isolated premature ventricular contractions and atrial arrhythmias will occur early after the onset of exercise. After continued exercise, runs of monomorphic or bi-directional ventricular tachycardia can degenerate into polymorphic ventricular tachycardia and, ultimately, ventricular fibrillation. The other consistent clinical feature is resting bradycardia. Patients with ARVD2 have very similar clinical features to CPVT. These patients also have stress-induced ventricular arrhythmias. The only potential clinical difference between the two conditions is that ARVD2 patients may have very mild pathologic changes in the right ventricle similar to other ARVD syndromes.

Diagnosis. The diagnosis of CPVT/FVT/ARVD2 depends on the demonstration of ventricular tachyarrhythmias that are elicited or exacerbated by exercise testing in a patient without either structural heart disease or LQTS. Exercise testing to evaluate for the possibility of CPVT/FVT/ARVD2 should be performed in any child with a history of exercise-induced syncope with normal electrocardiographic and echocardiographic studies. Exercise testing in asymptomatic children should also be performed in families where there is a history of one of these disorders or of exercise-induced syncope or sudden death of unknown etiology. Since the genotypes for these disorders are not yet fully defined, genetic testing is not yet available. In the future, genetic testing may be helpful in establishing the diagnosis and beginning treatment in non-symptomatic children from families with CPVT/FVT/ARVD2.

In families with CPVT/FVT/ARVD2, establishing the diagnosis prior to the onset of symptoms would be a major advance since the first manifestation of the disease may be sudden death in an otherwise healthy child.

Management. Treatment with beta-blockade has been highly effective in reducing morbidity and mortality. Some patients with persistent stress-induced arrhythmias despite beta-blockade may require ICD placement.

Familial Ventricular Fibrillation

Several families have been described with familial ventricular fibrillation (FVF). In one instance, the disorder was attributed to a mutation of the *SCN5A* gene but the patients did not demonstrate the electrocardiographic findings of BrS. Additional FVF genes likely remain to be identified since not all patients with the disorder have been demonstrated to have *SCN5A* defects.

Familial Atrial Fibrillation

Atrial fibrillation, the most common sustained cardiac-rhythm disturbance in adults, is very rare in children. It becomes increasingly prevalent with advancing age and is a common antecedent for stroke in patients over age 65. The rare incidences of familial AF have led to genetic mapping studies that have identified several genetic loci that may be responsible for this disorder. However, to date, no definite genetic causes of familial AF have been identified. As familial AF genes are identified, it is anticipated that their characterization will lead to improved treatment strategies for acquired AF.

CARDIOMYOPATHIES

Cardiomyopathies are due to abnormalities of structural components of the myocytes including changes in sarcomeric proteins, as well as in the proteins of the cytoskeleton. In many cases, these protein changes are due to mutations in the genes that encodes them. In patients with cardiomyopathies, there is often a predisposition to cardiac arrhythmias. There are various classification schema for the cardiomyopathies; however, the most common division is by clinical appearance, dividing the cardiomyopathies into hypertrophic, dilated, and dysplastic subtypes.

Hypertrophic Cardiomyopathy

Hypertrophic cardiomyopathy (HCM) is characterized by cardiac hypertrophy not attributable to other primary causes such as hypertension or a structural defect leading to hypertrophy. Abnormal hypertrophic growth results in enlarged and disordered myocytes that displace adjacent myocytes, causing the characteristic microscopic pathologic appearance of myocyte disarray. Despite the marked pathologic abnormalities, ventricular systolic function is usually normal or even supernormal. Due to the thickened stiff ventricular walls, diastolic function is often impaired, causing dyspnea on exertion, orthopnea, and angina. There is a high risk of atrial and ventricular arrhythmias, as well as sudden death, that does not always correlate with the degree of ventricular hypertrophy.

HCM can occur in isolation, as a primary disorder of the myocardium; in this case, it is the most common heritable cardiovascular disorder with an estimated incidence of 0.2% in the general population. This form of HCM is called familial HCM. HCM can also occur as a component of other medical syndromes including metabolic diseases, though this is much less common. Some syndromes that are commonly as associated with HCM include Noonan syndrome, LEOPARD syndrome, and Costello syndrome. Metabolic forms of HCM include lysosomal storage diseases, as well as mitochondrial diseases.

Based on genetic studies of familial HCM, it has been referred to as a "disease of the sarcomere." To date, mutations in ten different genes can cause familial HCM. Each of the genes identified thus far (Table 4) encodes for component of the sarcomere, the highly ordered contractile unit of striated muscle. The sarcomere can be divided into the thin filament (cardiac actin, the troponins, and α-tropomyosin), the thick filament (myosin heavy chain, the myosin light chains, and myosin binding protein C), and the structural elements (titin and muscle LIM protein).

Mutations in the genes that encode for β-myosin heavy chain, cardiac troponin T, and myosin-binding protein C account for the vast majority of the cases. The other genes, cardiac troponin I, regulatory and essential myosin light chains, titin, α-tropomyosin, α-actin, and α-myosin heavy chain, each account for a minority of HCM cases. Since a significant number of patients with HCM (approximately 40%) do not have a mutation of one of the previously-identified genes, additional HCM-causing genes remain to be identified.

The clinical manifestations of HCM vary widely even between patients that have identical mutations in the same gene. A variety of modulating factors, including systemic blood pressure and autonomic tone, can influence the degree of cardiac hypertrophy and diastolic dysfunction. Different mutations of the same gene can demonstrate great variability in clinical manifestations. The β-myosin heavy chain mutations Arg403Gln and Arg719Trp predispose to sudden death and heart failure while the mutations Gly256Glu, Phe513Cys, and Leu908Val are reported to be associated with lower morbidity and mortality. Each of the genes responsible for causing HCM presents a different disease spectrum. Mutations of myosin binding protein C and troponin T tend to be the most consistent in their disease presentation. Myosin binding protein C mutations, in general, cause cardiac hypertrophy later in life and are generally associated with a good prognosis, while troponin T mutations are associated with a high incidence of sudden death despite causing only very mild hypertrophy.

TABLE 4. Cardiomyopathies

Type	Locus	Gene product	Manifestations
Hypertrophic	1q32	Troponin T	Mild hypertrophy; high mortality
	2q31	Titin	
	3p31	Essential myosin light chain	Skeletal myopathy
	7q36	PRKAG2, Regulatory subunit of CAMP-activated protein kinase	With Wolff-Parkinson-White syndrome
	13p11	Myosin binding protein-C	Late onset; good prognosis
	12q23	Regulatory myosin light chain	Skeletal myopathy
	14q12	Cardiac β-myosin heavy chain	Variable, depends on mutation
	15q14	Cardiac actin	
	15q22	α-tropmyosin	
	19q13	Troponin T	
Dilated	1p1	Lamin A/C	Emery-Druifuss muscular dystrophy; conduction block
	1q32	Troponin T	
	2q22	?	
	2q31	Titin	
	2q35	Desmin	Skeletal myopathy
	3p35	?	
	5q33	δ-sarcoglycan	
	6q22	Photpholmaban	
	6q23	Desmoplakin	Skeletal myopathy
	9q22	?	
	10q22	Metavinculin	
	14q12	Cardiac β-myosin heavy chain	
	15q14	Cardiac actin	
	15q22	α-tropmyosin	
	19q13	?	
	Xp21	Dystrophin	Skeletal myopathy
	Xp28	Tafazzin	Barth syndrome; skeletal myopathy; neutropenia
Dysplastic (ARVD)	1q42	Ryanodine receptor 2	CPVT; minimal myopathy
	2q32	?	Localized LV involvement
	3p23	?	
	10p12	?	
	14q12	?	
	14q23	?	Naxos disease; woolly hair
	17q21	?	

Diagnosis. The diagnosis of HCM is made based on the echocardiographic finding of significant cardiac hypertrophy, greater than the 95th percentile for age, in the absence of other causes of hypertrophy such as hypertension or structural heart disease. The exact morphology of the hypertrophy may vary widely between patients but is often asymmetric with hypertrophy of the septum being much more prominent than that of the posterior free wall. This asymmetric septal hypertrophy can lead to subaortic narrowing and systolic anterior motion of the mitral valve (SAM). This process can result in significant left ventricular outflow tract obstruction (LVOTO), increased left ventricular systolic pressures and further acceleration of the hypertrophic process. In fact, the initially identified patients with HCM had LVOTO, and this finding led to earlier names of this condition such as hypertrophic obstructive cardiomyopathy (HOCM) or idiopathic hypertrophic subaortic stenosis (IHSS). It is now increasingly evident that most HCM patients do not demonstrate significant LVOTO.

The ECG pattern is abnormal in approximately 70% to 90% of the patients and differs widely depending on the nature of the HCM. Left atrial enlargement may be the first electrocardiographic sign of the diastolic dysfunction that accompanies HCM even before there is a significant increase in left ventricular forces consistent with left ventricular hypertrophy. Other electrocardiographic abnormalities that may be present include deep QS patterns in V1–V3, partial or complete bundle branch block, ST segment elevation/depression, or an abnormal T-wave axis or morphology.

If the hypertrophy is symmetric and the ECG is normal, the diagnosis of HCM can be very difficult to make. Trained athletes are known to have mild to moderate increases in their measured left ventricular wall thickness making the differentiation between "athlete's heart" and mild HCM complex yet critical (Table 5).

Furthermore, the risk of arrhythmia and sudden death has not been demonstrated to correlate strongly with the degree of cardiac hypertrophy, indicating that even HCM patients with mild hypertrophy may be at significant risk for sudden death. Family history may be helpful since some families appear to have a very high incidence of sudden death while in other families the affected individuals appear to have normal life spans.

Genetic testing has recently become available for HCM. The currently available test includes sequence analysis of the five most common genes know to cause HCM (*MYH7*, *MYBPC 3*, *TNNT2*, *TNNI3*, *TPM1*) as one panel; there is an additional panel with three additional genes tested (*ACTC*, *MYL2*, *MYL3*) also by sequence analysis. As is the case with genetic testing for Long QT and BrS, genetic testing does not identify mutations in all patients with clinically diagnosed HCM. The current genetic test for HCM is estimated to have an approximately 55% to 70% yield among patients clinically diagnosed with HCM who have both test panels performed. Thus, a positive genetic test (i.e., one that identifies a mutation), can confirm a clinical diagnosis of HCM, but a negative test is simply uninformative. As there is further experience with HCM genetic testing, as well as more detailed genotype/phenotype studies, genetic testing may become an important adjunct to clinical echo findings in establishing the diagnosis of HCM and determining the management recommendations for patients. Currently, genetic testing may be most helpful to determine mutation carriers in a family where a particular mutation can be identified. The family members that did not inherit the mutation would not need to be followed with serial echocardiograms, and those individuals that did inherit the mutation could be followed more closely to determine if and when they developed the clinical features of HCM.

TABLE 5. Diagnostic Criteria of HCM

HCM	Criteria	Athlete's Heart
Yes	Unusual Pattern LVH (Echo)	No
Yes	LV Cavity Size <45 mm (Echo)	No
No	LV Cavity Size >55 mm (Echo)	Yes
Yes	LA Enlargement (Echo)	No
Yes	Strange ECG Pattern	No
Yes	Abnormal LV Filling (Echo)	No
Yes	Female Gender	No
No	Decreased LV Thickening with Deconditioning	Yes
Yes	Family History HCM	No

Note: Adapted from Maron (2002).

Management. Current management of the patient with HCM depends upon the patient's symptoms, cardiac evaluation and family history. Clinical cardiac evaluation in all patients with suspected HCM should include an echocardiogram, an exercise test (if asymptomatic and non-obstructive), a 24-hour Holter ECG recording, and an extended family medical history. Treatment with beta-blockers or calcium channel blockers has been used to slow the progression of the cardiac hypertrophy through modification of the hypercontractile state that may have a role in the progression of the hypertrophy in some patients. While there are no strong data to support any affect on disease progression or symptoms, these agents may be beneficial in some patients. Treatment with antiarrhythmic agents has not proved to have a significant effect on the natural history of the disease and patients with symptomatic ventricular arrhythmias appear to be best treated with placement of an ICD. Indications for consideration of placement of an ICD include the following: aborted sudden death, unexplained syncope (particularly in childhood), strong family history of sudden death due to HCM, ventricular tachycardia during Holter monitoring or exercise test, absence of a systolic rise in blood pressure with exercise, and/or severe septal thickening (>30 mm).

Left ventricular outflow tract obstruction occurs in approximately 30% of the patients and can lead to myocardial ischemia, syncope, exercise intolerance, and acceleration of the hypertrophic process. The degree of hypertrophy in HCM patients does not correlate well with the incidence of sudden death, most likely due to the fact that certain HCM mutations are associated with minimal hypertrophy and a high incidence of sudden death. However, LVOTO-induced myocardial ischemia and syncope in a patient with a substrate for ventricular arrhythmia would be predicted to increase the risk of sudden death in an individual patient. Therefore, significant LVOTO is usually treated surgically with a myomectomy. More recently, intracoronary infusion of ethanol has been performed in the catheterization laboratory to induce septal ablation in adult HCM patients with LVOTO. Its ultimate place in treatment strategy remains to be defined.

HCM with Wolff-Parkinson-White Syndrome. Recently a syndrome of ventricular pre-excitation, early onset atrial fibrillation, and cardiac hypertrophy was identified in more than one family. In these families, a mutation was identified in the gamma-2 regulatory subunit (*PRKAG2*) of cAMP-activated protein kinase (AMPK). These observations suggest that AMPK may participate in the regulation of cardiac ion channels and may suggest a molecular basis of atrial fibrillation. Further evaluation of patients with *PRKAG2* mutations shows that these patients develop progressive hypertrophy, as well as progressive cardiac conduction disease often requiring pacemaker placement.

Dilated Cardiomyopathy

Dilated cardiomyopathy (DCM) is a common disorder that is characterized by dilation and decreased function of the left ventricle. It is the most frequent reason for heart failure and cardiac transplantation, and places patients at high risk for both atrial and ventricular tachyarrhythmias. The causes of DCM are myriad and include ischemic events, myocarditis, metabolic disorders, drug or toxin exposures (e.g., adriamycin), and infiltrative processes. In many cases, however, the cause of DCM can not be determined. In the last few years, it is clear that many of these idiopathic cases are due to mutations in a number of genes. A genetic cause is identified in at least 30% of cases of DCM and some studies suggest a genetic cause of DCM in up to 50% of cases. In familial cases, DCM has autosomal dominant transmission most commonly, but X-linked, autosomal recessive, and mitochondrial inheritance patterns have also been reported.

The search to identify the genes responsible for DCM has led to an improvement in the understanding of cardiac sarcomere structure-function relationship. To date, more than 15 genetic loci capable of causing DCM have been identified. Of these loci, 13 DCM-causing genes have been identified, including α-*tropomyosin*, *desmin*, β-*myosin heavy chain*, α-*cardiac actin*, δ-*sarcoglycan*, *dystrophin, troponin T, lamin A/C,* and *tafazzin* (Table 4). Each of the genes responsible for DCM can cause significantly different manifestations of the disease.

Isolated DCM. Four gene products responsible for DCM, α-*tropomyosin,* α-*cardiac actin,* β-*mMHC,* and *troponin T*, are sarcomeric proteins that have also been noted to be altered in some patients with hypertrophic cardiomyopathy. Therefore, different mutations of the same gene can lead to hypertrophy or dilatation, depending on how the mutation affects the physiologic properties of the sarcomere. However, DCM-causing alterations in these sarcomeric proteins usually result in isolated DCM without evidence of other systemic abnormalities.

DCM/Skeletal Myopathy. Other mutations resulting in DCM have been identified in the structural proteins that anchor the sarcomere to the cell membrane and the extracellular matrix such as *dystrophin,* δ-*sarcoglycan,* and *desmin*. These proteins stabilize the sarcomere and assist the transduction of force. Mutations in one of these genes often results in a skeletal myopathy in conjunction with a dilated cardiomyopathy. Whether the skeletal muscle or cardiac pathology predominates can depend on the site within the gene where the mutation occurs.

Another type of DCM associated with skeletal myopathy is Barth Syndrome. This syndrome is X-linked and thus only affected males have been seen clinically. Boys with Barth Syndrome have DCM and can present in infancy with congestive heart failure and hypotonia. Another common manifestation is cyclic neutropenia with associated infections. Affected boys can be severely ill as infants; however, if they survive infancy, the ventricular function often improves although there can be a return to symptomatic heart failure with or without significant arrhythmias in the years of puberty. The cause of Barth Syndrome is mutations in the *Tafazzin* gene. The diagnosis can be made by finding elevated 3-methylglutaconic acid in the urine. There is also genetic testing available for mutations in the *Tafazzin* gene.

DCM/Conduction System Disease. Emery-Dreifuss muscular dystrophy is characterized by early contractures of elbows and Achilles tendons, slowly progressive muscle wasting, and DCM with conduction block. The disorder can be inherited in an autosomal dominant, autosomal recessive, or X-linked pattern. At least some of the autosomal dominant and autosomal recessive cases have been determined to be due to mutations in the *lamin A/C* gene, a gene that encodes for two proteins of the nuclear lamina, lamin A and lamin C. The X-linked cases have been determined to be due to mutations in emerin, another nuclear lamina-associated protein.

Diagnosis. Individuals affected with DCM are defined as those that have a left ventricular ejection fraction of <50% on echocardiography, a regional fractional shortening of <27% on M-mode analysis, or both, in the presence of a left ventricular internal diastolic dimension of >2.7 cm/m^2 of body surface area in the absence of other common causes of DCM (coronary artery disease, myocarditis, hypertension, etc.). Due to the wide range of genetic causes of DCM, genetic testing has not been widely applied. It is just beginning to be available for a few of the DCM subtypes, particularly those associated with skeletal myopathy (desmin, dystrophin, lamin A/C, emerin) and other rare types of DCM such as *Tafazzin* gene testing in patients with DCM associated with Barth Syndrome. Molecular genetic characterization of families with the disorder at this time remains an important research tool to further understand this diverse disorder.

Management. Management strategies in patients with DCM are aimed at relief of symptoms as well as preservation of ventricular function since specific treatment for the underlying etiology of DCM is usually not possible. Medical treatment of the symptoms of congestive heart failure with diuretics, ACE inhibitors, and beta-blockade is a mainstay of therapy for symptomatic patients. Both atrial and ventricular arrhythmias occur frequently in patients with DCM. There is evidence that sustained ventricular arrhythmias can have an ominous prognosis in the setting of DCM with heart failure. In this case, placement of an ICD has been found to be superior to antiarrhythmic therapy in preventing sudden cardiac death in these patients.

Arrhythmogenic Right Ventricular Dysplasia

Arrhythmogenic right ventricular dysplasia (ARVD), characterized by fatty infiltration and fibrosis of the right ventricle, is the most common cause of sudden cardiac death in Italy, accounting for 17% of the cases, and an increasingly recognized cause in many other countries. It is transmitted as an autosomal dominant disorder. Genetic studies on ARVD families have determined that there are at least nine different genes that cause AVRD; specific genes identified thus far include *RYR2*, encoding the cardiac ryanodine receptor; *DSP*, encoding desmoplakin, and *PKP2*, encoding a cardiac desmosome protein. The other ARVD genes have been localized to human chromosomes 2q32.1-32.3, 3p23, 10p14-p12, 10q22, 14q12-q22 and 14q23-q24 but no specific genes at these ARVD loci have been identified to date. Thus, ARVD is a genetically heterogeneous disorder, making diagnosis and disease characterization difficult.

Diagnosis. ARVD can be extremely difficult to diagnose, even in families where there is a definite family history of the disorder. Unfortunately, the first symptom is often sudden cardiac death, although usually not during the childhood or adolescent years. Surface ECG abnormalities suggestive of the disorder include inverted T-waves in the right precordial leads and right ventricular arrhythmias with a LBBB configuration. The arrhythmias may be exacerbated by adrenergic stimulation or exercise and may be induced by electrophysiologic stimulation. Biopsy evidence of right ventricular fatty infiltration and fibrosis can establish the diagnosis. However, since the fibrosis usually begins in the RV free wall and then spreads to the interventricular septum where a biopsy would be performed, there are many "false negative" biopsies in patients with ARVD. The European Society of Cardiology and International Society and Federation of Cardiology established diagnostic criteria for ARVD that rely on structural and functional features of the right ventricle (based on MRI and echocardiographic assessment), fibrofatty infiltration of the right ventricle (based on biopsy), electrocardiographic evidence of depolarization and/or repolarization abnormalities, demonstration of ventricular arrhythmias of right ventricular origin, and family history. Major and minor criteria were established, the combination of which is used to categorize patients as having ARVD, probable ARVD, or unaffected (Table 6).

Management. Potential therapeutic options for patients with symptomatic ARVD include pharmacologic suppression of the ventricular arrhythmias, electrophysiologic radiofrequency ablation of the arrhythmogenic focus, and/or placement of an ICD. As with other autosomal dominant disorders, there is a 50% chance of passing the disorder to each offspring, so all first-degree family members (offspring, siblings, and parents) of every ARVD patient should be screened and followed for the disorder (or its potential emergence). Demonstration of sustained ventricular arrhythmias in response to electrophysiologic stimulation may be used to guide therapy in the older asymptomatic patient. Current practice for management of an affected family with children is to share with patients and families the current state of limited knowledge, especially regarding the natural history in the young, and encourage joining the recently formed national registry project.

TABLE 6. Suggested Criteria for the Diagnosis of ARVD

Feature	Major criteria	Minor criteria
Global and/or Regional Dysfunction and Structural Alterations	Severe RV dilatation and reduced RV ejection fraction with no (or mild) LV dysfunction Localized RV aneurysms	Minor global RV dilatation and/or reduction in ejection fraction with normal LV function Mild segmental dilatation of the RV Regional RV hypokinesia
Repolarization Abnormalities		Inverted T-waves in right precordial leads beyond V1 in the absence of right bundle branch block (patient age > 12 yrs)
Depolarization/Conduction	Epsilon waves or localized prolongation of the QRS complex (> 110 msec) in leads V1, V2, or V3	Late potentials (signal averaged ECG)
Arrhythmias		Left bundle branch block type ventricular tachycardia (sustained or non-sustained) on ECG, Holter or exercise testing Frequent ventricular extrasystoles with left bundle branch block morphology (> 1000/24 hrs) on Holter monitor
Family History	ARVD in family member confirmed at necropsy or surgery	Family history of premature sudden death (< age 35) due to suspected ARVD Family history of ARVD based on clinical criteria

Notes: Diagnosis of ARVD based on the presence of two major criteria, one major and two minor criteria or four minor criteria. Adapted from McKenna, et al. (1994) and Fontaine, et al. (1999), with permission from the Annual Review of Medicine 1999; 50 by Annual Reviews.

CONDUCTION SYSTEM ABNORMALITIES

While there are a variety of etiologies of cardiac conduction system abnormalities, there are a few instances where some hints toward an underlying genetic basis are known.

Complete Heart Block

The cardiac homeobox gene *Nkx2.5* is a transcription factor involved in the development of many different cardiac myocyte lineages. It is therefore not surprising that mutations of *Nkx2.5* have been associated with a range of cardiovascular defects that include atrial septal defects, tetralogy of Fallot and other conotruncal defects, Ebstein's anomaly, and anomalous pulmonary venous return. Not infrequently, mutations of *Nkx2.5* will lead not only to structural cardiac abnormalities but also to progressive atrioventricular block. Several families with structural heart disease, most commonly atrial septal defects and cardiac conduction abnormalities have been shown to have mutations of *Nkx2.5*. Mutations of *Nkx2.5* have also been noted in patients with isolated heart block and no structural heart defects. The exact cause of the conduction abnormality due to *Nkx2.5* mutation is not known but it has been proposed to be due to abnormalities of connexin expression. Connexin 40 and 43, two cardiac gap junction proteins responsible for establishing the electrochemical link between myocytes, are markedly down-regulated in mice that lack or have mutant *Nkx2.5* genes.

Kearns-Sayre Syndrome

Kearns-Sayre Syndrome is a multi-system disorder characterized by external opthalmoplegia, pigmentary degeneration of the retina, premature dementia, and a dilated cardiomyopathy, often with progressive conduction defect. The syndrome was first recognized by Kearns in 1965 and involves deletions of mitochondrial DNA. Most cases represent new deletions but there are reports of familial transmission of the disorder. Depending on the exact size and location of the mitochondrial DNA deletion, patients may also exhibit weakness of facial, pharyngeal, trunk and extremity muscles, deafness, short stature, and markedly increased cerebrospinal fluid protein.

Progressive Cardiac Conduction Defect

Progressive cardiac conduction defect (PCCD), also called Lenègre or Lev disease, is one of the most common cardiac conduction disturbances in adults. It is characterized by progressive slowing of conduction through the His-Purkinje system leading to right or left bundle branch block and, ultimately, to complete atrioventricular block, syncope, and sudden death. Several familial cases of PCCD have been described, and one gene responsible for the disorder has been localized to chromosome 19q13.3 but has not yet been identified. Recently, the cardiac sodium channel (*SCN5A*) was demonstrated to be mutated in two families with PCCD. Those families did not demonstrate features of either LQTS or BrS, again demonstrating that different mutations of the same gene can cause very different clinical manifestations.

Short QT Syndrome

While the dangers of prolongation of the QT interval have been recognized since the long QT syndrome was first described in the late 1950s and early 1960s, it is now becoming increasingly apparent that an abnormally short QT interval may also predispose to life-threatening arrhythmias and sudden death. The syndrome is characterized by a QTc (Bazett formula) of less than 320 msec and is frequently associated with atrial fibrillation. In fact, the diagnosis of short QT syndrome should be strongly considered in younger patients with isolated atrial fibrillation. Establishing the diagnosis is vital since patients with the disorder have a high incidence of syncope and sudden death due to ventricular tachyarrhythmias. As with long QT syndrome, short QT syndrome is a heterogeneous disorder with mutations in at least three different genes capable of causing the clinical manifestations. The mutations described to date are all gain-of-function mutations in genes in which loss of function can lead to the long QT or Andersen syndromes: *HERG* (I_{Kr}), *KvLQT1* (I_{Ks}), and *KCNJ2* (I_{Kt}). Treatment guidelines are still being established but case reports have recommended the use of ICD devices. Furthermore, quinidine may be of benefit, as adjunctive therapy in those patients with an ICD or primary therapy in those without, at least in patients with short QT syndrome due to mutations in the HERG channel.

Common Management Principles for Genetic Disorders of the Cardiac Impulse

When evaluating a patient with a newly diagnosed genetic arrhythmic disorder, it is very important to take a detailed and complete multiple generation family history. There is no substitute for a carefully drawn pedigree to help clarify if there may be other similarly affected family members and also to help clearly see the relatives who may also need clinical screening for these disorders. Those family members potentially at highest risk for being affected are the first-degree family members, meaning those family members related most closely to the proband. The most common inheritance pattern observed in these genetic disorders of the cardiac impulse is autosomal dominant, so first-degree family members have a 50% likelihood of being potentially affected.

All first-degree relatives of any patient with an autosomal dominant heritable disorder of the cardiac impulse should be examined including, in many cases, provocative testing. While those family members that demonstrate the disorder only during provocative testing may not need treatment, depending on the nature of the disorder, they are important to identify because they will help indicate which other family members will require testing. Examination within the family should continue antegrade and retrograde through the family pedigree until all potentially affected individuals are identified. In disorders that have inheritance patterns other than autosomal dominant, the potentially affected family members will be determined based on the specific inheritance pattern.

Genetic Testing. Genetic testing is just beginning to be clinically available for some of these disorders and the best clinical use of genetic testing is still being determined. Currently, because of its cost, the inability to exclude a given diagnosis (since not all of the potential genetic etiologies are known for any of these disorders) and the uncertainty of the predicted clinical course of a given mutation, genetic testing is something that should be considered carefully on a case by case basis to decide if it will be helpful in the management of a particular patient and family. When a mutation can be detected, this genetic information can be extremely helpful in confirming a clinical diagnosis and targeting screening among potentially affected family members. On the other hand, these tests have significant limitations and cannot be used to rule out a diagnosis of an inherited cardiac impulse disorder. As the technology advances and society develops and frames its understanding of the promise, limitations, and liability of genetic testing, techniques such as mutation analysis will be regularly performed on patients with LQTS, BrS Syndrome, ARVD, and the cardiomyopathies to confirm the diagnosis, to characterize the range of genetic defects that can cause these clinical syndromes, and to help define their pathophysiology. In the future, genetic testing will be used more extensively for diagnosis, particularly for those disorders such as LQTS and ARVD that are difficult to diagnose clinically. Because very similar clinical features may be related to very different underlying genetic defects, the results of testing may lead to different, more optimal treatment for these disorders, including gene-specific therapy.

SUGGESTED READING

1. Jervell A, Lange-Nielsen F. Congenital deaf-mutism, functional heart disease with prolongation of the QT interval and sudden death. *Am Heart J* 1957;54:59–68.
2. Romano C, Gemme G, Poginglione R. Aritmie cardiache rare dell'eta pediatrica. *Clin Pediatr* 1963;45:656–683.
3. Ward OC. A new familial cardiac syndrome in children. *J Irish Med Assoc* 1964;54:102–106.
4. Dumaine R, Antzelevitch C. Molecular mechanisms underlying the long QT syndrome. *Curr Opin Cardiol* 2002;17(1):36–42.
5. Zareba W, Moss AJ, Schwartz PJ, et al. Influence of genotype on the clinical course of the long-QT syndrome. International Long-QT Syndrome Registry Research Group. *N Engl J Med* 1998;339(14):960–965.
6. Schwartz PJ, Moss AJ, Vincent GM, Crampton RS. Diagnostic criteria for the long QT syndrome. An update. *Circulation* 1993;88(2):782–784.
7. Zhang L, Timothy KW, Vincent GM, et al. Spectrum of ST-T-wave patterns and repolarization parameters in congenital long-QT syndrome: ECG findings identify genotypes. *Circulation* 2000;102(23):2849–2855.
8. Brugada J, Brugada R, Brugada P. Right bundle-branch block and ST-segment elevation in leads V1 through V3: a marker for sudden death in patients

without demonstrable structural heart disease. *Circulation* 1998;97(5):457–460.
9. Wilde AA, Antzelevitch C, Borggrefe M, et al. Proposed diagnostic criteria for the Brugada syndrome: consensus report. *Circulation* 2002;106(19):2514–2519.
10. Priori S, Napolitano C, Tiso N, et al. Mutations in the cardiac ryanodine receptor gene (hRyR2) underlie catecholaminergic polymorphic ventricular tachycardia. *Circulation* 2000;r49–r53.
11. Maron BJ. Hypertrophic cardiomyopathy: a systematic review. *JAMA* 2002;287(10):1308–1310.
12. Varnava AM, Elliott PM, Baboonian C, et al. Hypertrophic cardiomyopathy: histopathological features of sudden death in cardiac troponin T disease. *Circulation* 2001;104(12):1380–1384.
13. Braunwald E, Seidman CE, Sigwart U. Contemporary evaluation and management of hypertrophic cardiomyopathy. *Circulation* 2002;106(11):1312–1316.
14. Gollob MH, Green MS, Tang ASL, et al. Identification of a gene responsible for familial Wolff-Parkinson-White syndrome. *N Engl J Med* 2001;344:1823–1831.
15. Towbin JA, Bowles NE. Molecular genetics of left ventricular dysfunction. *Curr Mol Med* 2001;1(1):81–90.
16. Fontaine G, Fontaliran F, Hebert, et al. Arrhythmogenic right ventricular dysplasia. *Annu Rev Med* 1999;50:17–35.
17. Schott JJ, Benson DW, Basson CT, et al. Congenital heart disease caused by mutations in the transcription factor NKX2-5. *Science* 1998;281:108–111.
18. McKenna WJ, Thiene G, Nava A, et al. Diagnosis of arrhythmogenic right ventricular dysplasia/cardiomyopathy. Task Force of the Working Group Myocardial and Pericardial Disease of the European Society of Cardiology and of the Scientific Council on Cardiomyopathies of the International Society and Federation of Cardiology. *Br Heart J* 1994;71:215–218.
19. Fontaine G, Fontaliran F, Hebert JL, et al. Arrhythmogenic right ventricular dysplasia. *Annu Rev Med* 1999;50:17–35.
20. Kasahara H, Wakimoto H, Liu M, et al. Progressive atrioventricular conduction defects and heart failure in mice expressing a mutant Csx/Nkx2.5 homeoprotein. *J Clin Invest* 2001;108(2): 189–201.
21. Kearns, TP. External ophthalmoplegia, pigmentary degeneration of the retina, and cardiomyopathy: a newly recognized syndrome. *Trans Ophthal Soc UK* 1965;63:559–625.
22. *Molecular Genetics of Cardiac Electrophysiology*, edited by Berul CI and Towbin JA. Kluwer Academic Press, Boston, 2000.
23. Gene Reviews for Long QT syndrome and Arrhythmogenic Right Ventricular Dysplasia/Cardiomyopathy at www.genetests.org. This website is also an excellent resource for updated information about genetic testing, available laboratories and specific testing methods.
24. Gussak I, Brugada P, Brugada J, et al. Idiopathic short QT interval: a new clinical syndrome? 2000, *Cardiology* 94, 99–102.
25. Schimpf R, Wolpert C, Gaita F, Giustetto C, Borggrefe M. Short QT syndrome. *Cardiovasc Res* 2005 (in press).
26. Tester DJ, Kopplin LJ, Creighton W, et al. Pathogenesis of unexplained drowning: new insights from a molecular autopsy. *Mayo Clin Proc* 2005;80(5):596–600.
27. Wehrens XH, Marks AR. Sudden unexplained death caused by cardiac ryanodine receptor (RyR2) mutations. *Mayo Clin Proc* 2004;79(11):1367–1371.

19

Fetal Arrhythmia

Elizabeth V. Saarel and Carlen Gomez

Clinical auscultation of fetal heart began in the late 1800s, and the first reports of fetal arrhythmia detection were published in the 1930s. The detection, diagnosis, and treatment of fetal arrhythmias have evolved considerably beyond these early observations.

FETAL ELECTROPHYSIOLOGY

Normal Sinus Fetal Heart Rate

The sinus node (SN), formed by the seventh week of gestation, generates the normal fetal rhythm. Important normal characteristics of fetal heart rate (FHR) include baseline rate, beat-to-beat variability, and periodic changes (transient decelerations or accelerations). Baseline rate is significantly higher early in gestation than at term. At a normal gestation of 20 weeks, the FHR is close to 160 bpm; by the mid-second trimester, baseline FHR ranges from 120 to 160 bpm; and at normal term, it is near 120 bpm. The gradual decline in baseline FHR is modulated by an increase in parasympathetic tone with resultant progressive vagal influence on the SN. If atropine is administered (to the mother), FHR reverts to the higher baseline of 160 bpm.

Beat-to-beat or short-term variability refer to changes in FHR (cycle length) between successive cardiac impulses. Periodic changes in FHR or long-term variability refer to changes in FHR between successive time intervals (seconds to minutes). A decrease in heart rate variability may be a sign of fetal distress. Heart rate variability is also modulated by the parasympathetic nervous system, and can be inhibited by atropine. Short- and long-term heart rate variability tends to be reduced or exaggerated in a parallel fashion, and there is no clear evidence that distinguishing between the two entities is clinically helpful.

A baseline FHR less than 120 bpm arising from the SN with normal conduction has been defined as sinus bradycardia. A baseline FHR greater than 160 bpm generated and propagated in a normal manner has been defined as sinus tachycardia. As in infants and children, sinus arrhythmia, evidenced by heart rate variability, is normal in the fetus.

OVERVIEW: FETAL ARRHYTHMIA

Consideration of fetal arrhythmia is usually triggered by auscultation of an irregular heart rate, bradycardia, or tachycardia during a prenatal visit. Abnormal FHR occurs in 0.2%–2% of pregnancies, with 10% of these having significant arrhythmia. Once the possibility of arrhythmia is raised, careful assessment of overall maternal and fetal health should be performed. If the arrhythmia is sustained at a markedly fast or slow rate, or if it is associated with structural congenital heart disease, fetal well-being may be compromised. Indeed, a less common presentation of fetal arrhythmia is hydrops fetalis—generalized edema that represents an end-stage fetal response to significant stress and insults. Moreover, several maternal conditions are associated with fetal arrhythmia, making maternal assessment essential in all cases of abnormal FHR.

When a fetal arrhythmia is suspected, a detailed prenatal echocardiogram should be obtained in an experienced pediatric prenatal echocardiographic unit. Although referral patterns vary by institution, abnormal FHR prompts roughly 25% of prenatal cardiology referrals. In addition to the evaluation of the fetal cardiac anatomy and function by echocardiogram during these visits, it is important to obtain a thorough maternal and family history. As many as 50% of fetuses with arrhythmia (particularly those with heart block) have associated structural cardiac malformations.

Optimal patient evaluation and treatment requires a team medical approach. The team should consist of obstetricians, perinatologists, pediatric cardiologists with fetal echocardiographic expertise, social workers, and nurses that work closely with pediatric electrophysiologists to care for affected fetuses. Other relevant ad hoc members may include neonatologists, anesthesiologists, geneticists, and endocrinologists. If significant structural heart disease is identified, pediatric cardiothoracic surgeons should be alerted near term. A team approach streamlines patient care, allows more accurate prognostication, improves communication between family and the health care team, and improves medical outcomes.

TECHNIQUES FOR DIAGNOSIS OF FETAL ARRHYTHMIA

Electrocardiography

Several non-invasive techniques are available for fetal arrhythmia analysis. In contrast to postnatal evaluation, electrocardiography is not practicable for diagnosis. Although fetal electrocardiograms can be recorded from electrodes placed on the maternal abdomen, signal detection is not reliable due to a low signal-to-noise ratio. Fetal cardiac electrical activity is low in amplitude, on the order of about 10 µV, $1/10^{th}$ the amplitude of maternal cardiac electrical activity. In addition, maternal abdominal wall musculature adds low amplitude noise to the electrocardiogram, often obscuring fetal myocardial electrical activity.

Magnetocardiography

A magnetocardiogram records the magnetic field generated by cardiac electrical activity. This technique has been applied to the fetus for the last decade. There have been multiple reports from several institutions describing magnetocardiograms that detail fetal cardiac electrical activity, including P-wave and QRS inscription, similar to a postnatal electrocardiogram. Magnetocardiographic equipment is complex and large, requires shielding and dedicated space, is very expensive, and is not yet commercially available; widespread use is not available. Fetal magnetocardiography would undoubtedly become more widespread if technical refinements decreased cost and increased accessibility.

Echocardiography

Because of limited access to the electrocardiographic forces of the fetal heart,

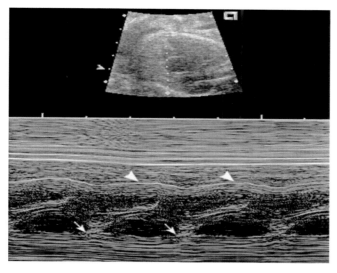

FIGURE 1. M-mode echocardiogram from a fetus in sinus rhythm. Upper image demonstrates beam position through the atrium and ventricle. Small arrows mark atrial contractions, and arrowheads mark ventricular septal motion. Note 1:1 ratio of atrial to ventricular contractions.

echocardiography is the current method for the diagnosis of fetal arrhythmias. Evaluation for arrhythmia begins by examining the timing and association of atrial and ventricular wall motion. This is best accomplished through the use of M-mode and Doppler echocardiography.

M-mode echocardiography displays motion of the cardiac tissue with respect to time. An M-mode tracing of the fetal heart chambers is obtained by placing the cursor line across both the atrial and ventricular walls simultaneously (Figure 1). The display shows movement of both chamber walls with time, reflecting atrial and ventricular systole.

Many factors influence the quality of the M-mode tracing. First, alignment with both heart chambers can be challenging, and several different probe positions and angles should be attempted to optimize the tracing. Second, a clearer tracing will be obtained where chamber walls have the greatest excursion with contraction, such as the atrial appendage, or lateral ventricular wall.

Doppler echocardiography utilizes spectral blood flow patterns during systole and diastole as representation of atrial and ventricular motion. An apical four-chamber image of the fetal heart is obtained, and the pulsed Doppler cursor is positioned with the gate spanning the mitral inflow as well as the aortic outflow (Figure 2).

Doppler tissue color M-mode imaging may aid in the diagnosis of fetal arrhythmias when standard M-mode and Doppler echocardiography are indeterminate. Gated pulsed Doppler may be used to measure fetal PR intervals. In addition, new techniques including tissue velocity imaging, with creation of a "fetal kinetocardiogram," as well as strain rate imaging, have recently been described. Tissue velocity and strain rate imaging may significantly enhance current fetal echocardiographic diagnostic capabilities.

In addition to determination of the arrhythmia mechanism, echocardiographic evaluation of the fetus should incorporate assessment for any hemodynamic and anatomic abnormality. Both tachyarrhythmias and bradyarrhythmias can cause heart failure, and ultimately hydrops fetalis. Complete evaluation for fetal heart failure includes assessment

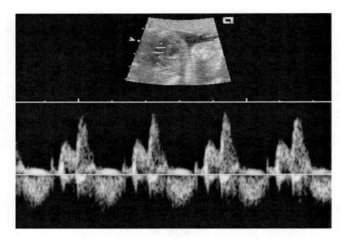

FIGURE 2. Doppler echocardiogram from fetus in sinus rhythm. Upper image demonstrates beam position in the ventricle. Note 1:1 relationship of normal mitral inflow (E- and A-waves above baseline) and normal aortic outflow (below baseline).

of heart size, ventricular systolic function, atrioventricular valve incompetence, venous Doppler patterns (increased reversal with atrial contraction) including umbilical venous Doppler (Figure 3), and documentation of the presence and size of pleural, pericardial, and abdominal effusions. All of these indices need to be reassessed for progression by serial echocardiographic studies along with ongoing obstetrical evaluation of fetal well being.

SPECIFIC FETAL ARRHYTHMIAS

Extrasystole

Extrasystoles account for 60% to 90% of fetal arrhythmia. Ectopic beats may arise in the atria, junctional tissue, or ventricle. Supraventricular ectopy (SVE) is most common. Ventricular ectopy (VE) comprises fewer than 10% of fetal extrasystole.

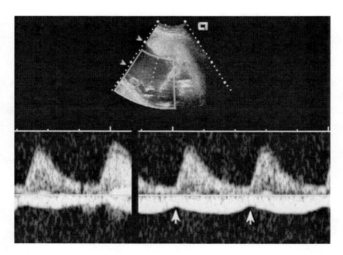

FIGURE 3. Pulsed spectral Doppler tracing from the umbilical vein in a fetus with hydrops fetalis. Upper panel demonstrates gate position in the umbilical vein. Arrows (lower panel) indicate pulsations with atrial systole, consistent with high atrial pressure.

Frequency of ectopic beats in the normal fetus has not been well established; however, the rate of extrasystoles in healthy premature infants is 20% to 30%, with a slightly lower frequency in term infants.

Ectopic beats present as an irregular FHR. As in older patients, very early SVE results in blocked atrioventricular conduction, either complete or partial (in the form of bundle branch block with aberrant depolarization). Hence, SVE can present as bradycardia. The majority of fetuses with premature beats are healthy, and the ectopy resolves over time.

Although the vast majority of fetuses with extrasystoles have a structurally normal heart, SVE and VE can be associated with anatomic congenital heart disease, cardiac tumors, and fetal genetic abnormalities such as Trisomy 18. In addition, SVE precedes supraventricular tachycardia (SVT) in 15%. Thus, all fetuses with premature beats should be monitored until delivery, or until resolution of the ectopy has been sustained.

Supraventricular Ectopy

Premature atrial and junctional depolarizations occur most often as single beats, but can present with bigeminy, trigeminy, or quadrigeminy. Diagnosis can be made with a combination of Doppler and M-mode echocardiography. With M-mode, simultaneous atrial and ventricular recording is performed, demonstrating a normal sequence of A-V contractions and an early atrial beat. The atrial beat after the premature contraction demonstrates an incomplete compensatory pause.

Doppler sampling at the junction of the mitral inflow and left ventricular outflow tract displays E- and A-waves (early, rapid ventricular filling followed by atrial contraction with active ventricular filling) and ventricular systole (with semilunar valve outflow). SVE will cause early active ventricular filling (A-wave), obscuring part of the early ventricular filling (E-wave) tracing with subsequent early ventricular systole (Figure 4). In cases of fully blocked SVE, no ventricular systole will follow the premature atrial contraction. The post-extrasystolic contraction will demonstrate a prolonged filling (diastolic) time interval. An additional Doppler finding is flow reversal in the IVC during early atrial contraction.

The fetal PR interval can be measured using a gated pulsed Doppler technique. Premature atrial contractions should demonstrate

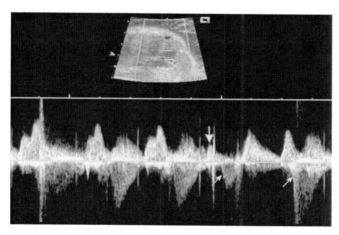

FIGURE 4. Doppler echocardiogram from a fetus with SVE. The large arrow marks the premature atrial contraction. The first small arrow (lower panel) marks the diminished aortic outflow volume with the SVE, and the second small arrow shows the increased aortic outflow volume during the post-extrasystolic contraction. Note the prolonged diastolic time interval following the SVE.

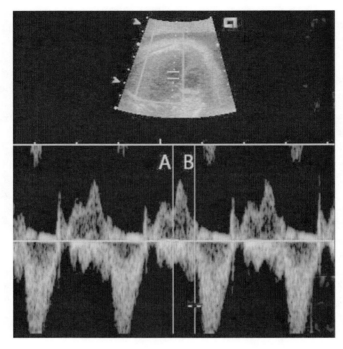

FIGURE 5. Pulsed spectral Doppler through the mitral inflow and aortic outflow in a normal fetus, demonstrating mechanical PR interval measurement. E- and A-wave components of mitral inflow are above the baseline, whereas the aortic outflow signal is below the baseline. Calipers measure from the beginning of the A-wave (line A) to the beginning of the aortic outflow signal (line B).

a normal (Figure 5) or prolonged PR interval (if slow conduction occurs).

The prognosis is favorable for the majority of fetuses with SVE. In addition to the conditions mentioned earlier, fetal SVE can be associated with maternal drug use or hyperthyroidism. For fetuses with associated maternal disease, structural congenital heart disease, tumors, or sustained tachyarrhythmias, the prognosis correlates with the associated condition. No treatment is indicated for isolated fetal SVE.

Ventricular Ectopy

M-mode echocardiography may demonstrate subtle distortion of normal ventricular contraction due to aberrant muscle depolarization (Figure 6). Also, a complete atrial compensatory pause is seen after most premature ventricular depolarizations. Doppler ultrasound demonstrates a characteristic AV valve inflow pattern with decreased diastolic antegrade flow. Marked retrograde flow in the IVC is seen during atrial contraction.

In addition to the conditions mentioned previously, VE has been associated with fetal myocarditis, cardiomyopathy, long QT syndrome, and complete AV block with a slow escape rate. Premature ventricular contractions also occur in healthy fetuses, for whom the prognosis is excellent. Prognosis is dependent on the associated condition. No treatment of isolated premature ventricular contractions is indicated.

Tachyarrhythmias

Most cases of elevated FHR are due to sinus tachycardia. SVT, atrial flutter, atrial fibrillation, and ventricular tachycardia (VT) are less common. In one large single-center

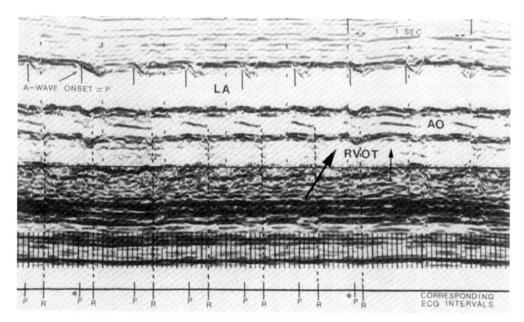

FIGURE 6. M-mode recording from a 36-week fetus with its back toward the transducer presenting an inverted view of the heart. Premature aortic valve opening (large arrow) can be clearly seen to be followed by atrial wall contraction but with no variation in P-P interval (A-wave interval), giving a compensatory pause (small arrow, allowing diagnosis of ventricular premature beat). LA = left atrium; Ao = aorta; RVOT = right ventricular outflow tract (5 MHz transducer). Reprinted from Clinical Cardiology 1985;8:1–10 with permission from Clinical Cardiology Publishing Company, Inc, Mahwah, NJ 07430 USA.

retrospective report, SVT and atrial flutter together accounted for 12% of fetal arrhythmia diagnoses.

Sinus Tachycardia

Fetal sinus rates rarely exceed 210 bpm, whereas fetal SVT rarely falls below 200 bpm. In sinus tachycardia, fetal M-mode echocardiography shows synchronous atrioventricular contractions. Sinus arrhythmia with varying cycle lengths may be present. Certain tachyarrhythmias that tend toward longer and more variable cycle lengths, such as persistent junctional reciprocating tachycardia and ectopic atrial tachycardia, are difficult to differentiate from sinus tachycardia using current ultrasound diagnostic techniques.

In sinus tachycardia, Doppler tracing of atrioventricular valve inflow often demonstrates amalgamation of the E- and A-waves. The mechanical PR interval, as measured by gated-Doppler ultrasound, should be constant and of normal duration.

Fetal sinus tachycardia is due to an underlying fetal or maternal abnormality such as drug exposure, hyperthyroidism, myocarditis, infection, hypoxia, or other causes of fetal distress. Treatment is directed at the primary cause of sinus tachycardia.

Supraventricular Tachycardia

Fetal supraventricular tachycardia (SVT) can be sustained or intermittent. Typical rates are 240–250 bpm, with a range from 200 to 320 bpm. Detection is most common after 15 weeks gestation, although earlier presentation has been reported. Fetal tolerance of SVT depends on the duration and rate of arrhythmia; intermittent and slower (≤260 bpm) rhythms are less malignant. Mechanisms of SVT in the fetus are similar to those in neonates (Figure 7).

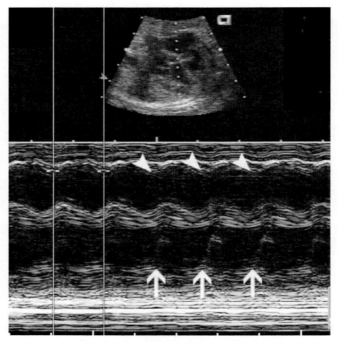

FIGURE 7. M-mode echocardiogram from a fetus with supraventricular tachycardia. Arrowheads indicate atrial contractions, and large arrows indicate ventricular contractions. Note the 1:1 relationship of atrial and ventricular contractions. The calipers measure 243 ms between successive ventricular contractions, indicating a heart rate of approximately 250 bpm.

Atrioventricular reentry tachycardia, utilizing an accessory connection between the atria and ventricles, is most common when examined post partum. AV node reentry tachycardia, ectopic atrial foci, persistent junctional reciprocating tachycardias, and junctional tachycardias are less common. Although an abrupt onset and termination of the tachycardia, if observed during the fetal echocardiogram, would support the diagnosis of a reentry tachycardia, it is not possible to distinguish with certainty between these entities using current ultrasound techniques.

Diagnosis of SVT is consistent with an M-mode tracing through the atria and ventricle showing sequential 1:1 contractions at retrograde time intervals of 80–120 ms. Similarly, Doppler ultrasound of ventricular inflow and outflow demonstrates sequential atrial and ventricular contractions. If measurable by gated-Doppler, the PR interval will be constant in reentry tachyarrhythmias.

Treatment is predicated on the effect of the tachyarrhythmia on fetal well being. If SVT is intermittent, slow, and late in pregnancy, fetal health is usually not jeopardized. In such patients, prognosis is generally excellent, and no treatment is indicated. If tachycardia is sustained at fast rates (>260 bpm), prenatal demise may be as high as 25%. It is therefore important to frequently (every 3–5 days) assess all fetal patients with SVT. Moreover, mothers should be educated to look for symptoms of fetal distress. Both obstetricians and cardiologists should follow patients. Of note, patients with structural congenital heart disease are at greater risk for tachycardia associated complications. Two of the earliest signs of fetal compromise are exaggerated umbilical venous or inferior vena cava flow reversal (greater than 30%) and cardiomegaly.

Other signs of fetal distress include decreased ventricular systolic function, atrioventricular valve regurgitation, and hydrops fetalis (pericardial effusion, pleural effusion, ascites, and/or skin edema). Ominous signs of distress include decreased fetal movement and abnormal umbilical artery pulsations.

Fetal treatment options include early delivery, transplacental (maternal) pharmacotherapy, or direct fetal pharmacotherapy. In addition, there is one report of fetal SVT conversion using transabdominal umbilical cord compression. Although not described in humans, there has been a single report of SVT conversion using transesophageal pacing in fetal sheep. Labor induction is the treatment of choice for term and near-term pregnancies with sustained fetal tachycardia or evidence of fetal compromise.

Although controversy exists, it is generally agreed upon that transplacental digoxin should be the first-line treatment of choice in pre-term pregnancies with sustained tachycardia or fetal compromise (Table 1). Digoxin treatment is safe and often effective. The drug can be administered to the mother in oral or IV form at relatively high maternal doses (up to 1 mg qd by mouth) in order to achieve an adequate level of fetal drug. SVT cessation is achieved in approximately three-quarters of cases with maternal oral therapy. If conversion has not been achieved after two weeks of therapy, a second antiarrhythmic agent may be added; flecainide, along with other drug choices are reasonable (Table 1). Other medications with reported efficacy include procainamide, verapamil, quinidine, amiodarone, or sotalol. There has been no definitive large, randomized published study comparing fetal antiarrhythmic agents.

In the case of significant fetal distress or rapid, sustained SVT, where prompt treatment is essential, digoxin can be given directly to the fetus through umbilical vein, intramuscular, or intraperitoneal administration; however, the later two routes are unreliable due to unpredictable drug absorption. Umbilical cordocentesis carries a significant risk in the setting of fetal distress. Maternal flecainide by mouth in cases where premature delivery of the fetus is too high a risk may be helpful. Cardioversion with flecainide is generally achieved within 3 to 4 days. Sotalol appears to be effective for atrial flutter, but probably should be avoided for SVT. Success with the combination of digoxin with amiodarone or the combination of digoxin with verapamil has been described.

Atrial flutter (intra-atrial reentry tachycardia) is responsible for approximately one-third of non-sinus fetal tachyarrhythmias. Atrial rates vary from 400 to 550 bpm. As in postnatal patients, ventricular response is variable, but rates are greater than 200 bpm in a majority of untreated fetuses. Variations in atrioventricular conduction frequently lead to an irregular FHR. When conduction is consistent, there is greater elevation of the FHR; 2:1 atrioventricular block is most common. Although atrial flutter is usually sustained, paroxysmal cases do occur.

The diagnosis of prenatal atrial flutter is confirmed by characteristic echocardiographic findings. M-mode tracings through the atria and ventricle demonstrate regular, fast atrial contractions with variable atrioventricular block, and thus less frequent ventricular depolarizations (Figure 8). Atrial rate is calculated after measurement of the time interval between successive atrial contractions (the mechanical P-P interval). Similarly, ventricular rate is calculated after measurement of the time interval between successive ventricular contractions (the mechanical R-R interval). Using this information, the degree of atrioventricular block can be determined. Doppler echocardiography of ventricular inflow or outflow can also be used to calculate the ventricular response rate.

Cardiac lesions reported in association with fetal atrial flutter include atrial septal defect, Ebstein's malformation of the tricuspid valve, atrioventricular septal defect with atrioventricular valvar regurgitation and atrioventricular block, right ventricular outflow tract obstruction with tricuspid regurgitation,

TABLE 1. Drugs for Treatment of Fetal Tachyarrhythmia

Drug	Route of Administration	Starting Dosage	Maximum Dosage	SVT	AFL	VT	Maternal/Fetal Adverse Effects
Adenosine	Fetal IV	0.1 mg/kg	0.2 mg/kg	Yes	No	No	Tachycardia commonly recurs requiring additional agent
Amiodarone	Maternal PO (Fetal IV; Fetal IM; Fetal IP)	800–1200 mg QD × 8–14 d load; then 200–400 mg QD maint.	800 mg QD maint.	Yes	No	Yes	Maternal or fetal hypothyroidism (common); bradycardia (common); IUGR; premature delivery; hepatotoxicity
Digoxin	Maternal PO; Maternal IV (Fetal IV; Fetal IM; Fetal IP)	PO: 0.25–1.5 mg load, then 0.125 BID maint. IV: 1–2 mg load (may repeat), then 0.5 mg QD maint.	PO: 0.25 mg BID maint. IV: 1.0 mg QD maint.	Yes[1]	Yes[1]	No	Maternal overdose; heart block; arrhythmia induction
Flecainide	Maternal PO	Start at 100 mg BID, advance to 200 mg BID-TID	Total daily dose not to exceed 600 mg; maternal blood level not to exceed 1 mg/L	Yes[2]	No	Yes	Dizziness; headache; paraesthesias; tremors; visual disturbances; nausea; vomiting; flushing; possible neonatal conjugated hyperbilirubinemia or neonatal prolonged QT
Lidocaine	Maternal IV	1 mg/kg load over 2 min, may repeat in 10–15 min × 2 (alternate 20–50 mg/kg/min)	3–5 mg/kg within first hour	No	No	Yes	Hypotension; asystole; seizures; respiratory arrest
Mexiletine	Maternal PO	200 mg PO q8 hrs; therapeutic level 0.5–2 mg/L	Max 1200 mg/d; maternal blood level not to exceed 2 mg/L	No	No	Yes	Atrial and ventricular arrhythmias; bradycardia; hypotension; confusion; dizziness; nervousness; tremor; ataxia; numbness of fingers or toes; weakness; blurred vision; tinnitus; nausea; hepatotoxicity
Procainamide	Maternal IV; Maternal PO	IV: 100 mg load over 2 min., then 1–6 mg/min PO: 1 g load then 200–500 mg q3–6 hrs	IV: 6 mg/min PO: 500 mg q3 hrs; plasma levels 4–10 mg/L	Yes	Yes	Yes	Hypotension; nausea; vomiting; blood dyscrasias; lupus-like syndrome; rash; confusion; prolonged QT
Quinidine	Maternal PO (sulfate form); Maternal IV (gluconate form)	PO: Test dose 200 mg; 800–1000 mg load, then 200 mg q4–6 hrs maint. IV: 0.3 mg/kg/min (infuse <10 mg/min)	PO: 600 mg q4–6 hrs IV: 0.5 mg/kg/min (infuse <10 mg/min)	Yes	No	Yes	Idiosyncratic reaction (prolonged QRS >0.02 sec); prolonged QT; torsades de pointes; nausea; vomiting; diarrhea; hypotension; tinnitus; confusion; blood dyscrasias; rash; heart block
Sotalol*	Maternal PO	80 mg BID	320 mg BID	No	Yes[2]	Yes[1]	Myocardial depression; arrhythmia induction; torsades de pointes
Verapamil	Maternal IV	5–10 mg over 3–5 min. (alternate 0.005 mg/kg/min)	May give second 10 mg dose after 30 min.	Yes	Yes	Yes	Myocardial depression; bradycardia; heart block; hepatotoxicity

[1] Recommended first-line agent
[2] Recommended second-line agent
* Use with extreme caution if hydrops fetalis is present

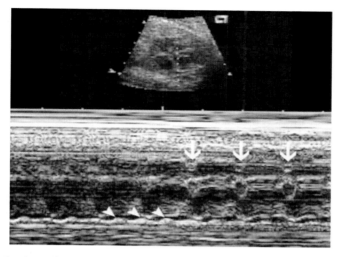

FIGURE 8. M-mode echocardiogram from fetus with atrial flutter. Arrowheads mark atrial flutter waves (400 bpm), and large arrows mark less frequent ventricular contractions (200 bpm), demonstrating 2:1 conduction.

hypoplastic left heart syndrome, cardiomyopathy, and endocardial fibroelastosis. As in older patients, it is likely that atrial dilatation due to a variety of causes will predispose to atrial flutter. In addition, atrioventricular reentry tachycardia and concealed accessory atrioventricular pathways are rarely associated with fetal atrial flutter.

Treatment of fetal atrial flutter is similar to that of SVT. Unfortunately, atrial flutter tends to be less responsive to drug therapy, in terms of cardioversion rates, as compared to other forms of reentry mediated SVT in the fetus. On the other hand, postnatal treatment of atrial flutter, via transesophageal overdrive pacing or DC cardioversion, is highly successful and postnatal recurrence is exceedingly rare.

Other Forms of Supraventricular Tachyarrhythmias

Other reported forms of fetal SVT include atrial fibrillation, junctional ectopic tachycardia, persistent junctional tachycardia, and chaotic atrial tachycardia. Prenatal definitive diagnosis of these arrhythmias is not possible using current echocardiographic techniques. All published reports of these unusual fetal arrhythmias rely on postnatal testing with an inferred mechanism of the prenatal arrhythmia. Currently, fetuses thought to have these tachyarrhythmias are evaluated and treated in the same manner as fetuses with other mechanisms of SVT.

Ventricular Tachyarrhythmias

Ventricular tachyarrhythmias (VT) occur in fetal patients, but as in infants and children, these arrhythmias are quite rare (Figure 9). In many reports, observed VT rates are slow, raising the possibility of an accelerated ventricular rhythm. Slow VT is usually well tolerated, and the prognosis for this group of patients is generally good. Fast ventricular tachycardia, however, is poorly tolerated in the fetus, and the prognosis is extremely poor.

M-mode and Doppler echocardiography show atrioventricular dissociation with a competing atrial sinus rhythm in all types of ventricular arrhythmia. Recently, Doppler tissue imaging has been reported useful for detection of atrioventricular dissociation. In addition, atrioventricular block has been reported in conjunction with VT.

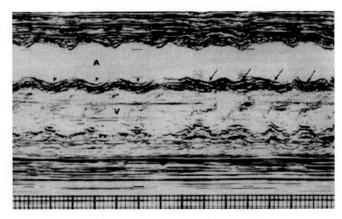

FIGURE 9. M-mode echocardiogram from fetus with intermittent ventricular tachycardia. When the fetus is in ventricular tachycardia (large arrows), the ventricular (V) rate exceeds the atrial (A) rate. When the fetus is in sinus rhythm (small arrows), the rates of the ventricular and atrial contractions are the same.

When treatment of a ventricular arrhythmia is indicated due to fetal compromise or persistence of a fast ventricular rate, the choice of a pharmacologic agent is based on postnatal infant data. Multiple fetal VT case reports, using an assortment of antiarrhythmic agents [all of which have been previously mentioned for treatment of SVT (Table 1)], have been published. In addition, the use of transplacental and transcordal lidocaine has been suggested. Sotalol is also reasonable for first-line treatment for prenatal VT.

Bradyarrhythmias

Sinus bradycardia (<120 bpm) is the most common cause of slow FHR. Other sources of fetal bradycardia include blocked premature atrial contractions (previously discussed) and atrioventricular block (AV block). The incidence of congenital complete heart block is approximately 1:20,000 live newborns. Fetal incidence of complete heart block is likely higher, given that some cases may result in fetal demise. AV block comprises about 5% of fetal arrhythmias.

Sinus Bradycardia

Transient fetal sinus bradycardia is very common, and is likely related to episodic parasympathetic stimulation. Sinus pauses have been documented in the healthy fetus, and prenatal sinus arrhythmia is common. Sequential atrioventricular 1:1 conduction can be observed with M-mode ultrasound through both the atria and ventricles when the fetus is in sinus rhythm (Figure 1). Doppler tracing of atrioventricular valve inflow shows normal E- and A-wave configuration and size (Figure 2). Doppler of ventricular outflow will be of normal velocity in patients with normal cardiovascular anatomy (Figure 2). The mechanical PR interval, measured with gated-Doppler ultrasound, will be of normal duration and constant length from beat to beat (Figure 5).

As with sinus tachycardia, sinus bradycardia is a reaction to fetal stimuli, not a primary cardiac arrhythmia. Sinus bradycardia can be triggered by stimuli that cause fetal distress. In addition, maternal medications, such as beta-blockers, may cross the placenta and decrease the FHR. In cases of severe hydrops fetalis, sinus bradycardia is an agonal rhythm.

Rarely, fetal sinus bradycardia has been a manifestation of prenatal long QT syndrome. The diagnosis of long QT syndrome cannot be proven definitively with ultrasound. Definitive diagnosis requires documentation of cardiac electrical activity with techniques such as magnetocardiography, or perhaps in the future with a transabdominal or fetal transesophageal ECG recording. A presumptive diagnosis can be made in cases of fetal bradycardia with a strong family history of long QT syndrome.

Treatment of sinus bradycardia is directed at correction of the underlying cause of fetal distress. If thorough investigation shows that the fetus is healthy, sinus bradycardia is a normal variant and no treatment is indicated.

Atrioventricular Conduction Block

First-Degree Atrioventricular Block

First-degree AV block may or may not be associated with sinus bradycardia. Diagnosis of first-degree AV block requires measurement of the fetal PR interval. M-mode and Doppler studies will show 1:1 atrioventricular conduction and normal Doppler tracings (Figure 5). The mechanical PR interval will be prolonged, and can be estimated using gated-Doppler ultrasound. The upper limit of normal for the fetal mechanical PR interval has been reported to be 13 ± 2 ms. Measurement of the true electrical PR interval requires magnetocardiography or another method of fetal ECG recording.

Second- and Third-Degree Atrioventricular Block

Second- and third-degree AV block are the most common sustained fetal bradyarrhythmias. In half of cases, the fetus has associated structural congenital heart disease, most commonly atrioventricular discordance or endocardial cushion defects (especially in conjunction with heterotaxy syndromes). In the majority of patients, a normal atrial rate is present. Ventricular rate may range from low normal to marked bradycardia (45–70 bpm).

In many cases of fetal AV block and normal cardiac anatomy, maternal autoimmune disease is present. Systemic lupus erythematosus (SLE) is the most commonly identified maternal condition, but other co-morbid maternal disorders, such as Sjorgren's syndrome, rheumatoid arthritis, Raynaud's syndrome, and mixed connective tissue disease, have been reported. Often the fetal conduction disturbance is the first presentation of maternal disease. Nearly all cases occur in the presence of circulating maternal Ro (Sjogren's syndrome A/SSA), and La (Sjogren's syndrome B/SSB) autoantibodies. These antibodies are directed against cellular ribonucleoproteins. Of note, not all fetuses exposed to anti-Ro/SSA and anti-La/SSB antibodies will develop heart block.

In cases of anti-Ro/SSA and anti-La/SSB associated fetal conduction disorders, antibodies deposit in the fetal cardiac tissue, instigating an inflammatory response. Inflammation leads to eventual fibrosis, calcification, and disruption of the normal cardiac conduction tissue. Severely affected fetuses will develop signs of generalized cardiac myocardial dysfunction and hydrops fetalis. In some severe cases, cardiac muscle and other fetal tissues are directly affected by local antibody deposition.

Second- and third-degree AV block present with fetal bradycardia in the second or third trimester. Although the overall rate of fetal demise is approximately 50% with complete AV block, most patients with normal cardiac anatomy do well. The presence of structural congenital heart disease, sinus bradycardia, with an atrial rate less than 120 bpm, and extreme bradycardia, with a ventricular rate less than 55 bpm, are indicators of poor prognosis. In addition, evidence of hydrops fetalis, including pericardial effusion and progressive slowing of the atrial rate, indicate impending fetal demise.

Progression from second- to third-degree AV block in the fetus has been documented. Moreover, return of sinus rhythm following second-degree AV block has been documented in fetuses exposed to maternal autoantibodies after treatment with transplacental corticosteroids. There has been no published report of resolution of persistent fetal third-degree block with or without treatment.

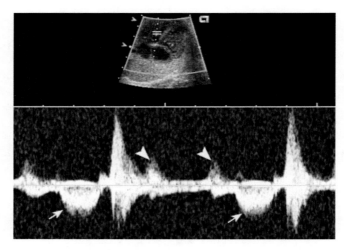

FIGURE 10. Doppler echocardiogram from fetus with third-degree heart block. Arrowheads mark atrial contractions above the baseline, and arrows mark aortic outflow signal below the baseline. Note complete A-V dissociation.

Diagnosis of second-degree AV block can be made using M-mode through the atria and ventricles. This technique demonstrates normally timed atrial contractions with intermittent AV block and missing ventricular contractions. If fetal mechanical PR intervals are measured using gated Doppler, they show progressive lengthening in Mobitz type I second-degree AV block. Constant PR intervals with intermittent loss of AV conduction will be seen in Mobitz type II AV block. Doppler of ventricular inflow will show abnormal E- and A-wave morphology during AV block.

The echo diagnosis of third-degree (complete) AV block can be made with M-mode or Doppler echo showing complete atrioventricular discordance (Figure 10). Atrial rate is calculated after measurement of the time interval between successive atrial contractions (the mechanical P-P interval). Similarly, ventricular rate is calculated after measurement of the time interval between successive ventricular contractions (the mechanical R-R interval). All affected fetuses show some degree of ventricular dilation.

Treatment goals for prenatal second- and third-degree AV block are enhanced patient survival and continuance of pregnancy until term. If anti-inflammatory medications, including corticosteroids, prove successful in ongoing prospective randomized trials, an additional goal of treatment for patients exposed to maternal anti-Ro/SSA and anti-La/SSB antibodies will be to prevent or improve the degree of heart block and signs of congestive heart failure.

As with other fetal arrhythmias, current treatment is determined by the effect of bradycardia on fetal well-being. If the fetus is healthy, showing no signs of distress, no treatment is indicated. Delivery of all fetuses with heart block should be performed at a facility capable of emergency postpartum pacing, even if the fetus appears to be healthy prior to birth.

One may institute prenatal empiric treatment with corticosteroids (maternal PO) under certain circumstances. In some patients, evidence of mild fetal compromise, decreasing fetal heart rate, or fetal bradycardia is present. Because there have been isolated reports documenting second-degree AV block reversal with corticosteroid therapy, one can strongly consider treatment using maternal oral corticosteroids for fetuses with this conduction abnormality. If the fetal ventricular rate is below 55 bpm, if the atrial rate is decreasing or below 120 bpm, or if signs of hydrops fetalis are present, early delivery is indicated in anticipation of postpartum pacing.

Other attempted treatment for fetal heart block and heart failure has included transplacental pharmacotherapy with digoxin, furosemide, and beta-receptor agonists. All of these drugs are of limited efficacy, and all have significant potential side effects. In case of severe symptoms with hydrops fetalis, early delivery of a preterm patient may be indicated. Initial pacing is best accomplished through temporary transvenous or epicardial leads. External pacing, if used at all, should be transient given the frequency of severe burns in neonates. Aggressive treatment of postpartum pericardial, pleural, and ascitic effusions should be performed. Standard methods of intensive systemic support, including mechanical ventilation, inotropic therapy, and correction of acidosis, should be utilized, as for any neonate with cardiac failure. Experimental in utero pacing has been described, but to date there has been no published report of an infant survivor.

FUTURE DIRECTIONS

Current understanding of fetal conduction system development and function is limited, but recently published molecular genetic studies have advanced our knowledge considerably. Molecular cardiology will eventually deliver a large impact on fetal diagnosis and therapeutics. Research on improved methods of diagnosis for fetal arrhythmia is ongoing. Tissue velocity imaging, strain rate imaging, magnetocardiography, and fetal transesophageal electrocardiography are emerging techniques for fetal arrhythmia diagnosis.

Finally, present treatment of fetal arrhythmias is imperfect. There is no consensus regarding fetal antiarrhythmic pharmacotherapy, largely due to the orphan status of the disease, and consequently, insufficient data. Moreover, all antiarrhythmic agents currently available have significant side effects for the fetus or mother. One area of ongoing study that may yield significant positive results in the near future is the prevention of immune-mediated heart block via transplacental anti-inflammatory agents. Minimally invasive fetoscopic techniques are now under intense study, and may offer improvements for fetal arrhythmia therapy in the future.

SUGGESTED READING

1. Anderson RH, Becker AE, Wenink AC, Janse MJ. The development of the cardiac specialized tissue. In: Wellens HJJ, Lie KI, Janse MJ, editors. The Conduction System of the Heart. Philadelphia, PA: Lea and Febiger, 1976:3–28.
2. Shenker L. Fetal cardiac arrhythmias. *Obstet Gynecol Surv* 1979;34(8):561–572.
3. Peters M, Crowe J, Piéri JF, et al. Monitoring the fetal heart non-invasively: a review of methods. *J Perinat Med* 2001;29(5):408–416.
4. Machado MV, Tynan MJ, Curry PV, Allan LD. Fetal complete heart block. *Br Heart J* 1988;60(6):512–515.
5. Schmidt KG, Ulmer HE, Silverman NH, et al. Perinatal outcome of fetal complete atrioventricular block: a multicenter experience. *J Am Coll Cardiol* 1991;17(6):1360–1366.
6. Cotton JL. Identification of fetal atrial flutter by Doppler tissue imaging. *Circulation* 2001;104(10):1206–1207.
7. Glickstein JS, Buyon J, Friedman D. Pulsed Doppler echocardiographic assessment of the fetal PR interval. *Am J Cardiol* 2000;86(2):236–239.
8. Rein AJJT, O'Donnell C, Geva T, et al. Use of tissue velocity imaging in the diagnosis of fetal cardiac arrhythmias. *Circulation* 2002;106(14):1827–1833.
9. Rein AJJT, Perles Z, Nir A, et al. Strain rate imaging is superior to tissue velocity imaging for measuring atrioventricular time interval in the fetus. *J Am Coll Cardiol* 2003;41(6 Suppl A):494A.
10. Quartero HW, Stinstra JG, Goldbach EG, et al. Clinical implications of fetal magnetocardiography. *Ultrasound Obstet Gynecol* 2002;20(2):142–153.
11. Ferrer PL. Fetal arrhythmias. In Deal BJ, Wolff GS, Gelband H, editors. Current Concepts in Diagnosis and Management of Arrhythmias in Infants and Children. Armonk, NY: Futura, 1998:17–63.
12. Wong SF, Chau KT, Ho LC. Fetal bradycardia in the first trimester: an unusual presentation of atrial extrasystoles. *Prenat Diagn* 2002;22(11):976–978.

13. Martin CB Jr, Nijhuis JG, Weijer AA. Correction of fetal supraventricular tachycardia by compression of the umbilical cord: report of a case. *Am J Obstet Gynecol* 1984;150(3):324–326.
14. Kohl T, Kirchoff PF, Gogarten W, et al. Fetoscopic transesophageal electrocardiography and stimulation in fetal sheep: a minimally invasive approach aimed at diagnosis and termination of therapy-refractory supraventricular tachycardias in human fetuses. *Circulation* 1999;100(7): 772–776.
15. Strasburger JF. Fetal arrhythmias. *Prog Pediatr Cardiol* 2000;11(1):1–17.
16. Friedman AH, Copel JA, Kleinman CS. Fetal echocardiography and fetal cardiology: indications, diagnosis and management. *Semin Perinatol* 1993;17(2):76–88.
17. Crowley DC, Dick M, Rayburn WF, Rosenthal A. Two-dimensional and M-mode echocardiographic evaluation of fetal arrhythmia. *Clin Cardiol* 1985;8(1):1–10.
18. Rein AJ, Levine JC, Nir A. Use of high-frame rate imaging and Doppler tissue echocardiography in the diagnosis of fetal ventricular tachycardia. *J Am Soc Echocardiogr* 2001;14(2):149–151.
19. Michaelsson M, Engle MA. Congenital complete heart block: an international study of the natural history. *Cardiovasc Clin* 1972;4(3):85–101.
20. Hosono T, Kawamata K, Chiba Y, et al. Prenatal diagnosis of long QT syndrome using magnetocardiography: a case report and review of the literature. *Prenat Diagn* 2002;22(3):198–200.
21. Buyon JP, Waltuck J, Kleinman C, Copel J. In utero identification and therapy of congenital heart block. *Lupus* 1995;4(2):116–121.
22. Serwer GA, Vermilion RP, Snider AR, et al. Prognostic indicators for fetuses with in utero diagnosed complete heart block. *Circulation* 1988;78(Suppl II):396.
23. Veille JC, Covtiz W. Fetal cardiovascular hemodynamics in the presence of complete atrioventricular block. *Am J Obstet Gynecol* 1994;170(5 Pt 1): 1258–1262.
24. Rentschler S, Zander J, Meyers K, et al. Neuroregulin-1 promotes formation of the murine cardiac conduction system. *Proc Natl Acad Sci USA* 2002;99(16):10464–10469.
25. Van Engelen AD, Weijtens O, Brenner JI, et al. Management outcome and follow-up of fetal tachycardia. *J Amer Coll Cardiol* 1994; 24:1371–1375.

20

Sudden Cardiac Death in the Young

Christopher B. Stefanelli

Sudden unexpected cardiac death (SCD) in the young is rare. Nonetheless, it has a devastating impact on families and communities. It is important to note that most pediatric cardiopulmonary arrests are of respiratory and trauma etiology. However, the death of several prominent young athletes, most from cardiac abnormalities, during the last two decades has led to an increase in public and media awareness of SCD as a medical, legal, and public health issue.

Sudden cardiac death can be defined as biologic death resulting from abrupt, unexpected cardiovascular collapse from which an individual does not recover or regain consciousness. *Sudden* usually implies an interval of less than one hour from the onset of symptoms to the time of death. Sudden arrhythmic (often presumed) and non-arrhythmic cardiac death in infants, children, and adolescents can occur in the absence or presence of known heart disease or other underlying medical condition. Sudden infant death syndrome (SIDS) represents a mechanistic spectrum of death in infants without suspected heart disease and in whom no cause can be found at standard necropsy. Recently, the use of "molecular autopsy" indicates that a minority of SIDS infants may exhibit one of the genetic cardiac ion channel mutations that can cause lethal arrhythmias. Resuscitated or aborted sudden cardiac death implies successful interruption of SCD by resuscitative efforts, and connotes a high recurrence risk.

The incidence of SCD in young persons is unknown. An estimated 4,000 to 8,000 children (≤ 18 years) in the United States die from sudden cardiac death annually, compared to more than 300,000 older individuals, most of whom die from ventricular fibrillation due to ischemic heart disease. In a review of nine reports, the rate of sudden death in children varied from 1.3 to 8.5 per 100,000 patient-years. Although much less common than in adults, SCD in children and young persons is more varied in both mechanism and cause. In contrast to adults, terminal bradycardia and ventricular fibrillation following profound bradycardia or asystole are not uncommon as a mechanism of SCD in newborns, infants, and young children (Figure 1).

SCD in athletic, competitive young persons is also rare. There are approximately ten million children and adolescents that participate in competitive athletics in the U.S. In the high school male athlete, the risk of SCD is less than 1 in 100,000 patient-years; the risk for female athletes is only a fraction of that. In most cases of SCD in children and young competitive athletes, a specific, often

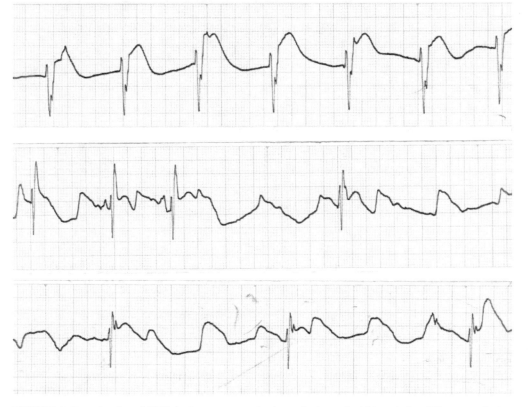

FIGURE 1. Several strips of ECG tracing over several minutes in 1-month-old boy during failing resuscitative efforts.

congenital, cardiovascular cause can be determined.

The causes of SCD in the young (Table 1) vary with age. In the infant (≤1 year), approximately one half may have coronary artery anomalies and in the other half, no structural cause can be identified, even after a thorough review of the history, death scene and a complete autopsy. This later group, by definition, is SIDS, and, as noted above, a "molecular autopsy" may identify a minority with a genetic cardiac ion channel mutations such as the long QT syndrome. In patients less than 21 years of age, the most frequent causes are hypertrophic cardiomyopathy, myocarditis, primary electrical disturbances, coronary artery abnormalities, and known, preexisting, structural congenital heart disease.

HYPERTROPHIC CARDIOMYOPATHY

Hypertrophic cardiomyopathy is a diverse disease in its genetic, biochemical, cellular, phenotypic, and clinical features. It is a result of different mutations in one of at least ten genes that encode protein elements of the cardiac sarcomere, and is inherited, as most defects in structural proteins, as an autosomal dominant trait (Chapter 18). Although it is relatively common, with an estimated prevalence of 1 in 500, it does not uniformly portend an unfavorable prognosis; overall, the annual mortality is about 1%, with a near normal life expectancy and little morbidity in most patients. There exists, however, a subset of patients, 10%–20% of those with hypertrophic

TABLE 1. Causes of SCD in the Young

Condition	Types
Myocardial Abnormalities	Hypertrophic Cardiomyopathy Dilated Cardiomyopathy Myocarditis Ventricular Noncompaction
Primary Electrical Disorders	Long QT Syndrome: Congenital and Acquired Ventricular Pre-excitation (WPW) Brugada Syndrome Catecholaminergic Ventricular Tachycardia Primary Ventricular Tachycardia/Fibrillation Complete Heart Block, Congenital and Surgical Arrhythmogenic Right Ventricular Dysplasia Commotio Cordis

Coronary Abnormalities	Congenital Anomalies
	    Anomalous origins from the aorta
	    Anomalous origins from the pulmonary artery
	    Ostial abnormality
	    Myocardial bridge (tunneled)
	Kawasaki Disease
	Transplant Vasculopathy
	Cocaine Vasospasm
	Coronary Atherosclerosis

Congenital Heart Disease	Aortic Stenosis Coarctation of the Aorta Tetralogy of Fallot (including variants) Transposition of the Great Arteries Atrioventricular Discordance Hypoplastic Left Heart Syndrome Single Ventricle/Fontan Ebstein's Anomaly Truncus Arteriosus
Miscellaneous	Primary Pulmonary Hypertension Marfan's Syndrome Eisenmenger Syndrome Mitral Valve Prolapse

cardiomyopathy, in whom the annual mortality from SCD exceeds 5%.

A sudden arrhythmia is the most frequent means of sudden death, too often as the first clinical manifestation of the disease, and can occur at rest or during mild or strenuous exertion. It occurs most commonly in young adults, adolescents, and children, a population with an estimated annual risk of mortality secondary to SCD of 2%–5%. Primary ventricular tachycardia and its degeneration to ventricular fibrillation is the predominant mechanism of SCD. Patchy and confluent myocardial scars, as a result of abnormal coronary architecture, diminished coronary flow reserve, coronary-myocardial mismatch and consequent myocardial ischemia, in addition to cellular disarray and collagen infiltration, serve as electrophysiologic substrates for the ventricular arrhythmias. Acute ischemia and vascular instability with hypotension are proposed triggers for ventricular arrhythmias.

Substantial effort has lead to the recognition of risk factors for SCD in individuals with hypertrophic cardiomyopathy (Table 2). Electrophysiology testing for inducible ventricular

TABLE 2. Risk Factors for SCD in Individuals with Hypertrophic Cardiomyopathy

Risk Factors

History of resuscitated cardiac arrest
Family history of SCD (particularly first-degree or multiple relatives)
History of sustained ventricular tachycardia
Nonsustained ventricular tachycardia
Syncope or presyncope (particularly if exertional, recurrent, or with documented arrhythmia)
Hypotensive exercise blood pressure response
Left ventricular wall thickness 30 mm or greater

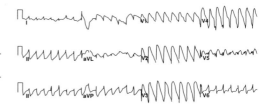

FIGURE 2. 12 lead electrocardiogram in a 7 month old boy with autopsy proven myocarditis. Note wide QRS tachycardia and the global ST elevations in the anterior lateral leads with reciprocal changes in the other leads.

arrhythmias is controversial; monomorphic ventricular tachycardia is infrequently induced and polymorphic tachycardia with aggressive extrastimulus protocols is a nonspecific finding.

Most patients are free of risk factors and SCD in such patients is rare, but those with two or more risk factors are at substantially increased risk. Patients with no risk factors have been shown to have a six-year SCD-free survival rate of 95%, while the corresponding rate for those with two or more risk factors was 72%. Implantable cardioverter defibrillator therapy is recommended for secondary prevention following events of resuscitated SCD and should be considered for primary prevention in those with two or more risk factors.

MYOCARDITIS

Acute and chronic myocarditis are associated with SCD. Acutely, myocyte necrosis and diffuse inflammatory infiltrates and, in the chronic form, patchy fibrosis and persistence of a low-grade inflammatory state, serve as an arrhythmogenic substrate for ventricular tachycardia (Figure 2). Ventricular ectopy or SCD may be the initial clinical manifestation of acute myocarditis and can occur in the presence of preserved ventricular function, confounding the diagnosis. Optimal treatment is unclear and the benefits of steroids and intravenous gamma globulin are unproven. Recovery can be complete, but is dependent upon the severity and persistence of the inflammatory process.

CONGENITAL LONG QT SYNDROME

The congenital Long QT syndrome represents a spectrum of genetic disorders (Chapter 18) that share in common the prolongation of ventricular repolarization due to abnormalities in potassium or sodium (LQT3) ion channels. The prolongation of repolarization can lead to delayed after-depolarization-mediated polymorphic ventricular tachycardia (torsades de pointes), which can degenerate into ventricular fibrillation and SCD.

Overall risk and environmental conditions and activities associated with ventricular arrhythmias and SCD have been shown, to an extent, to be gene specific. The risk of SCD is greatest in those with LQT3. SCD occurs most commonly during exercise among patients with LQT1, the most common subtype; there is also a tendency with LQT1 toward ventricular arrhythmias and SCD during recreational swimming. SCD occurs most commonly during emotional arousal or with auditory stimulus among those with LQT2 (Figure 1, Chapter 18), and during sleep or rest among patients with LQT3.

The genetic and phenotypic heterogeneity contributes to the difficulty in diagnosing and managing patients with the congenital long QT syndrome. The wide range of corrected QT interval values in both unaffected and affected individuals, as well as age-related variation in heart rate and sinus arrhythmia, also confound the diagnosis, particularly in the absence of symptoms or family history. Although a corrected QT interval of greater than 450–460 msec is a reasonable threshold to be concerned for an abnormality, the majority of individuals in this range are not affected.

Symptoms include syncope, seizures, and SCD. Patients with very prolonged event-free intervals or positive family history should be treated pharmacologically. In most cases, therapy involves β-blockade; the exception being individuals with LQT3 (Figure 3, Chapter 18), in whom ventricular arrhythmias are pause- or bradycardia-dependent and in whom permanent pacing may be beneficial. Mexilitine may be useful as well. In addition, secondary prevention of aborted SCD with an implantable defibrillator is indicated; the role of ICDs for primary prevention has not been established, but is a valid consideration, particularly for individuals with LQT3 and those in whom there is a strong family history of SCD. In the future, genetic testing will likely be instrumental in the diagnosis and guidance of therapy.

BRUGADA SYNDROME

The Brugada syndrome is characterized by the electrocardiographic findings of right bundle branch block and unusual ST elevation in leads V1–V3 (Figure 4, Chapter 18), polymorphic ventricular tachycardia, a structurally normal heart, and autosomal dominant inheritance with variable penetrance. It has emerged as an important cause of SCD worldwide and is the leading cause of SCD in South Asian males. Defects in the SCN5A human cardiac sodium channel gene result in loss of function producing epicardial action potential prolongation, dispersion of ventricular myocardial refractoriness, and susceptibility to early after-depolarization-mediated ventricular tachycardia (phase 2 reentry). This is in contradistinction to the gain of function defect in the same gene associated with LQT3, which results in phase 2 prolongation, rendering susceptibility to delayed after-depolarization. Like the congenital long QT syndrome, the Brugada syndrome is representative of several described and undetermined genetic defects. Unfortunately, the disorder is also difficult to diagnose, as the electrocardiographic features may manifest only occasionally. Provocative testing with ajmaline, flecainide or procainamide may uncover repolarization abnormalities in affected individuals. The use of implantable cardiovertor defibrillators is indicated for secondary prevention following resuscitated SCD and should be strongly considered for primary prevention as well since there is no preventive pharmacologic agent.

WOLFF-PARKINSON-WHITE SYNDROME

Patients with Wolff-Parkinson-White (WPW) syndrome have a very low risk of SCD, probably less than 1 per 1000 patient-years. The mechanism is rapid anterograde conduction across a fast-conducting accessory pathway(s) during atrial fibrillation, either spontaneous or initiated by atrioventricular reentrant tachycardia. The lack of sensitivity and specificity of noninvasive markers of SCD risk, such as intermittent loss of ventricular preexcitation or response to exercise testing or antiarrhythmic medications, limits their clinical usefulness. Patients at risk can be identified using electrophysiology study to determine the presence of multiple accessory connections, the inducibility of atrial fibrillation, the anterograde effective refractory period, and the shortest pre-excited RR interval during atrial fibrillation or rapid atrial pacing (Figure 3). Patients with slow anterograde accessory pathway conduction (accessory pathway effective refractory period >280 msec) are not at risk. On the other hand, the specificity for development of ventricular fibrillation with short pre-excited RR intervals during electrophysiology study is low.

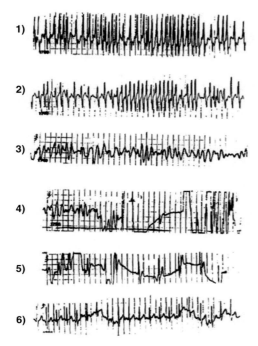

FIGURE 3. Panels 1-2: Atrial fibrillation in a 9-year-old boy with a very rapid ventricular response deteriorating into ventricular fibrillation (Panels 3 and 4) and cardiac arrest. He was successfully defibrillated (Panels 4 and 5) into sinus tachycardia with no preexcitation (Panel 6). At electrophysiologic study had an intermittent anterograde left lateral accessory pathway with an effective refractory period of 190 msec. It was successfully interrupted by radiofrequency ablation. Reprinted with permission from O'Connor, BK et al. What every pediatrician should know about supraventricular tachycardia. Pediatric Annals 1991;20(7):368–376.

It has been shown that arrhythmia inducibility and young age (adolescents and young adults) are associated with a significantly increased risk of arrhythmic events, including SCD, in previously asymptomatic patients. However, because of a small atrial mass, atrial fibrillation is very uncommon in infants and young children. The indications for radiofrequency ablation in patients with WPW and SVT are individualized, depending upon the frequency, duration, and associated symptoms of arrhythmias, as well as patient and family preference. The indications for electrophysiology testing to determine risk of SCD are also individualized.

Family history, occupation, and involvement in athletic and other risk-laden activities are important considerations in making an informed decision with patients and families. Guidelines for ablation in the young have recently been published by the Heart Rhythm Society (Chapter 22).

CONGENITAL CORONARY ARTERY ANOMALIES

Congenital coronary artery anomalies are diverse and occur with and without associated structural congenital heart disease. The congenital coronary anomaly most strongly associated with SCD is anomalous origin of the left coronary artery from the right sinus of Valsalva. If the anomalous artery courses between the great arteries particularly a long intramural segment, the risk of SCD is increased (Figure 4). Alternative courses include anterior to the pulmonary artery, posterior to the aorta and in the septal myocardium. Some hypothesize that in the hyperdynamic state, during or following exercise, the vessel may become compressed, resulting in acute ischemia and ventricular arrhythmias. Origin of the right coronary artery from the left coronary artery or sinus that courses between the great arteries is also associated with SCD.

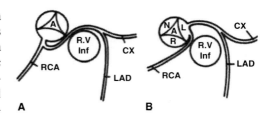

FIGURE 4. Panel A: Left coronary artery arising from the right coronary artery and coursing between the aorta and right ventricular outflow tract. Panel B: The right artery arises from the left coronary cusp and courses between the aorta and right ventricular outflow tract. Reprinted with permission from "Congenital Coronary Artery Anomalies" by Herlong JR in The Science and Practice of Pediatric Cardiology, Lippincott, Williams and Wilkins. 2nd Ed, 1998, p. 1657.

Other congenital coronary anomalies implicated in SCD include an acute angle of origin, ostial stenosis, transmural course and "myocardial bridge". These subtle abnormalities are more frequently diagnosed by experienced pathologists and have been implicated in a significant number of infantile cases of SCD. However, the majority of experts consider the "myocardial bridge" to be a normal variant and not predictably causative of SCD.

When abnormalities are discovered either fortuitously or following resuscitated SCD, surgery can be curative. Unfortunately, SCD is often the initial manifestation.

ACQUIRED CORONARY ARTERY ANOMALIES

Prior to the therapeutic implementation of intravenous immunoglobulin for Kawasaki disease, reports of SCD early in the disease process and with the development of aneurysmal and thrombotic complications were more common. SCD in association with Kawasaki disease is now a rarity. Transplant coronary vasculopathy, particularly in an accelerated form, is associated with both acute ischemia and chronic myocardial scarring and is, as such, associated with SCD.

STRUCTURAL CONGENITAL HEART DISEASE

Tetralogy of Fallot

Tetralogy of Fallot has long been associated with ventricular arrhythmias and SCD. The risk is small, however, probably 1–2 per 1000 patient-years. Risk factors include significant chronic pulmonary or conduit insufficiency and right ventricular dilation, and prolonged and *prolonging* (associated with worsening volume overload) QRS duration (>170 msec). Other potential risk factors, including residual right ventricular pressure load and non-sustained ventricular ectopy are in question. The risk of SCD among patients more recently repaired appears to be substantially less, as there has been a trend toward transatrial repairs, smaller or no ventriculotomy incisions and preservation of right ventricular outflow valve function.

Transposition of the Great Arteries

Late postoperative arrhythmias, most commonly intra-atrial reentrant tachycardia (IART, Chapter 8) and sick sinus syndrome (SSS), are common following atrial switch (Mustard or Senning) procedures, as is systemic right ventricular dysfunction; both are risk factors for SCD, which occurs in 5–10 per 1000 patient-years. Mechanisms include IART with rapid atrioventricular conduction and subsequent hemodynamic and metabolic instability, and pause dependent dispersion of ventricular refractoriness associated with sick sinus syndrome. The incidence and risk factors of late SCD associated with the arterial switch (Jatene) operation are unknown but are thought to be considerably less. The risk of SCD with L-TGA (ventricular inversion, atrioventricular discordance) is variable and dependent on the anatomy, surgical repair(s), and conduction system disease, but it is neither well established nor negligible.

Fontan Operation

Single ventricle palliation with variations of the Fontan operation (cavopulmonary or atriopulmonary anastamosis) is associated with a high incidence of IART (Chapter 8), and sinus node and ventricular dysfunction. Patients are thus at risk for SCD secondary to mechanisms similar to those following atrial switch procedures. Lateral tunnel and external cardiac conduit Fontan connections, as well as strategically-placed surgical incisions or cryoablation lesions, may lessen the incidence of atrial arrhythmias and SCD.

Left Heart Obstructive Lesions

SCD occurs in association with unrepaired aortic stenosis, almost exclusively with moderately severe or severe left ventricular outflow obstruction. Resultant left ventricular hypertrophy results in acute and chronic subendocardial ischemia, a substrate for ventricular arrhythmias. Embolism in association with endocarditis is a known cause of SCD with aortic stenosis. Symptoms, including chest pain, presyncope, syncope and palpitations, should not be taken lightly. Associated coronary artery abnormalities, most commonly ostial stenosis, contribute to risk. Transcatheter or surgical relief of the obstruction alleviates, but does not nullify the risk of SCD, particularly among those with aortic regurgitation, residual left ventricular dysfunction, or a mechanical aortic valve.

SCD occurs infrequently following repair for coarctation of the aorta. Mechanisms include aneurysm rupture associated with aortoplasty and those associated with progressive left ventricular hypertrophy, despite relief of repair.

Hypoplastic left heart syndrome in the newborn left untreated is lethal. SCD can occur following the three stages of repair. It occurs most often following the Norwood procedure, when coronary perfusion is tenuous and reliant on balance between the pulmonary and systemic vascular resistances. Diastolic "runoff" to the low-resistance pulmonary bed compromises coronary (particularly subendocardial) perfusion. The Sano operation (right ventricular to pulmonary artery conduit, i.e., "shunt") is designed and advanced as an antidote to the diastolic "run off." Similarly, the mechanism of diastolic "run off" can lead to congestive heart failure, ventricular arrhythmias, and SCD in infants with unrepaired truncus arteriosus.

COMMOTIO CORDIS

A blunt, non-penetrating chest wall blow to an otherwise normal child or adolescent that results in SCD, *commotio cordis*, is a rare and unfortunate event. The mechanism involves a precisely timed precordial impact, typically of a low-energy, low-velocity, solid core projectile object during a narrow window of ventricular repolarization, just prior to the T-wave peak. Ventricular tachycardia and fibrillation, unusually recalcitrant to resuscitative efforts, follows the impact. Prompt defibrillation is the major survival determinant. Commotio cordis occurs most often in school age, adolescent, and young adult males participating in competitive athletics, particularly baseball, and less often in ice hockey and lacrosse (presumably because of fewer participants). A recent comprehensive report, however, highlights several cases that occurred under unusual circumstances that were not regarded as threatening, such as a snowball or the head of a pet dog impacting the chest of a small child. The importance of protective equipment that completely covers the precordium of "at risk" athletes such as baseball catchers and hockey and lacrosse goalies is clear.

PULMONARY HYPERTENSION

The risk of SCD is high among patients with pulmonary hypertension, both primary and associated with congenital heart disease (Eisenmenger disease). The mechanism involves an acute imbalance in systemic and pulmonary vascular resistances, resulting in diminished cardiac output or increased right to left shunt. Resultant hypoxia and acidosis in myocardium already suffering from perfusion abnormalities provides a substrate for lethal arrhythmias. Risk factors include severely elevated pulmonary vascular resistance, previous symptoms such as syncope, and older age. Pregnancy, labor and the postnatal period, surgical procedures, and strenuous activity are particularly hazardous in at-risk patients. Therapeutic options include prostacyclins, Bosentan, and Sildenafil, "selective" pulmonary vasodilators which may promote beneficial remodeling, or, in the later stages, lung transplant. In the patient with Eisenmenger disease, there is no direct treatment.

SUGGESTED READING

1. Ackerman MJ, Tester DJ, Driscoll DJ. Molecular autopsy of sudden unexplained death in the young. *Am J Forensic Med Pathol* 2001;22(2):105–111.
2. Eggebrecht H, Mohlenkamp S. Images in clinical medicine. Myocardial bridging. *N Engl J Med* 2003;349(11):1047.
3. Humbert M, Sitbon O, Simonneau G. Treatment of pulmonary arterial hypertension. *N Engl J Med* 2004;30;351(14):1425–1436.
4. Liberthson RR. Sudden death from cardiac causes in children and young adults. *N Engl J Med* 1996;18;334(16):1039–1044.
5. Maron B, Pelliccia A, Spirito P. Cardiac disease in young trained athletes. Insights into the methods for distinguishing athlete's heart from structural heart disease, with particular emphasis on hypertrophic cardiomyopathy. *Circulation* 1995;91: 1596–1601.
6. Maron BJ. Sudden death in young athletes. *N Engl J Med* 2003;11;349(11):1064–1075.
7. Maron B, Poliac LC, Kaplan JA, Mueller FO. Blunt impact to the chest leading to sudden death from cardiac arrest during sports activity. *N Eng J Med* 1995;333:337–342.
8. Munger TM, Packer DL, Hammill SC, et al. A population study of the natural history of Wolff-Parkinson-White syndrome in Olmsted County, Minnesota, 1953-1989. *Circulation* 1993;87(3): 866–873.
9. Pappone C, Santinelli V, Manguso F, et al. A randomized study of prophylactic catheter ablation in asymptomatic patients with the Wolff-Parkinson-White syndrome. *N Engl J Med* 2003;349(19): 1803-1811.
10. Pappone C, Manguso F, Santinelli R, et al. Radiofrequency ablation in children with asymptomatic Wolff-Parkinson-White syndrome. *N Engl J Med* 2004;351(12):1197–1205.
11. Silka MJ, Hardy BG, Menashe VD, Morris CD. A population-based prospective evaluation of risk of sudden cardiac death after operation for common congenital heart defects. *J Am Coll Cardiol* 1998;32(1):245–251.
12. Silka MJ, Kron J, Walanc KCG, et al. Assessment and follow-up of pediatric survivors of sudden cardiac death. *Circulation* 1990;82(2):341–349.
13. Steinberger J, Lucas RV, Edwards JE, Titus JL. Causes of sudden unexpected cardiac death in the first 2 decades of life. *Am J Cardiol* 1996;77:992–995.
14. Tester DJ, Spoon DB, Valdivia HH, et al. Targeted mutational analysis of the RyR2-encoded cardiac ryanodine receptor in sudden unexplained death: a molecular autopsy of 49 medical examiner/coroner's cases. *Mayo Clin Proc* 2004;79(11): 1380–1384.
15. Young KD, Gausche-Hill M, McClung CD, Lewis RJ. A prospective, population-based study of the epidemiology and outcome of out-of-hospital pediatric cardiopulmonary arrest. *Pediatrics* 2004;114(1):157–164.
16. Wellens HJ. Catheter ablation for cardiac arrhythmias. *N Engl J Med* 2004;351(12):1172–1174.

21

Pharmacology of Antiarrhythmic Agents

Peter S. Fischbach

Cardiac arrhythmias result from alterations in the orderly sequence of depolarization and repolarization in the heart. The clinical severity of disordered cardiac activation range from asymptomatic palpitations to lethal arrhythmias. Clinicians have a number of therapeutic options from which to choose in an effort to suppress and/or eliminate the sources or structures that support the arrhythmias, as well as to revert the heart to normal rhythm if other therapies fail (Table 1). While recent technological advances have led to an increase in the use of nonpharmacological strategies including transcatheter radiofrequency or cryothermal ablation, intraoperative cryoablation as well as implantable pacemakers and defibrillators, pharmacological therapy remains a valuable tool for monotherapy or as adjunctive therapy in combination with device therapy. Pharmacological management of arrhythmias utilizes drugs that exert direct effects on cardiac cells by inhibiting the function of specific ion channels or ion pumps or by altering the autonomic input into the heart. Physicians caring for patients with arrhythmias therefore must understand and appreciate the benefits and risks provided by each therapeutic agent, what is the indication for each, and how they interact.

Antiarrhythmic drugs, like many drugs currently used in pediatric medicine, rely on data from adult studies for dosing and efficacy. There are relatively limited data concerning anti-arrhythmic drug use in children. Nearly all of the studies addressing the clinical efficacy of antiarrhythmic medications in children are retrospective. Large controlled studies comparing different drugs and dosing schedules are lacking.

The Vaughn-Williams system divides antiarrhythmic drugs into four classes based on their predominant mechanism of action. The classification system is, however, an oversimplification and does not address several drugs whose actions cross over multiple groups. Thus, although the grouping of antiarrhythmic agents into four classes is convenient, it should be understood that such a classification falls short of explaining the underlying mechanisms by which many drugs ultimately exert their therapeutic antiarrhythmic effect.

CLASS I

Class I antiarrhythmic drugs block the voltage gated sodium channel delaying phase 0 of the action potential and thereby slow conduction velocity in the tissue. These drugs act when the channel is either in the open or inactivated state rather than in the resting state.

TABLE 1. Pharmacological Therapy

	Acute therapy	Chronic therapy
Heart Rate	±	↓↓↓
Qtc	±	↑↑↑
AV node conduction	± to ↓	↓↓
Accessory pathway ERP	↑	↑↑↑
β–blockade	++	+++
Negative inotropic effect	+	±

Sodium channel inhibition also prolongs the effective refractory period of fast-response fibers by necessitating a more hyperpolarized membrane potential (more negative) be achieved prior to a return of excitability. Although many class I antiarrhythmic drugs possess local anesthetic actions and can depress myocardial contractile force, these effects are usually observed only at higher plasma concentrations. In addition to the effects on conduction velocity, class I drugs also suppress both normal Purkinje fiber and His bundle automaticity in addition to abnormal automaticity resulting from myocardial damage. Suppression of abnormal automaticity permits the sinoatrial (SA) node to resume the role of the dominant pacemaker.

Class I antiarrhythmic agents are subdivided into three groups: (1) Class IA drugs slow the rate of rise of phase 0 (V_{max}) of the action potential and prolong the refractory period; (2) Class IB drugs have a minimal effect on V_{max} and the refractory period of healthy myocardium while causing conduction block in diseased myocardium; and (3) Class IC drugs cause a marked depression in the conduction velocity with minimal effects on refractoriness in all cardiac tissue.

Class IA

Quinidine

Quinidine (*Quinidex*) is the dextro-isomer of quinine and was one of the first clinically used antiarrhythmic agents. Due to the high incidence of ventricular proarrhythmia and numerous equally efficacious agents, quinidine is now used sparingly. Quinidine shares all of the pharmacological properties of quinine, including anti-malarial, antipyretic, oxytocic, and skeletal muscle relaxant actions.

Electrophysiological Actions. Quinidine's effect depends on the parasympathetic tone and the dose. The anticholinergic actions of quinidine predominate at lower plasma concentrations and direct electrophysiological actions predominate at higher serum levels.

SA node and atrial tissue: At low concentrations a slight increase in heart rate results from the anticholinergic effects while at higher concentrations spontaneous diastolic depolarization is slowed. Quinidine slows the V_{max} of phase 0 slowing conduction through all tissue. Quinidine also has "local anesthetic" properties.

AV node: The anti-cholinergic effect of quinidine enhances conduction through the AV node. Quinidine's direct electrophysiological actions on the AV node decreases conduction velocity and increases the ERP.

His-Purkinje system and ventricular muscle: Quinidine decreases the slope of phase 4 depolarization, inhibiting automaticity. Depression of automaticity in the His-Purkinje system is more pronounced than depression of SA node pacemaker cells. Quinidine also prolongs repolarization in ventricular muscle resulting in an increase in the duration of the action potential and QT interval on ECG a result of blocking the delayed rectifier potassium channel (I_{Kr}).

Electrocardiographic Changes. Quinidine prolongs the PR, QRS, and QT intervals. QRS and QT prolongation is more pronounced than with other antiarrhythmic agents. The magnitude of prolongation is directly related to the plasma concentration.

Hemodynamic Effects. Myocardial depression is not a problem in patients with normal cardiac function while patients with compromised myocardial function may experience a decrease in cardiac function. Quinidine relaxes vascular smooth muscle directly as well as indirectly by inhibition of alpha-1-adrenoceptors.

Pharmacokinetics. Quinidine has nearly complete oral bioavailability with an onset of action within 1–3 hours, and peak effect within 1–2 hours. The plasma half-life is 6 hours with primarily hepatic metabolism. Therapeutic serum concentrations are 2–4 µg/mL.

Clinical Uses. The use of quinidine is limited by the poor side effect profile and the availability of equally or more efficacious agents. Quinidine may be used in combination with other agents such as mexilitine for the control of ventricular arrhythmias. Since the CAST study, the use of quinidine has declined. Currently, the inclusion of quinidine should be limited to patients with ICDs due to the significant risk of pro-arrhythmia. More recently, it may be useful in the patients with short QT syndrome (Chapter 18).

Adverse Effects. The most common adverse effects are diarrhea, upper-gastrointestinal distress, and light-headedness. Other relatively common adverse effects include fatigue, palpitations, headache, angina-like pain, and rash. These adverse effects are dose-related and reversible with cessation of therapy. Thrombocytopenia may also occur.

The cardiac toxicity of quinidine includes AV and intraventricular block, ventricular tachyarrhythmias, and depression of myocardial contractility. Ventricular proarrhythmia with loss of consciousness, referred to as "quinidine syncope," is more common in women and may occur at therapeutic or subtherapeutic plasma concentrations.

Large doses of quinidine can produce a syndrome known as cinchonism, which is characterized by ringing in the ears, headache, nausea, visual disturbances or blurred vision, disturbed auditory acuity, and vertigo. Larger doses can produce confusion, delirium, hallucinations, or psychoses. Quinidine can also cause hypoglycemia.

Contraindications. One absolute contraindication is complete AV block with a junctional or idioventricular escape rhythm that may be suppressed leading to cardiac arrest. Persons with congenital QT prolongation may develop torsades de pointes and should not be exposed to quinidine. Owing to the negative inotropic action of quinidine, the drug is contraindicated in congestive heart failure and hypotension. Digitalis intoxication and hyperkalemia accentuate the effect of quinidine on conduction velocity. The use of quinidine and quinine should be avoided in patients who previously have previously shown evidence of quinidine-induced thombocytopenia.

Drug Interactions. Quinidine increases the plasma concentrations of digoxin, requiring a downward adjustment in the digoxin dose. Drugs that inhibit the hepatic metabolism of quinidine and increase the serum concentration include acetazolamide, certain antacids (magnesium hydroxide and calcium carbonate), and cimetidine. Phenytoin, rifampin, and barbiturates increase the hepatic metabolism of quinidine and reduce its plasma concentrations.

Procainamide

Procainamide (*Pronestyl, Procan SR*) is a derivative of the local anesthetic agent procaine. Procainamide compared with procaine has a longer half-life, does not cause CNS toxicity at therapeutic plasma concentrations, and is effective orally. Procainamide is effective in the treatment of supraventricular, ventricular, and digitalis-induced arrhythmias. Its use is limited by its short serum half-life and frequent side effects when used chronically.

Electrophysiological Actions. Procainamide's direct electrophysiological effects are nearly identical to quinidine's, although it has a significantly weaker anti-cholinergic effect. The ECG changes are similar to quinidine.

TABLE 2. Class I Drugs

Drug	Class	Dose	$T_{1/2}$	Route of elimination	Therapeutic serum levels	ECG changes		
						PR	QRS	QTc
Quinidine gluconate	Ia	PO: 10–30 mg/kg/d ÷ bid-tid	6 hrs	H	2–6 mg/mL	±	↑↑	↑↑↑
Procainamide	Ia	PO: 15–50 mg/kg/d ÷ tid-qid IV: load: 7–15 mg/kg over 1 hr Infusion: 20–100 mcg/kg/min	2.5–4.5 hrs (6–8 hrs for SR)	HR	4–8 mcg/ml NAPA: 4–8 mcg/mL	±	↑	↑↑
Lidocaine	Ib	IV: Load = 1 mg/kg (may repeat × 3), infusion = 20–50 mcg/kg/min	1–2 hrs	H	1.5–6.0 mcg/mL	±	±	±
Mexiletine	Ib	PO: 6–15 mg/kg/d ÷ tid	10–12 hrs	HB	0.5–2.0 mcg/mL	±	±	±
Flecainide	Ic	PO: 4–6 mg/kg/d ÷ bid-tid	12–30 hrs	HR	0.2–1.0 mcg/mL	↑↑	↑↑	↑
Propafenone	Ic	PO: 8–10 mg/kg/d ÷ tid, may increase dose slowly to 20 mg/kg/d with careful monitoring	2–10 hrs	HR, 1/3 unchanged in urine	0.06–0.1 mcg/mL	↑↑	↑↑	±

H = Hepatic metabolism
HR = Hepatic metabolism with renal excretion
HB = Hepatic metabolism with biliary excretion
↑ = Increase
± = No significant change

Hemodynamic Effects. Hemodynamic compromise is less profound than with quinidine and seldom occurs after oral administration.

Pharmacokinetics. Procainamide is highly bio-available (75%–95%) with an onset of action of 5–10 minutes. The peak response following an oral dose is 60–90 minutes with a plasma half-life of 2.5–4.5 hours (6–8 hours for the sustained release preparation). The drug is metabolized hepatically and 50%–60% is excreted unchanged in the urine. The primary metabolite N-acetylprocainamide (NAPA) is cardioactive with class III properties and is eliminated unchanged in the urine. In patients who are rapid acetylators or have renal dysfunction, NAPA may accumulate more rapidly than procainamide. Therapeutic levels range from 4–8 µg/mL and may need to be slightly higher in neonates. NAPA levels should be considered separately from procainamide levels rather than combined and are also in the range of 4–8 µg/mL.

Clinical Uses. Procainamide is useful in the treatment of accessory pathway mediated tachycardia, atrial fibrillation of recent onset, all types of ventricular dysrhythmias, and combined with patient cooling for the treatment of post-operative junctional ectopic tachycardia.

Care should be used when initiating therapy in patients with atrial flutter or IART as procainamide may slow conduction in the flutter circuit allowing for 1:1 AV conduction and an increase in the ventricular rate. Additionally, procainamide may slow conduction velocity in other macrorentrant circuits (such as AVRT) and convert self-limited tachycardia into slower incessant tachycardia.

Intravenous administration for Brugada syndrome has emerged as a possible diagnostic test.

Adverse Effects. Acute cardiovascular reactions to procainamide administration include hypotension, AV block, intraventricular block, ventricular tachyarrhythmias, and complete heart block. The drug dosage must be reduced, or even stopped, if severe depression of conduction (severe prolongation of the QRS interval) or repolarization (severe prolongation of the QT interval) occurs. Long-term drug use may result in a clinical lupus like syndrome. The symptoms disappear within a few days of cessation of therapy.

Procainamide, unlike procaine, has little potential to produce CNS toxicity. Rarely, patients may experience mental confusion or hallucinations.

Contraindications. Contraindications are similar to those for quinidine. Procainamide should be administered with caution to patients with second-degree AV block and bundle branch block. The drug should not be administered to patients who have shown previous procaine or procainamide hypersensitivity. Prolonged administration should be accompanied by hematologic studies, since agranulocytosis may occur. Because of a potential hypotensive effect, intravenous administration should be titrated carefully monitoring blood pressure at no faster rate than 10 mg/kg/min.

Drug Interactions. Cimetidine inhibits the metabolism of procainamide. Simultaneous use of alcohol will increase the hepatic clearance of procainamide. The simultaneous administration of quinidine or amiodarone may increase the plasma concentration of procainamide.

Class IB

Lidocaine

Lidocaine (*Xylocaine*) is a local anesthetic that blocks sodium channels, binding to channels in both the open and inactivated state. Lidocaine, like other class 1B agents acts preferentially in diseased tissue causing conduction block and interrupting reentrant tachycardias.

Electrophysiological Actions

SA node and atrium: At therapeutic doses (1–5 mg/kg), lidocaine has no effect on the sinus rate and weak effects on atrial tissue.

AV node: Lidocaine has minimal effects on the conduction velocity and ERP of the AV node.

His-Purkinje system and ventricular muscle: Lidocaine reduces membrane responsiveness and decreases automaticity. Lidocaine in very low concentrations slows phase 4 depolarization in Purkinje fibers. In higher concentrations, automaticity may be suppressed, and phase 4 depolarization eliminated.

Electrocardiographic Changes. The PR, QRS, and QT intervals are usually unchanged, although the QT interval may be shortened in some patients. The paucity of electrocardiographic changes reflects lidocaine's lack of effect on healthy myocardium and conducting tissue.

Hemodynamic Effects. At usual doses, lidocaine does not depress myocardial function, even in the face of CHF.

Pharmacokinetics. Due to extensive first pass metabolism, lidocaine is not used orally. The onset of action is immediate when given intravenously with a plasma half-life of 1–2 hours. Elimination is primarily via the liver (90%) with the rest unchanged in the urine. Therapeutic serum levels range from 1.5–6.0 µg/mL. Lidocaine clearance is reduced by CHF, hepatic dysfunction, and concomitant treatment with cimetidine or beta-blockers.

Clinical Uses. Lidocaine is useful in the control of ventricular arrhythmias. It is not useful for the treatment of supraventricular arrhythmias. Lidocaine's use has decreased as amiodarone is frequently being used primarily for post-operative ventricular ectopy.

Adverse Effects. CNS toxicity is the most frequent adverse effect. Paresthesias, disorientation, and muscle twitching may forewarn of more serious deleterious effects, including psychosis, respiratory depression, and seizures. Myocardial depression may occur at very high doses.

Contraindications. Contraindications include hypersensitivity to local anesthetics of the amide type (a very rare occurrence), severe hepatic dysfunction or a previous history of grand mal seizures due to lidocaine. Care must be used in the presence of second- or third-degree heart block as it may increase the degree of block and abolish all idioventricular pacemakers.

Drug Interactions. The concurrent administration of lidocaine with cimetidine, but not ranitidine, may cause an increase in the plasma concentration of lidocaine. The myocardial depressant effect of lidocaine is enhanced by phenytoin administration.

Mexiletine

Mexiletine (*Mexitil*) is a structural analog of lidocaine altered to prevent first pass metabolism. Mexiletine has properties similar to lidocaine and is frequently combined with quinidine to increase efficacy while decreasing the risk of pro-arrhythmia.

Electrophysiological Actions. Mexiletine slows conduction velocity with a negligible effect on repolarization. Mexiletine demonstrates a rate-dependent blocking action on the sodium channel with rapid onset and recovery kinetics.

Hemodynamic Effects. Although its cardiovascular toxicity is minimal, the drug should be used with caution in patients who are hypotensive or who exhibit severe left ventricular dysfunction.

Pharmacokinetics. Mexiletine has an oral bioavailability of 90%. Its onset of action is 0.5–2.0 hours with a plasma half-life of 10–12 hours. Mexiletine is metabolized in the liver and excreted in the bile with 10% renal excretion. Therapeutic serum concentrations range from 0.5–2.0 µg/mL.

Clinical Uses. Mexiletine is useful in the management of both acute and chronic ventricular arrhythmias. While not currently an indication for use, there is interest in using mexiletine to treat the congenital long QT syndrome caused by a mutation in the SCN5A gene (*LQTS 3*).

Adverse Effects. A very narrow therapeutic window limits Mexiletine use. The first signs of toxicity are a fine tremor of the hands, followed by dizziness and blurred vision. Side effects include upper-gastrointestinal distress, tremor, light-headedness, and coordination difficulties. These effects generally are not serious and can be reduced by downward dose adjustment or administering the drug with meals. Cardiovascular-related adverse effects are less common and include palpitations, chest pain, and angina or angina-like pain.

Contraindications. Mexiletine is contraindicated in the presence of cardiogenic shock or preexisting second- or third-degree heart block in the absence of a cardiac pacemaker. Caution must be exercised in administration of the drug to patients with sinus node dysfunction or disturbances of intraventricular conduction.

Drug Interactions. An upward adjustment in dose may be required when mexiletine is administered with phenytoin or rifampin, due to increased hepatic metabolism of mexiletine.

Class IC

Flecainide

Flecainide (*Tambocor*) slows conduction throughout the heart, most notable in the His-Purkinje system and ventricular myocardium. Flecainide also weakly inhibits the delayed rectifier potassium channel (slightly prolonging repolarization) and inhibits abnormal automaticity.

Electrophysiological Actions

SA node and atrium: Flecainide causes a clinically insignificant decrease in heart rate. In the atrium, flecainide decreases the conduction velocity, shifts the membrane responsiveness curve to the right, and prolongs the action potential in a use-dependent fashion.

AV node: Atrioventricular conduction is prolonged.

His-Purkinje system and ventricular muscle: Flecainide slows conduction in the His-Purkinje system and ventricular muscle to a greater degree than in the atrium. Flecainide may also cause block in accessory AV connections, which is the principal mechanism for its effectiveness in treating atrioventricular reentrant tachycardia.

Electrocardiographic Changes. Flecainide increases the PR, QRS, and to a lesser extent, the QTc intervals. The rate of ventricular repolarization is not affected and the QT interval prolongation is caused by the increase in the QRS duration.

Hemodynamic Effects. Flecainide produces modest negative inotropic effects that may become significant in the subset of patients with compromised left ventricular function.

Pharmacokinetics. Flecainide is well absorbed with a bioavailability of 85%–90%. Oral absorption may be inhibited by milk and milk-based formulas. The onset of action is 1–2 hours with a serum half-life of 12–30 hours. The drug is primarily metabolized in the liver and excreted in the urine. Therapeutic serum concentrations are 0.2–1.0 µg/mL.

Clinical Uses. Flecainide is effective in treating atrial arrhythmias, particularly those supported by reentrant mechanisms, and is also used for life threatening ventricular arrhythmias. Based on the results of the CAST study in adults with ischemic heart disesase, and several reports of pro-arrhythmia in patients with repaired congenital heart disease, flecainide should be used with caution in patients with congenital heart disease. Flecainide crosses the placenta with fetal levels approximately 70% of maternal levels and in many centers is the second-line drug after digoxin for therapy of fetal arrhythmias. Flecainide is also the second line drug for SVT in children who are not well controlled on beta-blockers in many centers. Due to the possibility of proarrhythmia, initiation of therapy or significant increases in dosing should be performed as an inpatient.

Adverse Effects. Most adverse effects are observed within a few days of initial drug administration and include dizziness, visual disturbances, nausea, headache, and dyspnea. Worsening of heart failure and prolongation of the PR and QRS intervals may occur. The risk of pro-arrhythmia appears to be less than that observed in the adult population. The most frequent pro-arrhythmic effect is the occurrence of slow incessant SVT. Ventricular arrhythmias have been observed in patients following repair of congenital heart disease.

Contraindications. Flecainide is contraindicated in patients with preexisting second- or third-degree heart block unless a pacemaker is present to maintain ventricular rhythm. The drug should not be used in patients with cardiogenic shock.

Drug Interactions. Cimetidine may reduce the rate of flecainide's hepatic metabolism, thereby increasing the potential for toxicity. Flecainide may increase digoxin concentrations.

Propafenone

Propafenone (*Rythmol*) blocks the sodium channel, and like flecainide, propafenone weakly blocks potassium channels. Additionally propafenone is a weak β-receptor antagonist and L-type calcium channel blocker.

Electrophysiological Actions

SA node: Propafenone causes sinus node slowing.
Atrium: The action potential duration and effective refractory period are prolonged while the conduction velocity is decreased. Similar to other class I drugs, these effects may slow atrial flutter to a rate that allows for more rapid conduction into the ventricle.

AV node: Intravenous administration slows conduction through the AV node.
His-Purkinje system and ventricular muscle: Propafenone slows conduction and inhibits automatic foci.

Electrocardiographic Changes. Propafenone causes dose-dependent increases in the PR and QRS intervals.

Hemodynamic Effects. In the absence of cardiac abnormalities, propafenone has no significant effects on cardiac function. Intravenous administration may result in a decrement in decreased myocardial performance in patients with ventricular dysfunction.

Pharmacokinetics. Propafenone is nearly 100% absorbed following an oral dose. It has a serum half-life of 2–10 hours. It is metabolized in the liver with nearly one-third of the drug excreted unchanged in the urine. Therapeutic serum concentrations are 0.06–0.10 µg/mL.

Clinical Uses. Propafenone is useful for the treatment of supraventricular arrhythmias and life threatening ventricular arrhythmias in the absence of structural heart disease. Propafenone should be used with caution in patients with congenital heart disease due to the increased risk of ventricular pro-arrhythmia. Like flecainide, therapy should be initiated as an inpatient.

Adverse Effects and Drug Interactions. Concurrent administration of propafenone with digoxin, warfarin, propranolol or metoprolol increases the serum concentrations of the latter four drugs. Cimetidine slightly increases the propafenone serum concentrations. Additive pharmacological effects can occur when lidocaine, procainamide, and quinidine are combined with propafenone. As with other members of class IC, propafenone may interact in an unfavorable way with other agents that depress AV nodal function, intraventricular conduction, or myocardial contractility. The most common adverse effects are dizziness, light-headedness, a metallic taste, nausea, and vomiting.

Contraindications. Propafenone is contraindicated in the presence of severe congestive heart failure, cardiogenic shock, atrioventricular and intraventricular conduction disorders, and sick sinus syndrome. Other contraindications include severe bradycardia, hypotension, obstructive pulmonary disease, and hepatic and renal failure. Because of its weak β-blocking action, propafenone may cause dose-related bronchospasm.

CLASS II

Class II antiarrhythmic drugs competitively inhibit β-adrenoceptors. In addition, some members of the group (e.g., propranolol and acebutolol) cause electrophysiological alterations in Purkinje fibers that resemble those produced by class I antiarrhythmic drugs. The latter actions have been referred to as "membrane-stabilizing" effects.

Class II: β-blockers

Propranolol

Propranolol (*Inderal, Inderal LA*) is the prototype β-blocker. It decreases the effects of sympathetic stimulation by competitively binding to β-adrenergic receptors.

Electrophysiological Actions. Propranolol has two separate and distinct effects. The first is a consequence of the drug's β-adrenergic receptor blocking properties and the subsequent removal of adrenergic influences on the heart. The second is associated with the direct myocardial effects (membrane stabilization) of propranolol. The latter action, especially at the higher clinically employed doses, may account for its effectiveness against arrhythmias in which enhanced β-receptor stimulation does not play a significant role in the genesis of the rhythm disturbance.

SA node: Propranolol slows the spontaneous firing rate of nodal cells by decreasing the slope of phase 4 depolarization.

Atrium: Propranolol possesses local anesthetic properties and decreases action potential amplitude and excitability. The serum concentrations at which the membrane stabilizing effects are evident are similar to those that produce β-blockade; hence, it is impossible to determine whether the drug acts by specific receptor blockade or via a "membrane-stabilizing" effect.

AV node: Propranolol administration results in a decrease in AV conduction velocity and an increase in the AV nodal refractory period.

His-Purkinje system and ventricular muscle: Propranolol at usual therapeutic concentrations produces a depression of catecholamine-stimulated automaticity. At supra-normal concentrations, propranolol decreases Purkinje fiber membrane responsiveness and reduces action potential amplitude.

Electrocardiographic Changes. The PR interval is prolonged with no change in the QRS interval. The QT interval may be shortened by propranolol administration.

Hemodynamic Effects. β-adrenoceptor blockade leads to a decrease in the positive inotropic and chronotropic effects of catecholamines. Clinically, the heart rate and blood pressure falls and the myocardial oxygen consumption is decreased.

Pharmacokinetics. Table 3 gives the metabolic mechanisms of the class II β-blockers.

Clinical Uses. Propranolol is useful for a wide spectrum of arrhythmias. Propranolol or atenolol (see below) is usually the initial therapy for SVT in all age groups. It is also effective in several forms of ventricular ectopy/tachycardia including the suppression of symptomatic PVCs and catecholamine dependent idiopathic VT. Propranolol is the drug of choice for treating patients with the congenital long QT syndrome.

Adverse Effects. Cardiac adverse effects include bradycardia and hypotension. Propranolol may result in bronchospasm in patients with asthma, which may be life threatening. Propranolol crosses the blood–brain barrier and is associated with mood changes and

TABLE 3. Class II Drugs

Drug	Receptor	$T_{1/2}$	Metabolized	Excreted	Dose
Propranolol	$\beta_1 + \beta_2$	3–5 hrs	Hepatic	Renal	1–4 mg/kg/d ÷ qid
Atenolol	β_1	9–10 hrs	Minimal	Renal	1–2 mg/kg/d ÷ bid
Nadolol	$\beta_1 + \beta_2$	20–24 hrs	Minimal	Renal	1 mg/kg/d ÷ qd-bid
Esmolol	β_1	7–10 minutes	Esterases in erythrocytes		Load: 500 mcg/kg over 1 min., maintenance 200–800 mcg/kg

depression. School difficulties may be seen in children. Propranolol may also cause hypoglycemia in infants. Since propranolol crosses the placenta and enters the fetal circulation, fetal cardiac responses to the stresses of labor and delivery are blunted.

Contraindications. Propranolol should be used with caution in patients with depressed myocardial function. It may be contraindicated in the presence of digitalis toxicity because of the possibility of producing complete AV block and ventricular asystole. It should be used with extreme caution in patients with asthma. An up-regulation of β-receptors follows long-term therapy making abrupt withdrawal of β-blockers potentially dangerous.

Atenolol

Atenolol (*Tenormin*) is selective for the ß$_1$-receptor. Atenolol's advantages relative to propranolol are its longer serum half-life and limited diffusion across the blood–brain barrier leading to a marked reduction in CNS effects.

Electrophysiological Actions and Electrocardiographic Changes. Identical to propranolol.

Pharmacokinetics. See Table 3.

Clinical Uses. Atenolol has been used for all supraventricular tachycardias and for control of ventricular ectopy. In many centers it is the drug of choice for the initial therapy of SVT.

Adverse Effects. The side effect profile is favorable compared with propranolol due to the lack of penetration across the blood–brain barrier. Despite its relative selectivity for the β_1-receptor, a worsening of bronchospasm may result and therefore it should be used with caution in patients with a history of reactive airway disease.

Contraindications. Relatively contraindicated in patients with reactive airway disease.

Nadolol

Nadolol (*Corgard*) is a long-acting nonselective β-adrenergic antagonist without membrane-stabilizing or intrinsic sympathomimetic activity

Electrophysiological Actions and Hemodynamic Effects. Similar to propranolol.

Pharmacokinetics. See Table 3.

Clinical Uses. Nadolol has been used for the treatment of various forms of supraventricular tachycardia and for patients with the long QT syndrome. Like atenolol, its long serum half-life and reduced CNS effects make nadolol an attractive alternative to propranolol.

Adverse Effects and Contraindications. Similar to propranolol.

Esmolol

Esmolol (*Brevibloc*) is a short acting, intravenously administered β_1-selective adrenoceptor blocking agent. It does not possess membrane-stabilizing activity or sympathomimetic activity.

Electrophysiological Actions and Hemodynamic Effects. Similar to propranolol.

Pharmacokinetics. See Table 3.

Clinical Uses. Esmolol is useful for the acute treatment of supraventricular and ventricular tachyarrhythmias, as well as for acutely lowering blood pressure. Discontinuation of administration is followed by a rapid reversal of its pharmacological effects because of its rapid hydrolysis by plasma esterases.

Adverse Effects and Contraindications. The most frequently reported adverse effects are hypotension, nausea, dizziness, headache, and dyspnea. As with many β-blocking drugs, esmolol is contraindicated in patients with overt heart failure or for those in cardiogenic shock.

CLASS III

Class III antiarrhythmic drugs prolong the duration of the membrane action potential by delaying repolarization without altering depolarization or the resting membrane potential. Class III drugs have a significant risk of pro-arrhythmia due to action potential duration prolongation and the induction of torsades de pointes (Table 4).

Amiodarone

Amiodarone (*Cordarone, Pacerone*) is an iodine-containing benzofuran derivative identified as a class III agent due to its predominant action potential prolonging effects. Amiodarone also blocks sodium and calcium channels, as well as being a non-competitive β-receptor blocker (class I, II, and IV actions). Amiodarone is an effective agent for the treatment of most arrhythmias. Toxicity associated with amiodarone has led the FDA to recommend that the drug be reserved for use in patients with life-threatening arrhythmias.

Electrophysiological Actions. The electrophysiological effects of amiodarone are complex and not completely understood. Interestingly, the acute effects differ significantly from the chronic effects. The most notable electrophysiological effect of amiodarone after long-term administration is a prolongation of repolarization and refractoriness in all cardiac tissues, an action that is characteristic of class III antiarrhythmic agents.

SA node: Amiodarone and its metabolite desethylamiodarone inhibit nodal function. It may profoundly inhibit SA nodal activity in patients with underlying sick sinus syndrome and require permanent pacing due to hemodynamically significant bradycardia. This is a common problem in patients following the Fontan and atrial switch operations.

His-Purkinje system and ventricular muscle: Amiodarone and desethylamiodarone increase AV nodal conduction time and refractory period. The dominant effect on ventricular myocardium with chronic treatment is a prolongation in the action potential duration and increase in the refractory period with a modest decrease in conduction velocity.

Electrocardiographic Changes. The predominant electrocardiographic changes include a prolongation of the PR and QTc intervals, the development of U-waves, and changes in T-wave contour.

Hemodynamic Effects. Amiodarone relaxes vascular smooth muscle and improves regional myocardial blood flow. In addition, its effects on the peripheral vascular bed lead to a decrease in left ventricular stroke work and myocardial oxygen consumption. Intravenous administration may be associated with hypotension requiring volume expansion.

Pharmacokinetics. The pharmacokinetic characteristics of amiodarone are extremely complex. Absorption is slow and the oral bioavialabiltiy is low (35%–65%). The

TABLE 4. Class III Drugs

Drug	Dose	$T_{1/2}$	Therapeutic serum levels	Route of elimination	PR	QRS	QTc
Amiodarone	Loading: PO: 10–20 mg/kg/d bid × 5–10 days. IV: 5 mg/kg over 1/2 hour (may repeat ×1) Maintenance: PO: 5 mg/kg qd, IV: 10–15 mg/kg/d	26–107 days	0.5–2.5 mcg/mL	B	Acute: ± Chronic: ↑	± ↑	± ↑↑↑
Sotalol	PO: starting dose = 2 mg/kg/d bid, may increase slowly to 8 mg/kg/d	9.5 hrs (in children)	None established	HR	↑	±	↑↑↑
Dofetilide	No pediatric dosing. For creatinine clearance >60 mL/min = 500 mcg bid, for creatinine clearance 40–60 mL/min = 250 mcg bid	7–10 hrs	None established	R	±	±	↑↑↑
Ibutilide	>60 kg = 1 mg IV over 10 minutes, repeat ×1 if necessary 10 minutes after completion. <60 kg = 10 mcg/kg IV over 10, repeat ×1 if necessary 10 minutes after completion	6 hrs	Serum levels not related to clinical efficacy	HR	±	±	↑↑↑

B = Biliary
HR = Hepatic metabolism, renal excretion
R = Renal
± = No significant change
↑ = Increase

drug is almost completely protein bound and is concentrated in the myocardium (10–50x serum concentration), as well as in adipose tissue, the liver, and lungs (100–1000x serum concentration). The serum half-life ranges from 26–107 days with chronic administration. The primary route of metabolism is hepatic with excretion via the biliary tract. Therapeutic serum concentrations are 0.5–2.5 µg/mL.

Clinical Uses. Amiodarone is effective in a wide variety of cardiac rhythm disorders with minimal tendency for induction of torsades de pointes. The incidence of ventricular pro-arrhythmia is significantly less than other class III agents. Its use, however, is limited by the multiple and severe non-cardiac side effects.

Intravenous amiodarone has been used to treat a wide range of arrhythmias, particularly in the post-operative period including supraventricular tachycardia, atrial flutter, atrial fibrillation, intra-atrial reentrant tachycardia, junctional ectocpic tachycardia, and ventricular tachycardia. Chronic oral amiodarone administration is more efficacious than intravenous use. Oral amiodarone is effective in most forms of supraventricular and ventricular tachycardia with its use limited by the frequency and severity of its adverse effects. Because of its unknown effect on thyroid function and growth, use of amiodarone for treating SVT is reserved for patients who have failed several other medications and is time limited.

Adverse Effects. The most significant adverse effects include chemical hepatitis, worsening sinus node dysfunction, thyroid dysfunction (hypo or hyper), and pulmonary fibrosis (Table 5). Pulmonary fibrosis is frequently fatal and may not be reversed with discontinuation of therapy. Despite significant prolongation of the QT interval, the risk of torsades de pointes is relatively low.

Patients with underlying sinus node dysfunction tend to have significant worsening of nodal function, frequently requiring pacemaker implantation. Corneal microdeposits are common although complaints of halos or blurred vision are rare. The corneal microdeposits are reversible upon stopping the drug. Dermatological complaints are frequent including photosensitization and blue-gray discoloration. The risk is increased in patients of fair complexion. The discoloration of the skin regresses slowly, if at all, after discontinuation of amiodarone.

Amiodarone inhibits the peripheral and intra-pituitary conversion of thyroxine (T_4) to triiodothyronine (T_3) by inhibiting 5'-deiodination. The serum concentration of T_4 is increased by a decrease in its clearance, and increased synthesis due to a reduced suppression of the pituitary thyrotropin, T_3. The concentration of T_3 in the serum decreases and reverse T_3 appears in increased amounts. Despite these changes, the majority of patients appear to be maintained in a euthyroid state. Manifestations of both hypo- and hyperthyroidism have been reported.

Tremors of the hands and sleep disturbances in the form of vivid dreams, nightmares, and insomnia have been reported in association with the use of amiodarone. Ataxia, staggering, and impaired ambulation have also been noted. Peripheral sensory and motor neuropathy or severe proximal muscle weakness develops infrequently. Both neuropathic and myopathic changes are observed on biopsy. Neurological symptoms resolve or improve within several weeks of dosage reduction.

Contraindications. Amiodarone is contraindicated in patients with sick sinus syndrome and may cause severe bradycardia and second- and third-degree AV block. Amiodarone crosses the placenta causing fetal bradycardia and thyroid abnormalities. The drug is secreted in breast milk.

Drug Interactions. Amiodarone interferes with the metabolism of many drugs, most notably warfarin and digoxin. Patients receiving digoxin should have their dose decreased by 50%. Amiodarone also interferes

TABLE 5. Adverse Reactions

Adverse Reaction	Diagnosis	Incidence (%)	Screening	Management
Pulmonary	Cough, especially with local or diffuse infiltrates on CXR, suggesting interstitial pneumonitis; and decrease in DLCO from baseline.	1–20	Pulmonary function tests: Baseline and for any unexplained dyspnea, especially in patients with underlying lung disease; and if there are suggestive CXR abnormalities. CXR: Baseline and then yearly	Discontinue amiodarone. Consider corticosteroids.
GI	Nausea, anorexia, and constipation		History, physical exam	Symptoms may decrease with decreased dose
	AST or ALT elevations > 2 times normal	15–50	Liver function tests at baseline and every 6 months	Exclude other causes of hepatitis
	Hepatitis and cirrhosis	<3	Liver function tests at baseline and every 6 months	Discontinue amiodarone, consider liver bx to determine whether cirrhosis is present
Thyroid	Hypothyroidism	1–22	Thyroid function tests (T4 and TSH) at baseline and every 6 months	Thyroxine
	Hyperthyroidism	<3	Thyroid function tests (T4 and TSH) at baseline and every 6 months	Corticosteroids, propylthiouracil or methimaxole, may need thyroidectomy if cannot discontinue amiodarone
Skin	Blue discoloration	<10	Physical exam	Reassurance and sunblock
	Photosensitivity	25–75	Physical exam	Sunblock
CNS	Ataxia, paresthesia, peripheral polyneuropathy, sleep disturbances, impaired memory and tremor	3–30	Physical exam, history	Often dose dependent and improves or resolves with dose adjustment
Ocular	Halo vision, especially at night	<5	History, ophthalmologic exam at baseline if visual impairment is present or for symptoms	Corneal deposits are the norm; if optic neuritis occurs, discontinue amiodarone
	Optic neuritis	1	History, ophthalmologic exam at baseline if visual impairment is present or for symptoms	Discontinue amiodarone
Heart	Bradycardia and AV block (exaggerated effect in the face of existing sick sinus syndrome)	5	History, ECG (every 6 months), Holter for symptoms	May require permanent pacing
	Proarrhythmia (much less common than other class III agents)	<1	History, ECG (every 6 months), Holter for symptoms	Discontinue amiodarone
GU	Epididymitits and erectile dysfunction	<1	History and physical exam	Pain may resolve spontaneously

with the metabolism and elimination of flecainide, propafenone, procainamide, phenytoin, and quinidine.

Sotalol

Sotalol (*Betapace*) possesses nonselective β-adrenoceptor blocking properties in addition to class III actions via potassium channel blockade. The β-blocking effects are most evident at lower doses, with action potential prolonging effects predominating at higher doses.

Electrophysiological Actions

SA node and atrium: Pacemaker activity in the SA node is decreased and sotalol increases the refractory period of atrial muscle.

AV node: Sotalol decreases conduction velocity and prolongs the effective refractory period in the AV node.

His-Purkinje system and ventricular muscle: Sotalol's inhibition of the delayed rectifier potassium channel results in a prolongation of the effective refractory period in His-Purkinje tissue. Like other class III drugs, sotalol prolongs repolarization and increases the ERP of ventricular muscle.

Electrocardiographic Changes. Sotalol is associated with a dose and concentration-dependent decrease in heart rate and a prolongation of the PR and QTc intervals. The QRS duration is not affected with plasma concentrations within the therapeutic range.

Hemodynamic Effects. A modest reduction in systolic pressure and cardiac output may occur due to sotalol's β-adrenoceptor antagonist activity. Ventricular stroke volume is unaffected and the reduction in cardiac output is a consequence of the lowering of heart rate. In patients with normal ventricular function, cardiac output is maintained despite the decrease in heart rate due to a simultaneous increase in the stroke volume.

Pharmacokinetics. Sotalol has an oral bioavailability of 50% with an onset of action of 0.5 hours and a plasma half-life of 4 hours. The primary route of metabolism is hepatic with excretion primarily in the urine (20% unchanged and 40% as metabolite).

Clinical Uses. Sotalol possesses a broad spectrum of antiarrhythmic effects in ventricular and supraventricular arrhythmias. Use is limited by concerns for ventricular pro-arrhythmia. Sotalol is also used in many centers as second-line medication for fetal arrhythmias.

Adverse Effects. Side effects include those attributed to both β-adrenoceptor blockade and pro-arrhythmia. Other adverse effects of sotalol include, in decreasing order of frequency, fatigue, dyspnea, chest pain, headache, nausea, and vomiting.

Contraindications. Contraindications include severe heart failure or poor ventricular function. Use in patients with hypokalemia or prolonged QT intervals may be contraindicated, as they enhance the possibility of proarrhythmic events.

Drug Interactions. Drugs with inherent QT interval-prolonging activity (i.e., thiazide diuretics, and terfenadine) may enhance the class III effects of sotalol.

Dofetilide

Dofetilide (*Tikosyn*) is a "pure" class III drug. It prolongs the cardiac action potential and the refractory period by selectively inhibiting the rapid component of the delayed rectifier potassium current (I_{Kr}).

Electrophysiological Actions. Dofetilide blocks the cardiac ion channel carrying the rapid component of the delayed rectifier potassium current, I_{Kr}, over a wide range of concentrations with no significant effects on other repolarizing potassium currents.

The effects of dofetilide are exaggerated with hypokalemia and reduced with hyperkalemia. Dofetilide demonstrates reverse use dependence (i.e., less influence on the action potential at faster heart rates).

SA node and atrium: Dofetilide induces a minor slowing of the spontaneous discharge rate of the SA node via a reduction in the slope of the pacemaker potential and a hyperpolarization of the maximum diastolic potential. Dofetilide prolongs the plateau phase of the action potential thereby lengthening the refractory period of the myocardium. The effects on atrial tissue appear to be more profound than those observed in the ventricle. The reason for this is unclear.

AV node: There is no effect on the conduction through the AV node.

His-Purkinje system and ventricular muscle: Dofetilide increases the ERP of ventricular myocytes and Purkinje fibers. The ERP prolonging effects on the ventricular tissue is somewhat less than that in atrial tissue.

Electrocardiographic Changes. There are no changes in the PR or QRS intervals while the QT interval is prolonged. The increase in the QT interval is directly related to the dofetilide dose and plasma concentration.

Hemodynamic Effects. Dofetilide does not significantly alter the mean arterial blood pressure, cardiac output, cardiac index, stroke volume index, or systemic vascular resistance. There is a slight increase in the dP/dt in the ventricles.

Pharmacokinetics. The absorption of dofetilide is delayed by ingestion of food; however, the total bioavailability is not affected and is greater than 90%. The onset of action is a half hour with a plasma half-life of 7–10 hours. More than 60% of the drug is excreted unchanged in the urine with the remainder metabolized in the liver.

Clinical Uses. Dofetilide is approved for the treatment of atrial fibrillation and atrial flutter in adults. Due to the lack of significant hemodynamic effects, it may be useful in patients with CHF who are in need of therapy for supraventricular tachyarrhythmias. Dofetilide has been used in a few patients following Fontan operation with refractory IART with good results.

Adverse Effects. The incidence of non-cardiac adverse events is not different from that of placebo in controlled clinical trials. The principal cardiac adverse effect is the risk of torsades de pointes due to QT prolongation, which is approximately 3% in adult trials. Most pro-arrhythmic events are observed in the first 3 days. As such, initiation of therapy should be performed as an inpatient.

Contraindications. Contraindications include baseline prolongation of the QT interval or use of other QT prolonging drugs, history of torsades de pointes, creatinine clearance <20 mL/min, simultaneous use of verapamil, cimetidine, or ketoconazole, uncorrected hypokalemia (<4.0 mEq/100 mL), or hypomagnesemia and pregnancy or breast-feeding.

Drug Interactions. Verapamil increases serum dofetilide levels. Additionally, drugs that inhibit cationic renal secretion such as ketaconazole and cimetidine raise serum levels.

Ibutilide

Ibutilide (*Corvert*) is a structural analog of sotalol and produces cardiac electrophysiological effects similar to class III agents. Due to its significant first pass metabolism, ibutilide is only available as an intravenous preparation.

Electrophysiological Actions. Ibutilide prolongs action potential duration in isolated adult cardiac myocytes and increases both atrial and ventricular refractoriness in vivo. Ibutilide administration leads to activation of a slow, inward current (predominantly sodium) in addition to blocking the delayed rectifier potassium current. By prolonging the duration of sodium channel conductance during depolarization and by inhibiting outward potassium currents, the net effect is one of increasing the duration of atrial and ventricular action potentials and refractoriness.

SA node and atrium: There is no significant change in heart rate in healthy adult volunteers. Ibutilide causes an increase in the atrial effective refractory period with little if any reverse use dependence, which is different than most class III agents.

AV node: Experimental evidence suggests that ibutilide slows conduction through the AV node; however, there is no change in the PR interval on ECG.

His-Purkinje system and ventricular muscle: Ibutilide increases the ERP of ventricular myocytes and Purkinje fibers.

Electrocardiographic Changes. There are no changes in the PR or QRS intervals reflecting a lack of effect on the conduction velocity. Although there is no relationship between the plasma concentration of ibutilide and the antiarrhythmic effect, there is a dose-related prolongation of the QT interval. The maximum effect on the QT interval is a function of both the dose of ibutilide and the rate of infusion.

Hemodynamic Effects. Ibutilide has no significant effects on cardiac output, mean pulmonary arterial pressure, or pulmonary capillary wedge pressure in patients with and without compromised ventricular function.

Pharmacokinetics. The pharmacokinetics are highly variable between patients and due to extensive first pass metabolism, ibutilide is not suitable for oral administration. The drug is extensively metabolized by the liver and is excreted in the urine. It is 40% protein bound and has an elimination half-life of 6 hours (range: 2–12 hours).

Clinical Uses. Ibutilide is approved for the intravenous chemical cardioversion of recent onset atrial fibrillation and atrial flutter in adults. It appears to be more effective in terminating atrial flutter than atrial fibrillation. Ibutilide has also been demonstrated to lower the defibrillation threshold for atrial fibrillation resistant to chemical cardioversion. It has been used in a limited number of pediatric patients with congenital heart disease for the conversion of IART.

Adverse Effects. The major adverse effect associated with the use of ibutilide is the risk of torsades de pointes due to QT prolongation occurring in approximately 4% of adult patients usually within 40 minutes of initiating the infusion. Continuous ECG monitoring with the availability of equipment for urgent DC cardioversion is a necessity for up to 4–6 hours after administration. Other reported adverse cardiovascular events (all <2%) include hypo- and hypertension, brady- and tachycardia, and varying degrees of AV block. The incidence of non-cardiac adverse events with the exception of nausea did not differ from that of placebo in controlled clinical trials.

Contraindications. Contraindications include baseline prolongation of the QTc interval, use of other QT prolonging drugs, history of torsades de pointes, or hypersensitivity to ibutilide, uncorrected hypokalemia (<4.0 mEq/100mL) or hypomagnesemia, pregnancy or breast-feeding.

Drug Interactions. No significant drug interactions.

CLASS IV

Class IV drugs block the slow inward Ca^{2+} current (L-type calcium channel). The most pronounced electrophysiological effects are exerted on cardiac cells dependent on the Ca^{2+} channel for initiating the action potential, such as those found in the SA and AV nodes. The administration of class IV drugs slows conduction velocity and increases refractoriness in the AV node, thereby reducing the ability of the AV node to conduct rapid impulses to the ventricle. This action may terminate supraventricular tachycardias and can slow conduction during atrial flutter or fibrillation.

Class IV Calcium Channel Blockers

Verapamil

Verapamil (*Isoptin, Covera*) selectively inhibits the voltage gated calcium channel,

TABLE 6. Class IV Drugs

Drug	Dose	$T_{1/2}$	Metabolized
Verapamil	5–15 mg/kg ÷ tid-qid	3–7 hrs	Hepatic
Diltiazem	IV: bolus: 0.25 mg/kg over 5 min Infusion: 0.05–0.15 mg/kg/hr P.O.: 1.5–2.0 mg/kg/d ÷ tid	4 hrs	Hepatic

vital for action potential genesis in slow response myocytes such as those found in the SA and AV nodes (Table 6).

Electrophysiological Actions

SA node and atrium: Verapamil decreases the rate of SA nodal cells firing. Verapamil does not exert any significant electrophysiological effects on atrial muscle.

AV node: Verapamil slows conduction through the AV node and prolongs the AV nodal refractory period.

His-Purkinje system and ventricular muscle: Verapamil has no effect on intraatrial and intraventricular conduction. The predominant electrophysiological effect is on AV conduction proximal to the His bundle.

Hemodynamic Effects. Usual intravenous doses of verapamil are not associated with marked alterations in arterial blood pressure, peripheral vascular resistance, heart rate, left ventricular end-diastolic pressure, or contractility in adults and older children.

Pharmacokinetics. Verapamil is nearly completely absorbed but undergoes extensive first pass metabolism with only 10%–20% of an oral dose reaching the systemic circulation. The bioavailability is dramatically increased in patients with hepatic dysfunction and or decreased hepatic blood flow. It is metabolized by the p450 system with the majority eliminated in the urine. The serum half-life during chronic oral therapy is 3–7 hours. Therapeutic serum levels range between 0.125–0.4 µg/mL.

Clinical Uses. Verapamil is useful for slowing the ventricular response to atrial tachyarrhythmias such as atrial flutter and fibrillation. Verapamil is also effective in arrhythmias supported by enhanced automaticity such as ectopic atrial tachycardia and idiopathic LV-tachycardia. Verapamil is effective for the acute termination of supraventricular tachycardia that uses the AV node as a critical component such as AVNRT and accessory pathway mediated tachycardia.

Adverse Effects. Orally administered verapamil is well tolerated by the majority of patients. Most complaints are with respect to gastrointestinal side effects of constipation and gastric discomfort. Other complaints include vertigo, headache, nervousness, and pruritus.

Contraindications. Verapamil must be used with extreme caution or not at all in patients who are receiving β-adrenoceptor blocking agents due to exaggerating the depressant effects on heart rate, AV node conduction, and myocardial contractility. The use of verapamil in children less than 1 year of age, especially if in heart failure, is contraindicated due to the risks of cardiovascular collapse. Verapamil should be used with extreme caution in patients with ventricular dysfunction.

Diltiazem

The antiarrhythmic actions, electrophysiological effect and clinical uses of diltiazem (*Cardizem*) are similar to those of verapamil (Table 6).

Electrophysiological Actions. Similar to verapamil.

Hemodynamic Effects. Similar to verapamil.

Pharmacokinetics. Similar to verapamil, diltiazem is nearly completely absorbed but undergoes extensive first pass metabolism with only 45% of an oral dose reaching the systemic circulation. The serum half-life is 4–7 hours. Diltiazem is metabolized in the liver, but unlike verapamil, the majority is excreted via the GI tract (65%). Therapeutic serum levels range between 0.50–300 ng/mL.

Clinical Uses. Similar to verapamil. Experience in pediatrics is limited.

Adverse Effects. Similar to verapamil with perhaps less ventricular depression.

Contraindications. Same as verapamil.

MISCELLANEOUS ANTIARRHYTHMIC AGENTS

Digitalis Glycosides and Vagomimetic Drugs

Digitalis glycosides, especially digoxin (*Lanoxin*), due to their positive inotropic effects are widely used for treating patients with congestive heart failure. Additionally they continue to be used for the management of patients with supraventricular arrhythmias. Digoxin slows conduction through the AV node making it useful for use in reentrant arrhythmias that utilize the AV node as one limb of the circuit. It has fallen out of favor for limiting AV conduction during rapid atrial arrhythmias such as atrial fibrillation. Digitalis glycosides have theoretic advantages when compared with other medications that limit conduction through the AV node such as β-blockers and Ca^{2+} channel blockers by providing a positive rather than negative inotropic effect on the ventricles. The effects on the AV node are limited however in states of heightened sympathetic tone such as during advanced heart failure. Due to a potentially shortening effect on the effective refractory period of a manifest accessory pathways (probably very low incidence), digitalis should be avoided in older patients with Wolff-Parkinson-White syndrome. Its use in infants with WPW and supraventricular tachycardia has a long, apparently safe and successful record but the effect is not proven by randomized study.

Adenosine

Adenosine (*Adenocard*) is an endogenously occurring nucleoside that is an end product of the metabolism of adenosine triphosphate. It is used for the rapid termination of supraventricular arrhythmias following rapid bolus dosing.

Electrophysiological Actions. Adenosine receptors located on atrial myocytes and myocytes located in the SA and AV nodes act via a G-protein signaling cascade to open the same outward potassium current activated by acetylcholine. Adenosine stimulation leads to a hyperpolarization of the resting membrane potential, decrease in the slope of phase 4 depolarization, and shortening of the action potential duration. The effects on the AV node may result in complete conduction block with termination of tachycardias utilizing the AV node as a limb of a reentrant circuit. Adenosine does not affect the action potential of ventricular myocytes because the adenosine-stimulated potassium channel is absent in ventricular myocardium.

Electrocardiographic Changes. The most profound effect of adenosine is the induction of AV block (both antegrade and retrograde) within 10 to 20 secs of administration. Mild sinus slowing may initially be observed followed by sinus tachycardia resulting from mild vasodilation and hypotension. There is no effect on the QRS duration or QT interval.

Hemodynamic Effects. The administration of a bolus dose of adenosine is associated with a biphasic pressor response. There

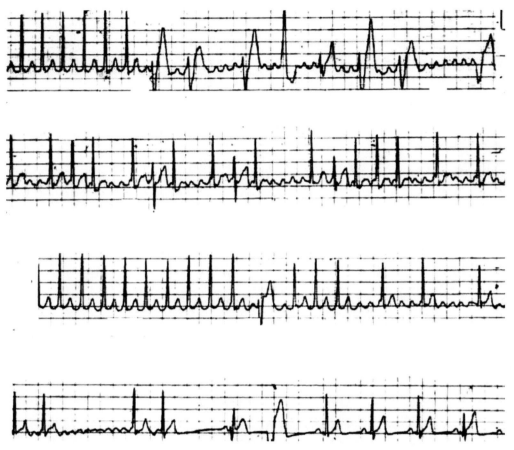

FIGURE 1. Continuous monitor electrocardiogram lead in 15-year-old boy. Adenosine unmasked atrial flutter (top tracing) by inducing AV block and underlying atrial flutter. Conduction was then variable. The flutter converted to atrial fibrillation (bottom tracing) which then spontaneously terminated and sinus rhythm returned.

is an initial brief increase in blood pressure followed by vasodilatation and a secondary tachycardia.

Pharmacokinetics. Adenosine has a nearly instantaneous onset of action and is rapidly metabolized by red blood cells with a plasma half-life of less than 10 seconds. Due to its rapid metabolism, there is no orally available form.

Clinical Uses. Adenosine is useful for the acute termination of supraventricular tachycardia that utilizes the AV node. Adenosine is also helpful for the diagnosis of narrow complex tachycardias by unmasking such as atrial flutter and ectopic atrial tachycardia (Figure 1).

Adverse Effects. Adverse reactions to the administration of adenosine are not uncommon; however, the short half-life of the drug limits the duration of such events. The most common adverse effects are flushing, chest pain, and dyspnea. Adenosine may induce profound bronchospasm in patients with known reactive airway disease. The mechanism for bronchospasm is unclear and the effect may last for up to 30 minutes despite the short half-life of the drug. Rarely, adenosine may induce atrial fibrillation (Figure 1) due to shortening of the atrial refractory period. This is potentially dangerous in the face of an accessory pathway that could rapidly conduct the atrial signal to the ventricles leading to ventricular arrhythmias.

Contraindications. As indicated previously, the use of adenosine in asthmatic patients may exacerbate the asthmatic symptoms. Known hypersensitivity to adenosine precludes its use.

Drug Interactions. Methylxanthines (such as theophylline) antagonize the effects of adenosine via blockade of the adenosine receptors and necessitate increased doses.

Magnesium Sulfate

Magnesium sulfate may be effective in terminating refractory ventricular tachyarrhythmias, particularly polymorphic ventricular tachycardia. Digitalis-induced arrhythmias are more likely in the presence of magnesium deficiency. There has also been a suggestion that hypomagnesemia increases the likelihood of post-operative junctional ectopic tachycardia. Magnesium sulfate can be administered orally, intramuscularly, or, preferably, intravenously, when a rapid response is intended. The loss of deep tendon reflexes is a sign of overdose.

DRUG–DEVICE INTERACTIONS

The use of ICDs in the pediatric population is expanding. Several large adult clinical trials performed in the 1990s demonstrated the superiority of ICDs compared with pharmacological therapy for secondary prevention of arrhythmic death (AVID, CASH, CIDS). Improvements in diagnostic techniques and

TABLE 7. Effects of Antiarrhythmic Drugs on Defibrillation Thresholds

No Change	Increase	Decrease
Quinidine	Amiodarone	Sotalol
Procainamide	Flecainide	Dofetilide
Disopyramide	Lidocaine	
Digitalis	Propafenone	
β-blockers	Mexiletine	

slow improvement in risk stratification have increased the use of ICDs for primary prevention of sudden arrhythmic death in pediatrics. Combination therapy employing both antiarrhythmic drugs and ICDs is becoming more common. Antiarrhythmic drugs continue to be an important component of therapy following device implantation to suppress both ventricular and supraventricular arrhythmias. While the antiarrhythmic drugs have multiple positive effects on the overall therapy, there are several possible deleterious effects. The principle adverse effects include an increase in the frequency of ICD discharges due to drug-induced proarrhythmia, a slowing of the VT rate to below the detection rate despite being hemodynamically unstable, and changing the electrogram morphology, which may affect the ability of the device to detect VT. An additional concern is the potential for a drug to increase the defibrillation threshold (DFT) thereby rendering the device ineffective.

The DFT is a statistical prediction of the amount of energy that is required to defibrillate the heart. The effects that antiarrhythmic drugs have on DFTs are somewhat inconsistent (Table 7). In general, drugs that block the sodium channel and shorten the action potential tend to increase the DFT. Drugs that prolong repolarization tend to decrease the DFT with the notable exception of amiodarone. Amiodarone appears to decrease the DFT acutely after intravenous administration; however, long-term therapy is associated with a significant increase in DFT energy requirements. These changes have obvious important ramifications for patients with ICDs.

SUGGESTED READING

1. Anonymous. The Sicilian gambit. A new approach to the classification of antiarrhythmic drugs based on their actions on arrhythmogenic mechanisms. Task Force of the Working Group on Arrhythmias of the European Society of Cardiology. *Circulation* 1991;84(4):1831–1851.
2. The Cardiac Arrhythmia Suppression Trial (CAST) Investigators. Preliminary report: Effect of encainide and flecainide on mortality in a randomized trial of arrhythmia suppression after myocardial infarction. *N Engl J Med* 1989;321:406.
3. Carmeliet E, Mubagwa K. Antiarrhythmic drugs and cardiac ion channels: mechanisms of action. *Prog Biophys Mol Biol* 1998;70:1–72.
4. Cavero I, Mestre M, Guillon JM, Crumb W. Drugs that prolong QT interval as an unwanted effect: assessing their likelihood of inducing hazardous cardiac dysrhythmias. *Expert Opinion Pharmacother* 2000;1:947–973.
5. Gillis AM. Effects of antiarrhythmic drugs on QT interval dispersion–relationship to antiarrhythmic action and proarrhythmia. *Prog Cardiovasc Dis* 2000;42:385–396.
6. Glassman AH, Bigger JT Jr. Antipsychotic drugs: prolonged QTc interval, torsades de pointes, and sudden death. *Am J Psych* 2001;158:1774–1782.
7. Hohnloser SH. Proarrhythmia with class III antiarrhythmic drugs: types, risks, and management. *Am J Cardiol* 1997;80(8A):82G–89G.
8. Huikuri HV, Castellanos A, Myerburg RJ. Sudden death due to cardiac arrhythmias. *N Engl J Med* 2001;345:1473–1482.
9. Link MS, Homound M, Foote CB, et al. Antiarrhythmic drug therapy for ventricular arrhythmias: current perspectives. *J Cardiovasc Electrophysio* 1996;7(7):653–670.
10. Nattel S. Singh BN. Evolution, mechanisms, and classification of antiarrhythmic drugs: focus on class III actions. *Am J of Cardio* 1999;84(9A):11R–19R.
11. Roden DM, George AL Jr. The cardiac ion channels: relevance to management of arrhythmias. *Ann Rev Med* 1996;47:135–148.
12. Schwartz PJ. Clinical applicability of molecular biology: the case of the long QT syndrome. *Curr Contr Trials Cardiovasc Med* 2000;1:88–91.
13. Vaughan Williams, EM. Significance of classifying antiarrhythmic actions since the cardiac arrhythmia suppression trial. *J Clin Pharmacol* 1991;31:123.

22

Transcatheter Ablation of Cardiac Arrhythmias in the Young

Macdonald Dick II, Peter S. Fischbach and Ian H. Law

Transcatheter ablation of cardiac arrhythmias emerged in the early 1990s as definitive treatment for many forms of tachyarrhythmias in adults; the technique was rapidly adapted for treatment of arrhythmias in children. A conference a decade later (2001) sponsored by the North American Society for Pacing and Electrophysiology (NASPE, later renamed the Heart Rhythm Society, HRS) led a discussion of current knowledge and experience regarding ablation of cardiac arrhythmias in children. From these presentations, indications for ablation for cardiac arrhythmias in children were distilled that were based on a consensus expert opinion among a selected number of the participants.

Class I indication is present when there is broad agreement among experts for ablation, where supportive data exist, and where the outcome is likely to be beneficial.
Class II indication is present when opinion regarding benefit is divergent.
    Class IIA is present when evidence/opinion favor benefit.
    Class IIB is present when evidence/opinion do not favor benefit.
Class III indication is present when there is agreement that radiofrequency ablation is not medically indicated.

Using these three classes, different clinical scenarios were devised as examples for suggested indications for radiofrequency ablation of arrhythmias in children.

Class I: RFCA

- WPW syndrome following aborted SCD
- WPW, syncope, AFib w/preexcited RR <250 msec or APERP <250 msec
- Chronic or recurrent SVT w/ventricular dysfunction
- Recurrent VT, Hemodynamic compromise, amenable to RFCA

Class IIA: RFCA

- Recurrent/symptomatic SVT, refractory to medical management, >4 yrs age
- CHD Surgery, chamber inaccessible
- Chronic (6–12 months) or incessant SVT w/normal ventricular function
- Recurrent, chronic IART
- Palpitations, w/SVT at EPS

Class IIB: RFCA

- Asymptomatic preexcitation (WPW), >5 years; risks/benefits of arrhythmia & ablation explained

- SVT, >5 yrs; replace effective AAD
- SVT, <5 yrs; ineffective AAD (Sol & Amio) or intolerable side effects
- IART—1–3/year
- AVN ablation for recurrent or intractable IART
- VT—one episode, hemodynamic compromise, amenable to RFCA

Class III: RFCA

- Asymptomatic preexcitation (WPW), <5 years
- SVT, <5 years; effective AAD
- NS-VT (not incessant – hours) no hemodynamic compromise
- NS-SVT no Rx required, minimally symptomatic

Abbreviations: ADD: Antiarrhythmic drugs; AFib: Atrial fibrillation; Amio: Amiodarone; APERP: Accessory pathway effective refractory period; Arrh: Arrhythmias; AVN: Atrioventricular node; CHD: Congenital heart disease; EPS: Electrophysiologic study; IART: Intra-atrial reentrant tachycardia; NS: non sustained; RFCA: Radiofrequency catheter ablation; RR: Successive R-waves; SCD: Sudden cardiac death; Sol: Sotalol; SVT: Supraventricular tachycardia; VT: Ventricular tachycardia; WPW: Wolff-Parkinson White Syndrome.

These recommendations are prudent and reasonable, particularly for Class III. However, for Class II, they are neither comprehensive nor absolute, and refer only to radiofrequency energy. A curious characteristic is dependence on age-related indications; a more rational recommendation would be based on the patient's size, a physiologic parameter, rather than the arbitrary chronological age. It is the occasional small child and even infant whose frequent trips to the emergency room for either pharmacologic or electrical cardioversion or who is refractory to the antiarrhythmic medications, including amiodarone, who will justify a deviation from these guidelines. Likewise a child may be physically larger and more robust than the chronologic age would suggest, meriting earlier catheter intervention consideration than suggested by the guidelines. In addition, symptoms attributable to the arrhythmia, as well as hemodynamic compromise, may warrant earlier intervention. Further, the recognized importance of the training and the experience of the operator, as well as that of the entire institutional cardiovascular team, not only with ablation but also in navigating the intracardiac space of infants and children with arrhythmias, including those with complex congenital heart defects, underscores the necessary critical attention to patient safety that governs all medical decision making. The development of cryotherapy as an alternative energy source significantly shifts the balance toward greater safety and perhaps earlier ablation. Because medical management with antiarrhythmic agents carries its own risks and side effects, families often elect a treatment that under usual conditions has been shown to be safe and which can be expected to eliminate the problem completely. Finally, after frank and open discussion regarding the potential risks and the expected benefits of all possible management strategies are outlined in detail with full understanding by the patient (if competent) and parents (legal guardian), the decision to manage the arrhythmias with ablation techniques outside the specific guidelines may be a reasonable course.

RADIOFREQUENCY ABLATION

Radiofrequency current is the primary, contemporary energy used to interrupt abnormally conducting pathways, ectopic atrial, or ventricular foci, and, on occasion, the atrioventricular (AV) node (Table 1). Although this technique had been employed to interrupt peripheral nerves and nerve roots for pain management by neurosurgeons for several decades, its first application for use in the human heart was in 1987. Practical, widespread use of radiofrequency energy for treatment of tachyarrhythmias grew rapidly in the 1990s.

TABLE 1. Biophysics of Radiofrequency Ablation

Biophysics of Radiofrequency Ablation

Alternating current delivered at 300–750 kHz
Resistive heating of tissue in contact with the catheter producing discrete lesion composed of desiccation and coagulation and necrosis 5–6 mm in diameter and 2–3 mm deep
Delivery through a 7 French catheter with a 4 mm electrode tip at 50–60 V for approximately 1 minute with ideal temperatures at 50°C–60°C
Mode (mono or bipolar) and waveform (modulated or not) of power output
Current frequency and density
Tissue temperature and impedance
Phase displacement
Convective loss and heat transfer to tissue
Tissue-tip contact—pressure, shape and size of electrode tip

Radiofrequency energy consists of alternating current delivered at 300–750 kHz between the oversized (4–8 mm) electrode tip mounted on a mapping/ablation 7 French electrode catheter and an indifferent electrode patch placed posteriorly on the patient's back behind the heart. This current produces, by way of resistive heating of the tissue in contact with the catheter tip electrode, irreversible tissue damage—dessication and coagulation necrosis—at temperatures approximately 50°C or greater.

High temperature (≥75°C–80°C) may result in blood coagulation and clot formation, indicated by a rise in the impedance and requiring cleaning of the catheter tip electrode. Temperatures of 100°C boil the liquid contents of the myocytes and may produce an audible "pop"; it results in larger areas of necrosis. Radiofrequency generators initially provided a continuous display of volts, impedance, and current as energy was being delivered. Newer models provide continuous graphic displays of impedance, watts (power) and, most importantly, temperature at the catheter tip. If there is good tissue contact, this provides a useful measure of tissue temperature—the chief indicator of tissue "burn." As observed in experimental animal models, the typical lesion is 5–6 mm in diameter and 2–3 mm deep. In some immature animal models, lesions placed on the epicardial surface of the ventricle tend to expand and thin with growth, but less so on the atrium and especially along the AV groove. Pathologically, desiccation and coagulation necrosis occur at the target site; the lesion heals to form a dense small scar. Factors that influence the effect of radiofrequency current on biologic tissue are the power output, the mode of delivery, the current frequency, current density, duration of current delivery, tissue temperature reached, convective heat loss by the passage of blood around the delivery site, the electrode tissue contact and pressure, and the shape and size of the electrode. Factors that can be controlled or directly influenced by the operator are electrode size, power delivered, duration of delivery, and tissue contact. For the majority of ablations of conducting pathways or ectopic foci, a 4-mm electrode at the tip of the catheter provides sufficient current density and temperature to achieve interruption of the target pathway or ectopic focus. In our experience, the majority of accessory pathways, if one is at the optimal site, are interrupted in about 4–5 seconds, the time it takes to reach maximal power and the pre-set auto-regulated temperature of 50°C–65°C, depending on the arrhythmia and tissue target (i.e., atrium, ventricle, accessory pathway, slow pathway). Routine procedure is to terminate the delivery of the current after about 10 seconds if the pathway is not ablated so as to minimize tissue injury. If there is successful interruption of the pathway, the energy is continued for 30–60 seconds. The operator can often improve contact between catheter tip and tissue by careful monitoring for low impedance, indicating sufficient current flow and satisfactory temperature rise, and by sensing tension along the catheter shaft as the tip rests firmly against the target site. Even with attention to these factors, intra-atrial reentry tachycardia in patients due to extensive atrial incisions and suture lines from extensive atrial surgery (Mustard, Senning, and Fontan operations) and complicated by thick fibrotic atrial walls, some ventricular arrhythmias, and isolation of pulmonary veins may benefit from a different size and configured electrode tip, greater power, and higher temperature.

CRYOABLATION

Delivery of radiofrequency energy to some locations within the heart (i.e., the anterior and mid septal area) may pose a direct risk to the AV node and His bundle by virtue of their proximity placing the patient at risk for complete heart block. In addition, delivery of radiofrequency energy to the right posterior septal area indirectly may injure the AV node through damage to the AV artery. Cryotherapy, which is the application of hypothermia for therapeutic purposes, has long been a mainstay of pain control associated with trauma, in the surgical management of congenital heart disease in infants, and more recently the intraoperative treatment of arrhythmias during open heart surgery. With the development of cryocatheters that deliver a refrigerant fluid (NO_2) through a catheter-housed closed capillary circuit to the catheter electrode tip (4 mm), transcatheter cryotherapy has emerged as an alternative energy source to ablate right anterior septal, right mid septal, and posterior septal accessory pathways that are adjacent to the normal conducting tissues or its blood supply. As the liquid-to-gas phase of the cryogenic fluid passes into the tip that is in good contact with the myocardial tissue, the tip temperature, monitored through a thermocouple sensor, rapidly drops to a preset level of $-25°C$ to $-30°C$. The gas is conducted away through a vacuum return tube. When the temperature reaches approximately $-20°C$, an ice ball forms around the catheter tip and abutting myocardial tissue, resulting in adherence of the tissue to the catheter tip. Because the catheter tip is fixed to the target tissue, there is total stability of the catheter, even during pacing protocols. Cryomapping is the first step; here the temperature is reduced to $-30°C$ for 30 seconds. This cooling produces a transient loss in cellular electrophysiologic function such as loss of conductivity in the AV node. However, this effect, whether from the cooling or from manipulation of the catheter, is reversible. If during the mapping procedure the desired effect is achieved without an adverse response (i.e., producing AV delay or block), the temperature is rapidly reduced ("dive") to $-70°C$ to achieve cryoablation. At this temperature the early change in the cell membranes' physical state (less fluid), seen at about $-30°C$, is accelerated and the membrane ion pumps loose their transport capacity, with decrease in the amplitude and duration of the action potential and prolongation of repolarization. Both extracellular and intracellular ice formation disrupts the cell in a catheter tip-to-cell inverse manner—the closer the cell to the tip and ice ball, the greater the disruption and chance of lethality. Time exposure to the ice formation also affects lethality—the longer the exposure, the more lethal and the less reversible the effect (Table 2). The resultant cryogenic lesion is homogenous, acellular, and well demarcated, with no endocardial disruption and full of collagen fibers—fibrous stroma.

During delivery of the cryothermal energy careful continuous monitoring of the activity of the target is mandatory. If the target is potentially near the AV node or His bundle, such as in slow pathway ablation (Chapter 5) or in mid septal or anterior septal accessory

TABLE 2. Biophysics of Cryotherapy

Characteristics

Cryomapping: $-30°C$ for 30 seconds
Cryoablation: $-70°C$ for 3–4 minutes
Choice of cryogen
Tip contact
Cooling rate
Temperature of tip
Duration of freeze
Frequency and interval between freeze–thaw cycles
Tissue vascularity
Depth of target

pathways (Chapter 4), monitoring of AV conduction (PR interval) can alert one to an increase or even AV block during delivery of cryotherapy. At the moment of an interruption of AV conduction, the application is terminated.

The cryothermal technique offers the advantage of "testing" or "mapping" the site of ablation by inducing transient cellular injury (temperatures of -20 to -30 C) prior to performing the ablation (-70 C). If undesirable effects are seen during cryomapping (e.g. changes in atrioventricular conduction) discontinuing the cooling reverses the untoward effect. Cryothermal technology also offers the advantage of catheter stability during ablation as the catheter adheres to the endocardial surface at temperatures less the -25 C. This cooling technology is ideally suited for parahisian pathways or other arrhythmias in which damage to the surrounding tissue is undesirable.

The technique of cryothermal ablation is similar to that of radiofrequency ablation with regard to mapping the accessory pathway. When the ablation site is found cryomapping is performed, observing for the desired affect (loss of pre-excitation, loss of eccentric retrograde conduction). Testing, e.g. ventricular pacing or atrial extrastimuli, can be performed during cryomapping to confirm success. If the cryomapping is successful within the first 10–20 seconds, the temperature is rapidly dropped to -70 C for 4 minutes. In contrast to radiofrequency ablation, application of cryothermal energy can be applied safely during supraventricular tachycardia, without fear of catheter movement upon tachycardia termination, due to the adherence of the catheter tip during cooling.

TECHNIQUES OF ABLATION IN CHILDREN

Radiofrequency ablation was first introduced to treat tachyarrhythmias in adults. However, adoption of this technique for the treatment of supraventricular tachycardia in children was rapid. Nonetheless, a number of conditions and factors require a modified approach in children, especially those approximately 20 kg in weight or less. A number of anatomic and physiologic factors (Chapter 3) influence the approach and technique of electrophysiologic study in children as well as radiofrequency ablation. Because the procedure is often long, requires a stable patient state, and is uncomfortable, general anesthesia is often used in children less than 12 years of age. The increase in patient comfort and stability during radiofrequency delivery, as well as the ability to closely monitor the patient, offsets the very small risk of general anesthesia. In adolescents and young adults, conscious sedation with very close cardio-respiratory monitoring is usually sufficient.

The variability in patient size necessitates comparable differences in dimensions of intravascular devices. Adolescents are usually of sufficient size to allow use of catheters and sheaths of the length and caliber of those used in adults. However, smaller children require smaller catheters and sheaths, especially in the diameter of the operator-controlled curve. Typically, for a child 20 kg or less, catheters should not exceed 6 French. For an infant under 10 kg, 4 and 5 French catheters are necessary, including the mapping/ablation catheter (Chapter 3). Because of the variable dimension of the cardiac chambers in different size children, steerable catheters with different shapes and different sized curves are also necessary to appropriately accommodate each patient (Figure 1).

In an infant, an additional atrial recording and stimulation site is the esophagus (Chapter 3, Figure 8), behind the left atrium.

Due to the vigorous contractile state and the faster heart rates in small children compared to adults, both during sinus rhythm and tachycardia, catheter stability is at risk. There is always a tradeoff between catheter size (and stiffness) and catheter stability. Using the His-RVA catheter adds stability to the His catheter since it is, in part, secured by the placement

 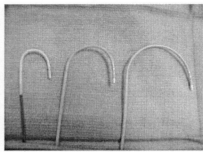

FIGURE 1. Examples of mapping and ablation 7 French catheters with an operator controlled steerable or flexible property. The arc radius is shown in three sizes in the fully flexed (left panel) and the open positions (right panel). The catheter tip can move in the reverse direction as well over a full 180–220° in a one-dimensional plane. Catheters are also available in 5 French and 6 French sizes.

of the catheter tip in the right ventricular apex (Chapter 3, Figure 9). Multiple catheters produce artifacts when the electrodes from different catheters strike one another, a confounder increased in the smaller volume of a child's heart (≤20 kg).

When radiofrequency ablation was first introduced, access to accessory pathways or ventricular ectopic foci located in the left side of the heart was achieved by retrograde passage of the catheter from the femoral artery around the aortic arch into the ascending aorta and left ventricle (Figure 2). Although this route provided access for simultaneous arterial monitoring (albeit attenuated with a catheter in it), it clearly was not applicable to small children (≤30 kg). Transseptal entry into the left atrium has virtually replaced retrograde arterial catheterization for ablation of tachycardias arising from left-sided structures (Chapter 3, Figure 13) in children.

Several features associated with successful ablation of either conduction pathways or ectopic foci have been identified. For a patient with manifest preexcitation (Chapter 4), local preexcitation (i.e., earliest ventricular activation recorded through the intracardiac mapping electrode) that is ≥15 ms earlier than the earliest onset of the preexcited QRS complex on the electrocardiogram suggests a favorable site for ablation of an accessory pathway (Figure 3). Similarly, success of ablation of ectopic foci is associated with the intracardiac electrogram that precedes the surface ECG P-wave or QRS complex, depending on either an atrial or ventricular origin of the tachyarrhythmia (Chapter 10, Figure 4). Another characteristic associated with successful ablation of either a concealed or manifest accessory pathway is the shortest ventriculo-atrial (VA; i.e., retrograde) conduction time attainable during mapping of the AV ring during tachycardia (Figure 4).

In contrast, amplitude of the atrial or ventricular electrograms reflects only relative position across the AV ring; due to the dominant ventricular mass relative to the atrium, the ratio of the amplitude of the atrium to that of the ventricle, if correctly placed across the AV junction, is usually less than 0.5. Thus, the amplitude of the atrial and ventricular electrograms confirms optimal proximity to the AV ring whereas the retrograde conduction time conveys information regarding exact proximity to the putative accessory connection. Much of the inference that is contained in the amplitude of the electrograms and in the VA conduction times is dependent on operator integration of the anatomic and electrical data on the fly; more recently, computer-assisted mapping techniques produce an electro-anatomic map correlating in a three-dimensional image the activation sequence and the anatomic configuration (Chapter 10, Figure 5).

Currently, we place an arterial line (4 French pigtail catheter) for continuous

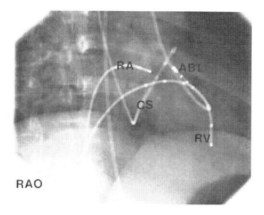

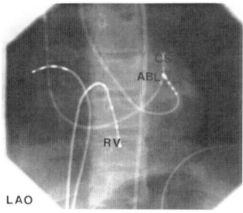

FIGURE 2. Top: Right anterior oblique view. Bottom: Left anterior oblique view. Note the retrograde passage of the ablation catheter around the aortic arch into the left ventricle oriented along the AV (mitral) ring. Note also the antegrade passage of the coronary sinus catheter into the left subclavian vein, through the right superior vena cava and right atrium into the coronary sinus. These catheter routes are rarely used today in children. Reprinted with permission from Dick, M. et al. Use of radiofrequency energy to ablate accessory connections in children. Circulation 1991;84:2318–24.

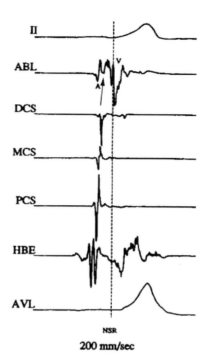

FIGURE 3. Single beat with ventricular preexcitation (Leads I, avL) and intracardiac electrograms; His bundle electrogram (HBE, no His potential because it is buried in the ventricular electrogram), The distal coronary sinus (DCS), the mid-coronary sinus (MCS), the proximal coronary sinus (PCS), and the distal ablation/mapping electrogram (ABL). Note the local ventricular preexcitation in the ablation electrogram (arrow) precedes the global preexcitation expressed in the two surface leads. A = atrial electrogram; V = ventricular electrogram.

blood pressure monitoring in those individuals who are ≤20 kg or who are undergoing electrophysiologic study and ablation for ventricular arrhythmias. This approach minimizes risk to the femoral arteries while preserving intra-arterial cardiac monitoring in those individuals most at risk. Thus, it is clear that the size, site, and number of catheters and sheaths placed in children of varying size is dependent on the specific anatomic and physiologic features of the individual undergoing electrophysiologic study and radiofrequency ablation.

Fluoroscopy is critical for transseptal entry into the left atrium. In addition, fluoroscopy is used for rapid and precise placement of catheters for both intracardiac recording and stimulation and remains the benchmark imaging technique to assess anatomic landmarks, to perform endocardial mapping, and to identify initial catheter positions. Due to their 90° orientation one to another, the left anterior oblique (LAO) and right anterior oblique (RAO) provide different but equally important information regarding

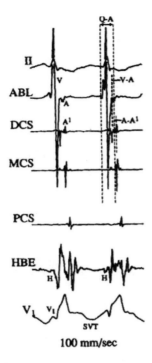

FIGURE 4. Two beats of supraventricular tachycardia with two electrocardiographic leads (leads II and V1 with right bundle branch block), an ablation/mapping electrogram (ABL), distal coronary sinus (DCS), mid coronary sinus (MCS), proximal coronary sinus (PCS), and His bundle electrogram (HBE). Dashed vertical lines indicate the retrograde conduction time from ventricular activation to the mapped site recorded through the ABL bipolar electrodes (Q-A) and the interval between the earliest mapped site recorded through the coronary sinus electrodes (DCS) and the earliest mapped site recorded through the ABL electrodes. Note the shortest interval through the ABL catheter.

catheter positions. The RAO projection places both the mitral orifice and tricuspid orifice *in profile*. This projection provides, in conjunction with the atrial and ventricular electrogram configurations, information with regard to catheter tip position relative to proximity to the AV groove (i.e., the AV valve), as well as to the immediate upstream atrium and downstream ventricle. In contrast, the LAO projection images the mitral and tricuspid orifices *en face* and provides information regarding the position of the catheter tip along the AV groove. The mapping locations on the left side are recognized by the coronary sinus catheter with its multiple electrode array. On the right, viewing from the right ventricular apex, a clock face is superimposed on the tricuspid orifice beginning at the His bundle at approximately 1 o'clock; the annotation of the sites by the hour hand then moves clockwise (Figure 5). The development of non-radiologic navigation systems (Biosence Webster's CARTOMERGE™ Image Integration Module, which uses a magnetic field, and Localisa® and Navix®, which use a small electric field) that can detect and display in a 3-D image the position of the catheter(s) within the cardiac chamber of interest, has reduced fluoroscopy exposure markedly, particularly the radiation exposure necessary for delivery of ablation energy. Using these systems, the target site for ablation as well as vital anatomic landmarks—AV node, His bundle, tricuspid valve, coronary sinus os and body, and atrial septum—can be effectively mapped (Figure 5), allowing a safe ablation by either radiofrequency current or cryotherapy.

Because of the catheter and the sheath in, as well as the delivery of radiofrequency current to, the systemic side of circulation (left atrium and left ventricle), maintenance of anticoagulation to prevent the introduction and development of thrombi is of special concern. Once entry into the left atrium is achieved, it is useful and prudent to verify entry into the left atrium by pressure and oxygen saturation determinations. Bolus and continuous heparin infusion is delivered to maintain an activated clotting time of ≥ 300 seconds. Following the procedure, the patient is placed on 81 mg of aspirin a day for 6 weeks.

APPROACH TO ACCESSORY PATHWAYS

Left-Sided Pathways

Approximately 60%–65% of accessory pathways are located on the left side

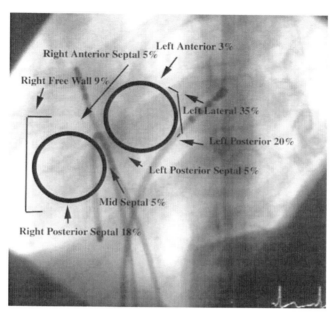

FIGURE 5. Left anterior lateral oblique radiograph depicting position of tricuspid (left circle) and mitral (right circle) valves and the distribution of accessory pathways around the AV groove.

(Figure 5). This left-sided dominance is higher in infants and small children. As the infant or child grows, functional or anatomic loss of the pathway may occur, perhaps due to molding of the AV groove, which, at least theoretically, may interrupt these pathways leftward and lateral to the crux, or fulcrum, of the heart. Such a natural attrition would reduce the absolute incidence of left-sided pathways. In contrast, right-sided pathways may theoretically be somewhat protected from this "molding effect" by their closer proximity to the crux and by the fixation of the right side by the vena cavae compared to the relatively less tethered left lateral aspect of the heart.

Access to the left side is by the transseptal technique into the left atrium (Chapter 3). Once in the left atrium, there should be free movement of the steerable catheter tip so that it can sweep (Chapter 3, Figure 8) in an arch from near the top of the left atrium down to the mitral posterior rim for mapping and ablation of left lateral accessory. Left posterior pathways require a slightly tighter curl to the catheter curve, achieved by gently decreasing the radius of the curve and moving the curved tip inferiorly by slightly withdrawing the catheter and sheath, and thus positioning the electrode–ablation tip slightly more medial on the posterior mitral ring (Chapter 3, Figure 5). Further tightening the curve and slightly withdrawing slightly the catheter and sheath will bring the catheter tip to the left posterior septal region. All three of these mapping and ablation applications require continued clockwise torque on the catheter–sheath apparatus to ensure the catheter remains on the posterior mitral ring. The anterior septal region, the last site on the mitral ring, is a particularly difficult location to secure a stable position of the catheter tip. In larger children, this is one site where using a specialized curved sheath, which directs the tip toward the targeted area and conveys firm stability to the catheter tip, is helpful. Delivery of radiofrequency energy with the catheter tip at the mapped site of the accessory pathway often will ablate the pathway and terminate the tachycardia (Figure 6).

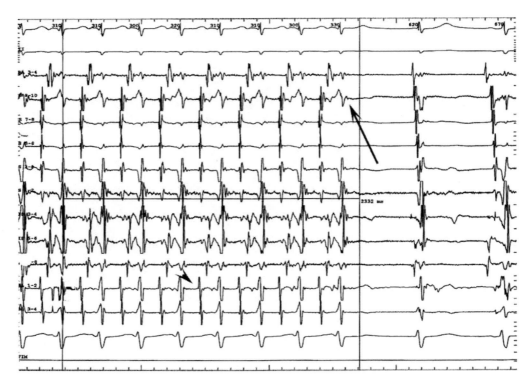

FIGURE 6. Radiofrequency ablation during supraventricular tachycardia. Top 2 tracings and bottom tracings are leads I, II, V1. The third tracing from the top is recorded from the high right atrium. The middle 4th to the 14th tracings are recorded respectively from the proximal, three middle, distal coronary sinus sites, three His bundle sites, and a distal and proximal mapping/ablation site. The arrowhead indicates earliest retrograde activation recorded through the distal pair of electrodes on the ablation catheter. Radiofrequency energy delivered at that site terminated the tachycardia (large arrow) within 3 seconds.

Catheter stability is a critical component to achieve successful ablation of a tachyarrhythmia. Because the tachycardia of small children is often faster than that of adults, catheter stability is often difficult to achieve, particularly at the moment of successful ablation when the tachyarrhythmia is terminated and sinus rhythm ensues. At that instant, the catheter tip, while relatively stable during the tachyarrhythmia, can be abruptly dislodged from the mapped ablation site, losing sufficient contact for complete ablation of the targeted pathway. There are two strategies to improve stability. The first is to map the site of earliest local preexcitation (Figures 3 and 7) in patients with manifest pathways during sinus rhythm or an atrial pacing; local preexcitation at an optimal ablation site is usually ≥ 15 before global excitation is detected on the ECG. The second is to locate and isolate the accessory pathway by mapping the retrograde activation sequence during ventricular pacing (Figure 8) in patients with either manifest or concealed pathways. One must pace at a sufficiently fast rate (usually 150 bpm or greater) to ensure that retrograde conduction through the accessory pathway is not masked by the robust retrograde AV nodal–His-Purkinje conduction often seen in young patients. Energy is then delivered during ventricular pacing, thus avoiding abrupt displacement of the catheter when successful ablation is achieved and allowing delivery of radiofrequency current beyond the point of observed interruption of anomalous conduction (Figure 8).

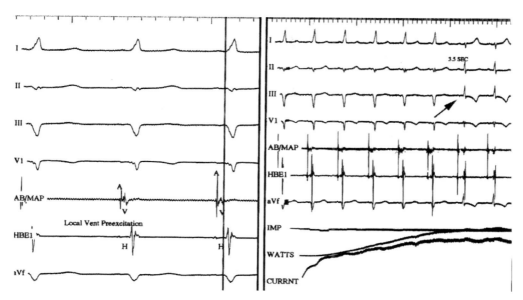

FIGURE 7. Left-hand panel: Sinus rhythm with preexcitation in a 12-year-old girl. Leads I, II, III, V1, and aVF are due to a right-sided pathway. Ablation/mapping catheter at site of maximal local ventricular preexcitation (vertical line denotes onset of global ventricular excitation. Note the onset of local ventricular preexcitation (v in front of the vertical line) indicating the earliest site of ventricular activation. Right-hand panel: Leads I, II, II, V1, aVF, ablation catheter (ABLMAP), and His bundle catheter (HBE) and continuous tracings of impedance (IMP) Watts, Current. Note the loss of preexcitation (arrow) when the radiofrequency power reaches the highest level, indicating interruption of the manifest accessory pathway.

If supraventricular tachycardia cannot be induced in the electrophysiology laboratory, retrograde conduction through left-sided concealed pathways may be difficult to map and isolate. Two solutions are available. The first is the bolus administration of adenosine (Figure 9). This technique will usually block the AV node retrograde, but usually not the accessory pathway, particularly if it is a fast conductor. However, its transient effect limits its usefulness. The second solution is more stable but requires placing a catheter in the left ventricle. The left ventricle is stimulated at a fixed cycle length (600–500 msec, 100–120 bpm). Due to the proximity of the left ventricular pacing site to the suspected left-sided pathway, this technique more selectively engages the left accessory pathway than does pacing from the right ventricular apex, thus unmasking conduction in the left-sided pathway (Figure 10).

Right-Sided Pathways

Access to right-sided structures is clearly less complicated than access to the left. On the other hand, catheter stability on the AV ring, especially along the right free wall and anterior areas, is less secure. In addition, delivery of radiofrequency energy to the tricuspid AV ring in the septal area places the AV node, His-Purkinje system, and AV nodal artery at greater risk. Right-sided accessory pathways are depicted on the clock face of the tricuspid valve in Figure 5. Right posterior septal pathways are the most common and are ablated by placement of the catheter tip at the AV groove determined in the right anterior oblique projection and at the 5 o'clock position determined in the left anterior oblique projection. Occasionally, these pathways can be found in the mouth of the coronary sinus. Difficulty arises when mapping indicates

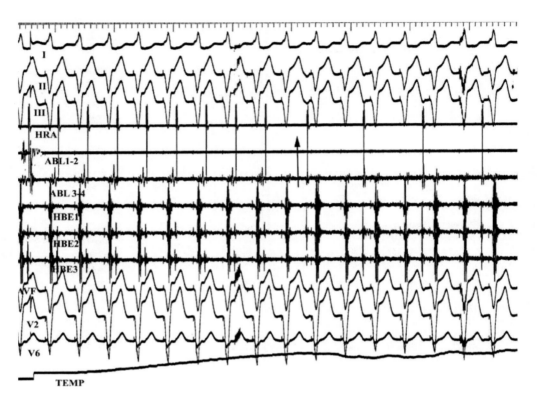

FIGURE 8. Ventricular pacing in a 15-year-old girl with a left concealed accessory pathway. (lead II, low septal right atrium in the region of the His bundle (HBE 1-3). Note the retrograde activation is earlier in the left atrium (ABL 3-4) reached by the transseptal approach rather than on the right compatible with a left-sided accessory pathway. Radiofrequency energy is applied (ABL 1-2) and retrograde conduction is interrupted (arrow) just as the temperature reaches its preset peak (55°C). The pacing is maintained, preserving stability of the ablation electrode at the target site.

early activation (in either antegrade or retrograde direction) at both the right posterior site (5 o'clock) and the left posterior site (mid coronary sinus site). Application of energy is necessary at both sites. Proximity to the AV nodal artery can be determined by right coronary angiography; if the artery is within 1–2 mm of the mapped site of the pathway, cryoablation should be considered.

Ablation of right free wall pathways is confounded by the instability of the catheter tip at the target site on the AV groove. There are no intracardiac structures to stabilize the catheter tip. It is often necessary to utilize a long sheath with specially directed curves that orient the catheter tip toward the right AV groove. In small children with small atrial dimensions, a long sheath with a minimal curve is sufficient to stabilize the catheter tip (Figure 11). If preexcitation is present, ablation should be delivered during sinus rhythm. If the pathway is concealed, catheter stability is enhanced by ventricular pacing, as opposed to SVT, during delivery of radiofrequency current. Not infrequently, the atrial and ventricular electrograms, particularly after a few unsuccessful applications of the radiofrequency current, become attenuated and fragmented, falsely suggesting a poor site for the catheter tip for successful ablation.

Since this observation may incorrectly imply that the catheter tip is not on the AV

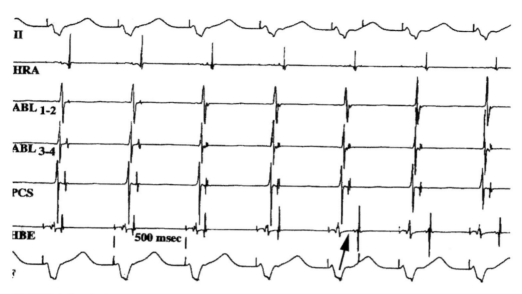

FIGURE 9. Ventricular pacing in an 8-year-old girl with a concealed left lateral pathway. Adenosine is administered (bolus 200 mcg/kg). Note the prolongation of retrograde conduction (in the HBE tracing at the arrow) in the right low septal and His bundle area (HBE), but not in the electrograms recorded from the left atrium (ABL 1-2 and ABL 3-4), indicating block in the AV node but not in the left-sided accessory pathway unmasking its location.

groove, the catheter position should be reviewed by fluoroscopy in the right anterior oblique projection. In fact, as noted above, the amplitudes of the atrial and ventricular electrograms are not decisive in terms of actual target site. The strongest indicators of correct location, which predicts successful ablation, are maximal local preexcitation if the pathway(s) is manifest (Figure 3), and the shortest ventriculo-atrial conduction time during either supraventricular tachycardia or ventricular pacing if the pathway is concealed (Figure 4).

The right anterior septal and the right mid septal sites are closest to the AV node and His bundle. Accordingly, application of radiofrequency ablation at these sites may jeopardize the normal conduction system. Inspection of the electrogram signal recorded through the ablating catheter's distal bipolar pair of electrodes for a His bundle potential, and then correlating the site with fluoroscopy, can be used to prevent untoward energy delivery too close to the normal conducting pathway. Application during sinus rhythm with low power allows the operator to monitor closely the appearance of first-degree heart block, which could lead to abrupt termination of current delivery. A stabilizing sheath of the appropriate size increases operator control of the catheter tip (Figure 11). Rarely, access from the right internal jugular may assist in delivering the tightly curved catheter to an anterior site by placing the tip at the target and then opening the steerable curve (rather than increasing the tightness of the curve). This maneuver applies pressure to the catheter tip as the curve opens up and helps hold the tip at the target site on the anterior AV groove. Nonetheless, as noted previously, delivery of radiofrequency energy to the anterior and mid septal pathways pose a direct risk to the AV node and His bundle by virtue of their proximity, whereas delivery of radiofrequency energy to the right posterior septal area indirectly may injure the AV node through damage to the AV artery. Cryotherapy is an alternative energy source in these situations.

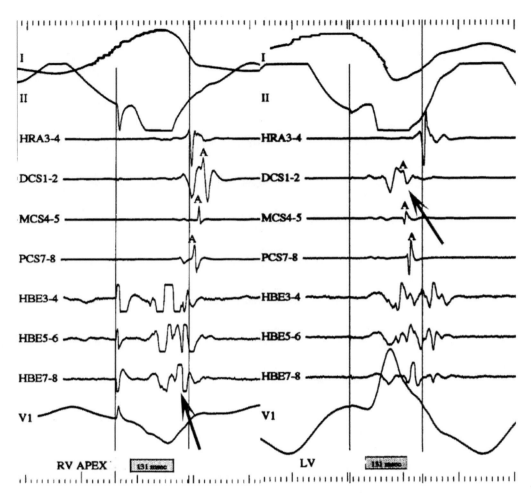

FIGURE 10. Left-hand panel: Pacing from the right ventricular apex. Right-hand panel: Pacing from the left ventricle—posterior wall. Note the much shorter VA interval in the distal coronary site (DCS, arrow) when pacing from the left ventricle is compared to the shortest site in the His bundle area (HBE, arrow) when pacing from the right ventricle. This difference unmasks the left-sided pathway. Surface ECG leads I, II, V1, and intracardiac tracings high right atrium (HRA), mid coronary sinus (MCS), and proximal coronary sinus (PCS).

Criteria for Successful Ablation of an Accessory Pathway

Criteria for successful interruption of an accessory pathway are outlined in Table 3. Acute success is greater than 98% if these criteria are met and persist for greater than 30 minutes. Our practice is to wait in the laboratory for 45–60 minutes after the last application. Recurrence (resolution of the local injury and return of functionality) of the pathway can be expected in 5%–10% of those on the left and 15%–20% of those on the right. These recurrences can be safely and permanently ablated at a second session.

APPROACH TO ATRIAL ECTOPIC (AUTOMATIC) TACHYCARDIAS

Left atrial foci are often near the orifices of the pulmonary veins. The typical ECG pattern consists of a deformed (bimodal and long) P-wave that is purely positive in V1

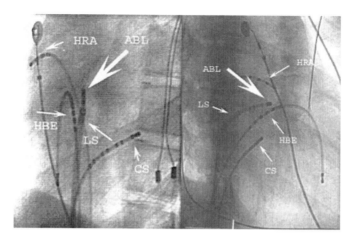

FIGURE 11. Left-hand panel: Left anterior oblique projection. Note the ablation catheter tip (large arrow, ABL) in the 1 o'clock position against the septum. The small arrow indicates the long sheath (LS) containing the ablation catheter. Right-hand panel: Right anterior oblique projection. The ablation catheter tip electrode is on the AV ring. HRA = high right atrium; HBE = His bundle electrogram; CS = coronary sinus.

and usually lead 1. Right atrial foci maybe scattered throughout the right atrium. Their P-wave axis varies depending on the origin high or low in the atrium. An ectopic focus in the orifice of the right atrial appendage near the sinus node exhibits a normal axis but is often markedly deformed (bimodal and long).

The approach to atrial ectopic (automatic) tachycardias depends on their location. Left-sided ectopic foci require the transseptal approach whereas those on the right side are immediately accessible. Special consideration for the form of analgesia and anesthesia are important as deep anesthesia may suppress the ectopic focus. Conscious sedation is often the best approach to preserve the atrial ectopic rhythm. Often the infusion of isoproterenol will bring out an otherwise suppressed rhythm. Endocardial mapping with an electrode catheter is directed toward identifying the earliest site of atrial activation prior to the onset on the P-wave (Chapter 10, Figure 4).

Because of the heat generated during application of energy that excites the ectopic focus, the tachycardia rate usually transiently accelerates before extinguishing completely Chapter 10, Figure 4). Because the atrium is thin and thus requires less energy, it is wise to start at low power, monitoring the temperature rise until a temperature of 50°C–60°C is reached; application should last 30–60 seconds. Termination of the tachycardia is

TABLE 3. Criteria for Successful Ablation of an Accessory Pathway

Criteria
Rapid (≤ 4 sec) loss of preexcitation, termination of SVT or VA conduction through the pathway with ventricular pacing during application of radiofrequency energy.
Absence of preexcitation in sinus rhythm (if present prior to ablation) or with atrial pacing.
Pharmacological (adenosine IV bolus) induced antegrade delay/block in the AV node without emergence of preexcitation after ablation.
No inducible tachycardia after ablation (present prior to ablation).
VA block or no eccentric retrograde atrial activation with ventricular pacing after ablation.
VA block with adenosine bolus infusion after ablation but not prior to ablation during ventricular pacing.

the only confirming marker of success; therefore, delivery of radiofrequency energy during tachycardia is necessary. Dislodgement of the catheter tip from the target site is less frequent in this group. Enhanced mapping techniques such as electro-anatomic mapping and non-contact mapping can facilitate precise anatomic location (Chapter 10, Figure 5).

PERICARDIAL APPROACH TO EPICARDIAL SITES

Unsuccessful ablation may occur (<10% of the time) for a number of reasons, including inadequate mapping and localization of the pathway (or ectopic focus), inadequate energy delivery, and movement and displacement of the electrode tip from the targeted site. In addition, the target—pathway or focus—may be located closer to or on the epicardial surface of the heart rather than on or near the endocardial surface. In this case, percutaneous pericardial entry from a sub-xiphoid or sub-costal site permits access to the epicardial surface—particularly the AV groove for pathway-dependent and atrial tachycardias and the ventricular surface for ventricular tachycardias. A critical part of this technique is to define by angiography the caliber and course of the right and left coronary arteries, both before and after energy delivery. This information, in conjunction with the recorded electrograms, serves as a roadmap to the appropriate ablation site. Furthermore, it assesses the effect of radiofrequency energy, if any, on the coronary circulation. However, the pericardial approach is rarely necessary in children and has not yet been used in children.

APPROACH TO ATRIOVENTRICULAR NODAL REENTRY TACHYCARDIA

Atrial ventricular nodal reentrant tachycardia (AVNRT) is infrequent in infants and children but increases with age. After identification of this form of arrhythmia (Chapter 5), radiofrequency current is delivered to the area of slow conduction (slow pathway) in the triangle of Koch. This anatomic region is bounded by the tendon of Todaro, the posterior leaflet of tricuspid valve, and the line drawn from the posterior margin of the tricuspid valve to the posterior rim of the coronary sinus os. The His bundle catheter in conjunction with the coronary sinus catheter provides fluoroscopic landmarks for the triangle of Koch and aids in guiding the mapping and ablating catheter to the target site within the triangle (Chapter 5, Figure 8). Although the dimensions of the triangle of Koch are relatively similar in adults of various sizes, the area of the triangle of Koch increases with growth in children (Figure 12). At less than 20 kg, the area is equal to or less than 50 mm^2. This small dimension along with the proximity of the AV node–His bundle axis suggests either deferring ablation in infants and small children if AVNRT is the mechanism of the tachyarrhythmia, minimizing ($\leq$5) the number and duration of radiofrequency energy applications or, more recently, employing cryotherapy. The use of a long sheath to enhance catheter stability is an important adjunct to the technique for ablation of this arrhythmia. However, in this age group, supraventricular tachycardia is most often dependent on a manifest or concealed accessory pathway and is only rarely dependent on functional changes in the triangle of Koch, which contains the slow pathway.

In contrast to the ablation of the accessory pathways, amplitudes of the atrial and ventricular electrograms are critical. The ratio between the amplitude of the atrial and ventricular (A–V ratio) electrograms is optimally less than 1:2. The morphology of the atrial electrogram is helpful; an m-shaped (or w-shaped), fragmented atrial electrogram suggests an area of slow conduction and is a supporting marker of a successful site for interruption of the "slow" conducting pathway. Atrial electrogram morphology and stabilization of the catheter tip can be

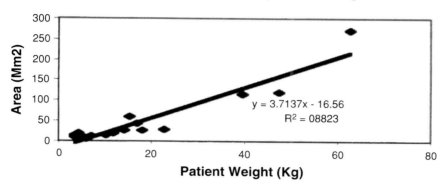

FIGURE 12. Plot of area of triangle of Koch versus patient weight.

improved by examining the effect of the respiratory cycle on these measures. Extended expiration usually stabilizes the catheter and provides the optimal electrogram shape. This technique, however, requires general anesthesia, endotracheal intubation, and controlled mechanical ventilation by an anesthesiologist. Radiofrequency current is delivered during sinus rhythm for 60–120 seconds in the area of slow conduction with the triangle of Koch, with careful attention to atrial-junctional association, as well as intact AV conduction (PR interval) during delivery. Intermittent pacing during application may be helpful in monitoring antegrade conduction if junctional tachycardia appears. Indicators for successful ablation are the appearance of junctional rhythm with constant junctional–atrial association during current delivery. The junctional rhythm observed during current delivery should be somewhat faster than the sinus rhythm ($\leq 125\%$ of the baseline) and demonstrate junctional–atrial association (fixed constant simultaneous atrial electrograms and QRS complexes). An acceleration of the junctional rhythm beyond that level indicates excessive heat is being delivered to the AV node; current delivery should be immediately terminated. Repeat applications can be attempted at adjacent, usually more posterior (dorsally) positions, if slow pathway conduction persists.

Optimal temperature should not exceed 55°C and often 47°C–48°C is sufficient. The absence of echo beats or inducible supraventricular tachycardia following ablation is not a definitive indicator of success, but is helpful if the tachycardia was not easily induced prior to ablation. The strongest indicator for successful ablation is elimination of inducible tachycardia. In addition, the elimination of slow pathway conduction determined by repeat atrial extrastimulation and noting the absence of a "jump" in conduction from the "fast" pathway to the "slow" pathway (Chapter 5, Figure 5) is another surrogate for successful ablation. Due to the proximity of the "slow" pathway to the AV node and normal conduction pathway, cryotherapy is emerging as an alternate and often performed energy source in these patients (Chapter 5).

INTERRUPTION OF THE ATRIOVENTRICULAR NODE BY RADIOFREQUENCY ABLATION

Although interruption of the AV node with radiofrequency ablation followed by pacemaker implant is an accepted treatment for symptomatic, drug resistant chronic atrial fibrillation in adults, it is rarely necessary in children and young adults. When employed, it is usually for intractable atrial arrhythmias

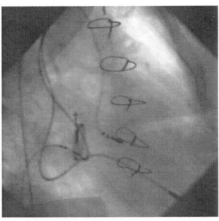

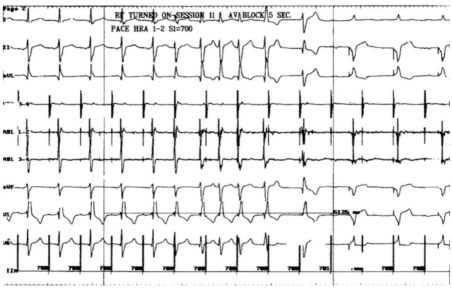

FIGURE 13. This patient had recurrent atrial fibrillation with a rapid ventricular response, as well as repaired AV septal defect and a mitral valve replacement. The AVN–His conduction system, is as often the case in this malformation, could not be detected on the right side. Top panel: Radiograph in the right anterior oblique projection with the retrograde ablation catheter in the left ventricle (arrow). Other leads are for the implanted transvenous pacemaker. Bottom panel: Atrial pacing during delivery of radiofrequency current through the ablation catheter (arrow, top panel). After 6 seconds, AV block is produced and the ventricular pacemaker escapes and captures the ventricles.

in the setting of complex heart disease. Although the standard approach to the AV node is antegrade through the right atrium to the low septal area at the apex of the triangle of Koch (the home of the node), in patients with abnormal anatomy such as transposition of the renal arteries and AV septal defect, the only or best access to the conduction tissue may be retrograde through the aorta (Figure 13).

ABLATION OF ATRIAL FLUTTER AND INTRA-ATRIAL REENTRY TACHYCARDIA: ENTRAINMENT

Typical atrial flutter, characterized by negative flutter waves (300 bpm) in 2, 3, AVF, and V1 on the surface of the electrocardiogram, is rare in children and adolescents. On the other hand, intra-atrial reentrant

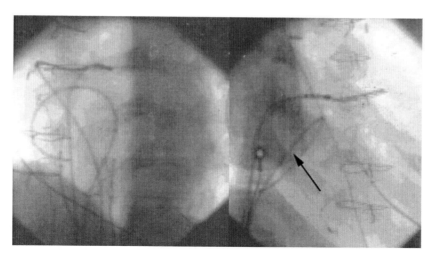

FIGURE 14. Radiographs in the left anterior oblique (left-hand panel) and right anterior obliged (right-hand panel) projections of the 20-electrode "halo" catheter deployed around the tricuspid orifice (circular catheter). Note the *en face* orientation on the left and in profile orientation on the right around the tricuspid valve orifice. A typical atrial flutter activation wavefront moves in a counter-clock wise direction as viewed from the right ventricular apex.

tachycardia, a form of atrial flutter seen in individuals who have undergone extensive atrial surgery for repair of complex congenital heart disease, is not infrequent. The clinical, electrocardiographic, and electrophysiologic characteristics, mechanisms, and complex management of these tachyarrhythmias are extensively discussed in Chapter 8. When sufficiently frequent and long in duration, especially in the face of several failed antiarrhythmic drug trials, and especially with symptoms, radiofrequency ablation of the reentrant circuit of these various forms of atrial flutter is a reasonable and feasible therapeutic approach. In typical atrial flutter, the reentrant circuit circumscribes the tricuspid valve in a counter-clockwise rotation (looking from the right ventricular apex). This circuit can be discerned by deploying a steerable curved 20 electrode "halo" catheter around the tricuspid orifice (Figure 14) and note the atrial activation sequence (Figure 15).

Suspicion that the tachycardia is reentrant is confirmed by entrainment mapping (Table 4). Entrainment requires pacing the atria during the intra-arterial reentrant tachycardia at a cycle length slightly shorter than the tachycardia cycle length for several seconds of atrial capture. If, on termination of pacing, the tachycardia resumes and the post pacing interval (PPI) is equal to the tachycardia cycle length (TCL), the surface ECG P-waves during pacing and tachycardia are identical and the activation sequence of the atrium is identical to the activation sequence during the tachycardia, one could infer that concealed entrainment was achieved and the tachycardia is reentrant (Figure 15). Manifest entrainment indicates that the pacing catheter electrode tip is distant from the reentrant circuit, whereas with concealed entrainment, the catheter tip is in the circuit (Table 4).

Termination of the tachycardia upon cessation of pacing indicates that the pacing cycle length is too short for concealed entrainment. In this instance, the pacing impulses enter both the orthodromic and antidromic limbs of the reentrant circuit and collide within the circuit, self-extinguishing and interrupting the tachycardia when the pacing is terminated. Although not entrainment, termination with overdrive pacing strongly suggests that the arrhythmia is reentrant (Figure 16).

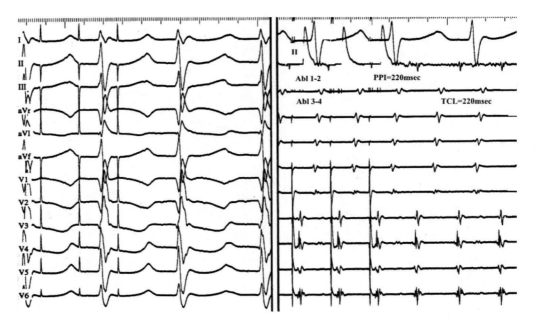

FIGURE 15. Concealed entrainment in the systemic atrium in a 7-year-old boy with intra-atrial reentrant tachycardia 5 years after the Fontan operation (atrio-pulmonary connection). Left-hand panel: 12-Lead ECG at cessation of atrial pacing. Note the similarly shaped) low amplitude (barely discernable—typical for flutter in a Fontan patient) P-waves during pacing and tachycardia, compatible with concealed entrainment. Right-hand panel: Lead II, ablation catheter (Abl 1-2, 3-4) and multiple intra-atrial electrograms (8 bottom tracings). Note the same atrial activation pattern within the 8 bottom tracings during both pacing and the tachycardia. Note also the match in msec between the post pacing interval (PPI; interval from last pacing stimulus to first local electrogram at paced site) and the tachycardia cycle length (TCL). These findings indicate concealed entrainment and place the ablation catheter tip within the reentrant circuit.

In typical atrial flutter, the course of the reentrant circuit traverses the isthmus of excitable tissue, often the site of slow conduction, between the tricuspid valve (and coronary sinus orifice) and the inferior vena cava, both of which are electrically inert. This isthmus is a patient size-dependent, narrow corridor of excitable tissue in the triangle of Koch, bound by two electrically silent barriers (the tricuspid valve orifice and the inferior vena cava). Placing a line of radiofrequency lesions between these two sites will create a line of block, interrupting the reentrant circuit. Such a line of block is confirmed by pacing

TABLE 4. Entrainment of Tachyarrhythmias

Entrainment of Tachyarrhythmias

The appearance of identical P-wave (or QRS if the tachycardia is ventricular and one is pacing the ventricle) morphology (12-lead ECG) during both the tachycardia and atrial pacing (pace mapping).

Manifest (Overt) Entrainment: Acceleration of the tachycardia rate to the pacing cycle length with constant (or progressive) P (between the flutter wave and the wave produced by the stimulus pulse) or QRS fusion and without termination of the tachycardia upon cessation of pacing.

Concealed (Covert) Entrainment: Acceleration of the tachycardia rate to the pacing cycle length with no fusion (i.e., no alteration in the P- or QRS-wave morphology or endocardial activation sequence during pacing) and without termination of the tachycardia upon cessation of pacing.

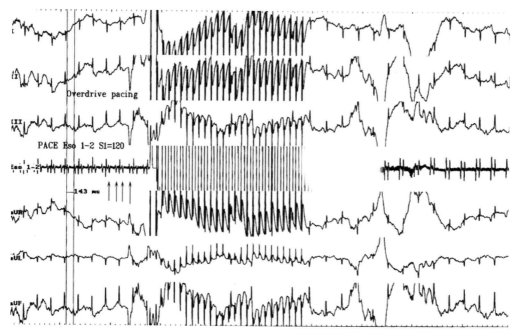

FIGURE 16. Left-hand side: 1-day-old boy with atrial flutter (cycle length = 120 ms). Middle: Overdrive transesophageal atrial pacing (cycle length 100 ms) for 2.9 sec). Right-hand side: Normal sinus rhythm. Leads I, II, III, vVR, aVL, aVF, and transesophageal lead (middle tracing—Eso).

on both sides of the isthmus (i.e., the coronary sinus os on the left and the low right atrial wall on the right) and noting failure of conduction in the pacing impulse across the isthmus.

The problem is more complicated in patients following congenital heart disease surgery (Chapter 8). In this group of patients with reentrant circuits within an atrium, the primary problem is the multiple incisions and suture lines rendering localization of the circuit difficult. Rather than the typical reentrant counter-clockwise circuit around the tricuspid valve, due to existing suture lines, incisions and pathologic changes in the atrial tissue, the multiple reentrant wavefronts can circulate around different scars, suture lines, and the usual anatomic obstacles, can traverse different circuits, and display different, usually slower, cycle lengths. Detecting these circuits is challenging. Creating a line of block across the isthmus of slow conduction of the reentrant circuit is the strongest indicator of successful ablation (Figure 17).

Because of the complexity of the atrial anatomy (postoperative Fontan patients with single ventricle physiology and anatomy, and s/p Mustard or Senning patients for transposition of the great arteries), the complexity of the arrhythmias and their underlining substrate including multiple circuits, scars, and thick atrial walls, radiofrequency ablation in these patients is only 70% successful with a high incidence of reoccurrence. Because of these post-operative conditions within the atria, it is not unusual, nor unreasonable, to have to return to the electrophysiology laboratory for further sessions. Computer-assisted approaches utilizing electro-anatomic mapping and/or non-contact mapping are useful in identifying non-conducting tissue (i.e., scar or anatomic obstacles), as well as the propagating wave front as it traverses the critical isthmus (Chapter 8).

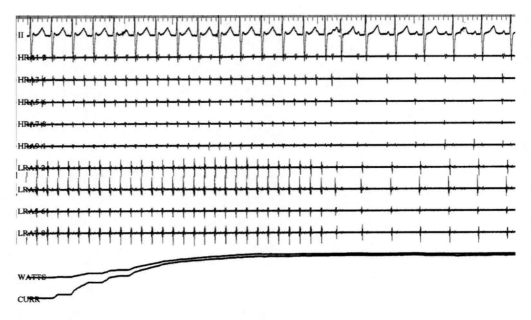

FIGURE 17. In the same patient as in Figure 16, radiofrequency current is delivered as the ablation catheter is drawn from one electrically silent barrier (in this case a scar created at surgery) to another (inferior vena cava) creating a line of block interrupting the reentrant circuit and the thus the tachycardia.

RADIOFREQUENCY ABLATION OF ATRIAL FIBRILLATION

In contrast to the incidence of atrial fibrillation in adults, which has reached public health concern, atrial fibrillation in the young is exceedingly rare. Rarely, teenagers will experience paroxysmal ("lone", i.e., no associated heart disease) atrial fibrillation. Potentially underlying factors increasing the risk of atrial fibrillation such as hyperthyroidism cardiomyopathy should be eliminated in this age group. However, ablation is not usually recommended. For the consideration of more definitive therapy for paroxysmal atrial fibrillation, referral to a medical electrophysiologist is appropriate.

ABLATION OF NON-ISCHEMIC VENTRICULAR TACHYCARDIA

Idiopathic ventricular tachycardia can arise from either the right or left ventricle (Chapter 12). Because this tachycardia appears to be an ectopic mechanism, general anesthesia, which can suppress the tachyarrhythmia, is avoided. Under conscious sedation, the tachycardia must be precisely located using either pace mapping or endocardial mapping during the tachycardia (with a localized ventricular electrogram preceding the QRS morphology, Figure 18). New computer-dependent mapping systems, such as electro-anatomic mapping and non-contact mapping, add correlation between virtual anatomic images and the pattern of excitation; however, they usually require larger size catheters and are therefore less applicable in small children (<30 kg).

Some tachycardias, which appear to arise from the right ventricular outflow tract, may be within the conal septum or on the left side of the conal septum underneath the aortic valve. Ablation above the right coronary cusp, within the coronary cusp or within the ventricle beneath the right coronary cusp may localize these tachyarrhythmias and lead to successful ablation. Right ventricular tachycardia can occur from the muscular septum, as well as the

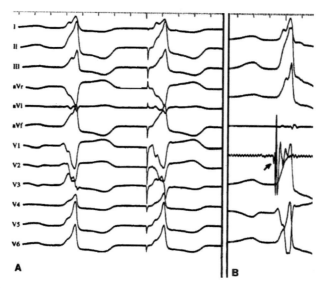

FIGURE 18. Panel A: 12-Lead ECG demonstrating single beat (left-hand QRS complex) of the right ventricular outflow tract ventricular tachycardia (left bundle branch pattern and inferior leftward frontal plane axis.) and a single identical beat (right-hand QRS complex) paced from the right ventricular outflow tract indicating a successful pace map. Panel B: Single spontaneous beat from a run of tachycardia. The top three tracings are Leads I, II, III. The fourth tracing is recorded from the atrium. The bottom 3 tracings are VF, V1, V6. The fifth tracing from the top is recorded from the right ventricular outflow tract; the ventricular electrogram is very early and preceded by a fast action deflection (arrow) in advance of the surface QRS complex, illustrating effective endocardial mapping, localizing the origin of the tachycardia.

inferior wall; these tachyarrhythmias are more difficult to ablate.

Left-sided structures giving rise to ventricular tachycardia can be accessed either through the transseptal route with passage of the mapping/ablating catheter from the left atrium into the left ventricle or by way of the retrograde route around the aortic arch, across the aortic valve into the left ventricle. In small children, the former is favored. Left ventricular tachycardia often arises from downstream specialized conduction tissue of the left bundle and Purkinje network. Sharp fast action deflections indicating activation of specialized conduction tissue may often be recorded, preceding the QRS complex and the ventricular electrogram of these tachycardias (Figure 19). Ablation at those sites is generally successful.

In is important to remember that many young patients with these arrhythmias may be asymptomatic. In addition, the arrhythmia is usually not associated with a risk of sudden cardiac death. Thus, indications for ablation falls within Class III and can often be deferred.

In older patients, especially those with chronic heart disease, the site of the tachycardia may arise from the epicardial surface and benefit from a pericardial approach. This technique has not generally been necessary in young patients with idiopathic ventricular tachycardia.

ABLATION OF ARRHYTHMIAS IN PATIENTS WITH CONGENITAL HEART DISEASE

Several cardiac malformations are associated with accessory pathways. These are Ebstein's anomaly of the tricuspid valve, congenitally corrected transposition of the great arteries (L-TGA), some forms of hypertrophic cardiomyopathy, and some forms

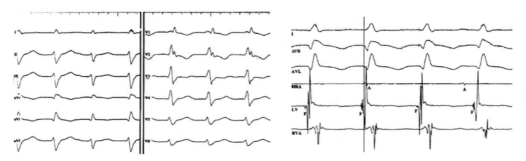

FIGURE 19. Left-hand panel: 12-Lead ECG of left ventricular tachycardia. Right-hand panel: Third tracing from the top demonstrates an early ventricular electrogram relative to the ECG QRS complex preceded by a fast action electrogram (F) indicating the origin of the tachycardia. This site, by fluoroscopy, was located in the region of the left posterior fascicle (f) of the left bundle. Reprinted with permission form Dorostkar, PC., et al. The use of radiofrequency energy in pediatric cardiology. *J Interv Cardiol* 1995;8(5): 557–68.

of single ventricle, including heterotaxy syndromes. When these defects are detected, the presence of a possible manifest or concealed pathway should be considered. Other congenital defects, such as ventricular septal defects and atrial septal defects coarctation, may be complicated by accessory pathways, but do not have a sufficiently strong association nor theoretical developmental link that would warrant undertaking their investigation (in the absence of symptoms).

The accessory pathway in patients with Ebstein's anomaly is in our experience on the right side. However, because of the underdevelopment of the tricuspid valve and the associated often severe tricuspid regurgitation, placing the catheter on the AV groove is difficult, and can be facilitated by a long stabilizing sheath. At the same time, presumably due to the abnormally formed right-sided AV junction, the pathway seems to be, by mapping techniques, broad and diffuse. Placement of the ablation catheter beyond the "true" AV groove deeper into the "atrialized" [indicated by low pressure (Figure 20)] ventricle to locate a possibly constricted insertion end of the pathway may lead to a favorable ablation site.

If patients with congenitally corrected transposition have an accessory pathway, it is usually on the anatomic left side (relative to the thorax), but morphologic right (ventricular inversion) relative to the ventricles. Ablation, therefore, requires a transseptal approach (Chapter 3, Figure 14).

Although not frequently found in patients with AV septal defects, ablation of these pathways may present a challenge, particularly if right-sided. In the presence of an AV defect, the normal location and course of the AV node-His-Purkinje axis is displaced posteriorly. If the pathway is located in the right posterior septal region, delivery of radiofrequency energy at that site may involve the posteriorly placed AV node and result in heart block. Close monitoring of the indicators of proximity to the AV node and attention to AV conduction during energy application is critical. It is not unusual to produce at least first-degree heart block during this application. The absence of clinical or inducible tachycardia, as well as the absence of a short preexcited RR interval during atrial fibrillation, suggests that ablation can de deferred.

COMPLICATIONS OF RADIOFREQUENCY ABLATION

The complication rates are directly dependent on patient weight, patient age, and operator experience. For patients under 15 kg and 3 years of age, the reported incidence of complications by the Pediatric Radiofrequency Ablation Registry is higher than for

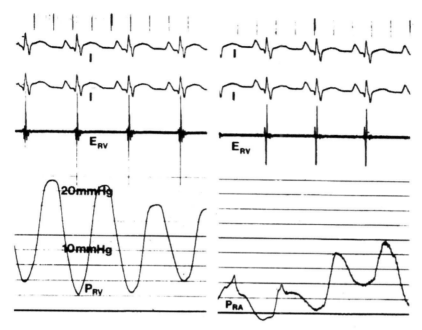

FIGURE 20. Lead I (2). Intracardiac unipolar electrogram (E$_{RV}$) and intracardiac pressure tracing (P$_{RV}$) recorded through a single catheter with a lumen for pressure and a single electrode at the catheter tip .in a patient with Ebstein's anomaly. Left-hand panel: Note simultaneous right ventricular pressure (20 mmHg) and ventricular electrogram. Right-hand panel: The catheter has been withdrawn slightly. Note that the pressure is now low at trail level while the electrogram is still ventricular in origin, indicating the "atrialized portion of the right ventricle."

older patients. Likewise operator experience decreases the complication rate. Complications can be broken into both major (requiring intervention) and minor (resolve spontaneously with no consequence) types. In a recent Pediatric Radiofrequency Ablation Registry analysis, inadvertent ablation of the AV node with complete heart block, perforation, systemic thrombosis with central nervous system embolization, and death (1 in 3187 patients without other heart disease) constitute the most significant complications. However, the occurrence of these adverse events has been significantly reduced since the early 1990s.

SUMMARY

Transcatheter ablation offers the possibility of complete, safe correction of tachyarrhythmias, particularly supraventricular, in children. Although radiofrequency energy is the predominant form of energy delivery to ablate abnormal pathways and ectopic foci in the human heart, cryotherapy, long used in the operating room but now delivered through a catheter, introduces a new technique that is, prior to full application, reversible. Thus, as one is applying a discrete, well-localized cryotherapy lesion, one can see its effects, whether salutary or deleterious, and terminate the delivery, allowing reversal of an adverse cryo effect. Early experience suggests that transcatheter cryoablation is a useful addition to the treatment of a number of tachyarrhythmias—especially AV nodal reentry tachycardia. Other advances in the understanding of arrhythmias, in catheter design (smaller, more flexible), in navigation techniques, in imaging systems (real time magnetic resonance imaging), and in ablation energy sources promise to bring even further innovation in the future.

SUGGESTED READING

1. Friedman RA, Walsh EP, Silka MJ, et al. NASPE Expert Consensus Conference: Radiofrequency catheter ablation in children with and without congenital heart disease. Report of the writing committee. North American Society of Pacing and Electrophysiology. *Pacing Clin Electrophysiol* 2002;25(6):1000–1017.
2. Van Hare GF, Lesh MD, Scheinman M, Langberg JJ. Percutaneous radiofrequency catheter ablation for supraventricular arrhythmias in children. *J Am Coll Cardiol* 1991;17(7):1613–1620.
3. Dick M 2nd, O'Connor BK, Serwer GA, et al. Use of radiofrequency current to ablate accessory connections in children. *Circulation* 1991;84(6):2318–2324.
4. Van Hare GF, Witherell CL, Lesh MD. Follow-up of radiofrequency catheter ablation in children: results in 100 consecutive patients. *J Am Coll Cardiol* 1994;23(7):1651–1659.
5. Erickson CC, Walsh EP, Triedman JK, Saul JP. Efficacy and safety of radiofrequency ablation in infants and young children <18 months of age. *Am J Cardiol* 1994;74(9):944–947.
6. Erickson CC, Carr D, Greer GS, et al. Emergent radiofrequency ablation of the AV node in a neonate with unstable, refractory supraventricular tachycardia. *Pacing Clin Electrophysiol* 1995;18(10): 1959–1962.
7. Aiyagari R, Bradley DJ, Etheridge S, et al. *Pediatric Cardiology* 2005, in press.
8. Blaufox AD, Felix GL, Saul JP. Pediatric Catheter Ablation Registry. Radiofrequency catheter ablation in infants ≤18 months old: when is it done and how do they fare?: short-term data from the pediatric ablation registry. *Circulation* 2001;104(23): 2803–2808.
9. Danford DA, Kugler JD, Deal B, et al. The learning curve for radiofrequency ablation of tachyarrhythmias in pediatric patients. Participating members of the Pediatric Electrophysiology Society. *Am J Cardiol* 1995;75(8):587–590.
10. Kugler JD. Radiofrequency catheter ablation for supraventricular tachycardia. Should it be used in infants and small children? *Circulation* 1994; 90(1):639–641.
11. Van Hare GF, Lesh MD, Stanger P. Radiofrequency catheter ablation of supraventricular arrhythmias in patients with congenital heart disease: results and technical considerations. *J Am Coll Cardiol* 1993;22(3):883–890.
12. Kanter RJ, Papagiannis J, Carboni MP, et al. Radiofrequency catheter ablation of supraventricular tachycardia substrates after Mustard and Senning operations for d-transposition of the great arteries. *J Am Coll Cardiol* 2000;35(2):428–441.
13. Friedman PL, Dubuc M, Green MS, et al. Catheter cryoablation of supraventricular tachycardia: results of the multicenter prospective "frosty" trial. *Heart Rhythm* 2004;1(2):129–138.
14. Fischbach PS, Saarel EV, Dick M 2nd. Transient atrioventricular conduction block with cryoablation following normal cryomapping. *Heart Rhythm* 2004;1(5):554–557.
15. Estes NA 3rd. Catheter cryoablation of supraventricular tachycardia: Quo vadis? *Heart Rhythm* 2004;1(2):139–140.
16. Miyazaki A, Blaufox AD, Fairbrother DL, Saul JP. Cyro-ablation for septal tachycardia substrates in pediatric patients: mid-term results. *J Am Coll Cardiol* 2005 15;45(4)581–588.

23

Nursing Management of Arrhythmias in the Young

Sarah Leroy

The nursing care of children and adolescents with arrhythmias has significantly changed in response to major medical advances over the last decade. These medical advancements include catheter ablation as first-line therapy for many childhood arrhythmias, the proven efficacy of implantable cardiovertor-defibrillators (ICD) for life-threatening ventricular arrhythmias, and the growing population of adolescents and young adults who have survived complex congenital heart surgery with chronic arrhythmias. In addition, recent advances in understanding the genetic etiology of many arrhythmias have significant implications for nursing roles. These rapid technological advances and the complexity involved in the diagnosis and treatment of pediatric arrhythmias necessitates a multidisciplinary approach and skilled, knowledgeable nursing personnel are critical to optimal, patient-centered safe care.

SCOPE OF NURSING PRACTICE

The scope of nursing practice (i.e., what roles nurses may legally assume) is regulated by state legislation via nurse practice acts and other legislation with oversight provided via the state board of nursing. Nonetheless, there is considerable ambiguity and variation in specific roles across institutions. Because there are no formal nursing educational programs that are specific for arrhythmia management, the Heart Rhythm Society (www.hrsonline.org) has published comprehensive educational guidelines that outline core curriculum and standards of practice for "allied health professionals" in the field of pacing and electrophysiology (EP). The society offers two certification review examinations: one that assesses knowledge of pacemakers and ICDs and the other focused on basic electrophysiology and electrophysiologic studies (EPS). These examinations are important resources but the materials are primarily oriented to the care of the adult patient. Although there is considerable overlap, they do not address issues specific to the pediatric population. Expert nursing care of these patients is demanding as it requires integrating the core knowledge of pediatric nursing with that of cardiac arrhythmias. Additional specialized knowledge and skills are necessary for the EP nurse caring for the patient undergoing specific therapies (e.g., radiofrequency ablation, administration of conscious sedation, device implantation).

Nursing: Definitions and Processes

Nursing goals focus on the diagnosis and treatment of human responses to illness. Generic goals include restoration of health, optimal functioning with chronic illness, or a peaceful death. The nursing process is used to develop a comprehensive plan of care for each patient family (Figure 1). Optimally, goals are set in collaboration with the family (Table 1). The nursing plan is achieved through provision of physical, emotional, and spiritual care to both the child and the family.

The initial phase of the nursing process is an in-depth assessment. This assessment can be based on either a longstanding relationship between the patient/family and the electrophysiology team or a newly developed relationship that is primarily procedure/hospitalization focused.

A child's developmental level, particularly the cognitive level, is key to providing effective, age-appropriate nursing care. Cognitive development determines the timing and content of information given, appropriate language to be used (particularly medical terminology), optimal methods for pre-procedure preparation, and may influence choice of general anesthesia or conscious sedation for children undergoing electrophysiologic study or radiofrequency ablation.

Child and parent responses to the diagnosis can be understood from a conceptual framework of stress and coping. For most families, the diagnosis of a cardiac arrhythmia represents a health crisis causing varying degrees of child and family stress. Stress occurs when the relationship between a person and his/her environment is or is perceived as taxing or threatening to personal well-being, eliciting a variety of physical, cognitive, and emotional outcomes ("flight or fight" sympathetic nervous system response). Coping refers to the set of cognitive, emotional, and behavioral techniques employed to manage stress. An individual's coping style is dependent on a number of interacting factors including problem-solving skills, social skills, social support, health and energy, beliefs, temperament, developmental level, family coping patterns, and material resources. Successful coping is not simply the absence of distress, but rather a perception by the child or parent that the event, although unpleasant, is manageable.

Nursing Roles for the Child with an Arrhythmia

Care of the child with an arrhythmia is frequently provided in large tertiary care centers that cross traditional hospital boundaries. While some arrhythmias are self-limited or carry only minor health implications, others represent chronic problems requiring highly specialized lifetime treatments. Fortunately, few pediatric arrhythmias are life-threatening. In response to patient needs, nursing roles span the continuum of care from the outpatient clinic, to the electrophysiologic laboratory, surgery, inpatient units, and long-term follow-up. In addition to the traditional nursing roles, EP nursing roles are specialized and are summarized in Table 2. The specially trained advanced practice nurse (APN) is ideally positioned to provide care from pre-admission assessment through post-discharge follow-up, including pre-procedure education and preparation, telephone triage, pre-procedure history and physical, education, and post-procedure management and follow-up.

Nursing Care during the Diagnostic Phase

The time from onset of symptoms until diagnosis can be a particularly stressful time for children and their parents. Many families have personal experience with family members or friends who have experienced heart attacks or other serious heart problems. Parents of young athletes are particularly anxious due the extensive media coverage of sudden

Nursing Assessment

Child Factors
1. Developmental stage
2. Temperament/coping style
3. Previous healthcare experiences
4. Understanding of situation/procedure
5. Fears
6. Learning style

Family Factors
1. Family composition/roles
2. Support networks
3. Other family stressors
4. Parental coping
5. Knowledge of arrhythmia/treatment /treatment options
6. Fears
7. Cultural/religious factors
8. Previous hospital experiences (particularly if aversive)
9. Decision-making
10. Learning style

Healthcare Factors
1. System resources
2. Procedure planned
   a. EPS
   b. Surgery
   c. Device implant
   d. Catheter ablation

Nursing Plan of Care

1. Patient goals and expected outcomes
2. Plan: pre-procedural preparation, education
   a. Optimal timing
   b. Methods
   c. Personnel
3. Interdisciplinary communication and coordination
4. Documentation
5. Referrals as indicated: e.g. language/cultural factors Special needs (deaf/blind/autistic)
6. How care delivery is evaluated/ follow-up

Implementation

1. Performance of non-invasive testing
2. Assist invasive testing
3. Support and education throughout hospitalization
4. Documentation

Evaluation

1. Were expected outcomes achieved for patient/family?
2. Are there opportunities to improve system of care delivery?

FIGURE 1. Nursing Care of Children and Adolescents with Arrhythmias.

TABLE 1. Nursing Plan

1. Nursing care that ensures optimal, safe, and effective use of diagnostic and treatment procedures.
2. Child and family understanding of the diagnosis, treatment options, and treatment plan.
3. Reduced anxiety and effective coping for patients and family members/caretakers during specific stress points such as during the diagnosis process, initiation of treatment, and surrounding an invasive procedure.
4. Increased sense of mastery of the environment and maintenance of personal self-control by patients and family members/caretakers.
5. Enhanced trust between patients, family members/caretakers, and healthcare providers.
6. Positive long-term emotional and behavioral adjustments in patients and family members/caregivers.
7. Transition of care from childhood to self-sufficiency as young adult able to take personal responsibility for own healthcare needs

cardiac death in athletes. Fortunately, reassurance is often effective during the diagnostic process. Nursing roles during this phase are outlined in Table 3.

Anxiety during these tests tends to be more of a problem for younger children (<5 years) who often respond well to distraction techniques such as puppet play or video movies. Once reassured that the tests are not painful, older school age children generally tolerate non-invasive testing well and are often interested in the equipment and output (e.g., ultrasound pictures of their heart).

TABLE 2. Specialized EP Nursing Roles

1. Pre-procedure preparation for invasive procedures
2. Administration and monitoring of intravenous sedation
3. Assisting in the electrophysiology laboratory during EP study, radiofrequency ablation, electrical cardioversion/defibrillation, device implantation and testing, inpatient telemetry monitoring
4. Outpatient device follow-up
5. Transtelephonic monitoring
6. Non-invasive cardiac testing
7. Phone triage

TABLE 3. Nursing Roles–Diagnostic Phase

1. Preparing children for cardiac testing
2. Providing information about the heart, the conduction system and, the arrhythmia.
3. Vagal maneuvers
   Expiring against a closed glottis ("blow on one's thumbas if it were a trumpet")
   Gag reflex
   Headstand
   Immersion of face into ice water
4. Preparation regarding non-invasive testing such as electrocardiograms, echocardiograms, exercise treadmill testing, and transtelephonic monitoring

Nursing of the Patient Diagnosed with an Inheritable Arrhythmia

Since the mid 1990s, knowledge regarding the heritable nature of a number of pediatric arrhythmias has increased dramatically (Chapter 18). The diagnosis of a hereditary illness imposes significant additional family stress as siblings and/or other family members may be affected. At a time when parents are already coping with the stress of one child's diagnosis, they may be informed that other family members are also affected. Parents may experience feelings of guilt due to an interpretation that "I may have given this to my child." Although usually not initially a concern, inheritable diagnoses also raise questions about future childbearing.

For many of these diseases, genetic testing at this time may not be definitive and is very expensive, limiting the value of molecular testing. In addition, there are no practice guidelines to direct the choice or timing of genetic or non-invasive testing. Younger family members may require serial testing over years as new mutations are discovered. This uncertainty leads to considerable ambiguity and stress, particularly since many of these diagnoses involve life-threatening arrhythmias. Thus, in addition to providing family support, education regarding the genetics of the disease, which is constantly being updated, screening recommendations, and available supportive resources is required. While referral for genetic counseling may be desired these services may be difficult to schedule on a timely basis and may not be accessible to all family members. However, the role in genetic counseling particularly requires specific skills and knowledge so nurses who assume this role require additional training. Nonetheless, when genetic information about the diagnosis and genetic implications is unambiguous and given in writing, family comprehension is improved and distribution of information to the extended family facilitated.

Nursing Interventions for Pre-procedure Preparation

Pre-procedure preparation remains a critical role for nurses caring for children undergoing invasive electrophysiologic procedures. The importance of pre-procedure preparation in decreasing child and parent anxiety as well as decreasing the long-term ill effects of potentially aversive hospital experiences that persist after hospital discharge is well known. During the initial phases of diagnoses and treatment, nurses assess child and family stressors and coping responses. A phone call one to two weeks prior to the procedure provides an excellent opportunity to initiate assessment of child and family needs and to provide anticipatory information about the proposed procedure. This phone call also allows the nurse to begin to establish rapport with the family which is particularly important if they are new to the institution. Assessment of social needs such as transportation, hotel, and insurance during this call facilitates timely referral to social work.

Early studies document the efficacy of providing information about anticipated medical procedures and this continues to provide the foundation for most pre-procedure preparation programs. Booklets, videotapes, structured medical play, peer counseling or tours of the electrophysiology lab are useful methods to increase understanding of the proposed procedures. Use of these materials is enhanced by creative verbal communication with health professionals. In one of the earliest studies regarding preparation of children for heart catheterization, the researchers showed a positive effect on children's behavioral stress through information delivered to children via a puppet show. Several studies have shown that information-giving is significantly enhanced when sensory information (what the patient will see, feel, smell, taste) is provided. Cues from the family are used to guide the interventions as health information can actually increase anxiety for some patients.

The timing of information-giving is based on the child's cognitive level (Table 4). Young children (<5–7 years of age) respond best if information is given the day before the procedure, in a short conversation using

TABLE 4. Pre-procedure Screening Phone Call

Medical screening	Psychosocial screening	Anticipatory guidance
Update health history	Child/family response to planned procedure	Ask parent/guardian to identify concerns and then address these concerns
Identify co-morbidities Develop plan as needed Consultation with specialists as needed	Assess family composition/support systems/coping methods	Describe procedure, post procedure care, discharge plan, restrictions
Obtain brief family health history (anesthesia reaction, bleeding problems, seafood/iodine allergies	Identify any current psychosocial problems needing plan or referral	Describe recovery period (when child can return to usual activities including sports
Identify and obtain any missing medical information (esp tracing of arrhythmia)	Identify social needs; financial, insurance, transportation, housing Refer to social work PRN	Offer information re: talking with child about procedure: timing and content. Refer to child life PRN

TABLE 5. Common Stressors

EPS/RFA	Device implantation
Blood draws, IV start, separation from parents, if IV sedation may be partially awake and feel arrhythmia induction, long procedure, post procedure cardiac monitoring.	Anesthesia induction, separation from parents, IV start, blood draw, cardiac monitoring, incision care, body image changes, if ICD: induction of VT/VF to test device, social changes

age appropriate terminology. Approaches involving mothers have proven effective for younger children including a pre-admission home-based program taught to mothers. Older school age children seem to respond optimally when information is provided one to two weeks prior to the procedure, while adolescents are often involved in the initial and ongoing decision-making process.

Due to the length and potential discomfort associated with most invasive electrophysiologic procedures, pharmacological sedation or general anesthesia is required. However, children and families still experience various stress points (Table 5). Comprehensive stress management programs that combine information-giving and emotional support with cognitive behavioral intervention are associated with positive short- and long-term coping by children and parents. These include conscious breathing, guided imagery, biofeedback, and refocusing. Overall, the critical dimension of pre-procedure preparation may not be the actual intervention used but the extent to which the children or adolescents have a "plan" for dealing with a procedure. For children who are highly anxious, particularly if there is a history of prior aversive hospital experiences, additional pre-procedure preparation with referral to a psychologist, child life therapist, or social worker may be indicated.

Assessment of parental anxiety and stress levels is also important as both can affect the child's stress. Accurate information given to parents should dispel any misconceptions from previous cardiac experiences and, as with children, should be individualized based on cultural and intellectual background, prior hospital experience and knowledge, as well as emotional needs. Parental understanding about what to expect during the child's hospital stay allows them to anticipate the physical and behavioral changes that may occur and develop strategies to prepare the child.

Conscious Sedation versus General Anesthesia

Goals of pre-medication and sedation (or anesthesia) are to decrease anxiety, ensure patient comfort during the procedure, promote amnesia, and facilitate the performance of the procedure so that it may be undertaken as safely as possible. The choice of sedation is based on clinical factors, such as the age of the patient, the estimated length of the procedure, and the expected inducibility of the arrhythmia. In some patients, general anesthesia may not be an option as it may affect successful induction of the clinical arrhythmia.

Administration of intravenous (IV) sedation requires specialized nursing skills. Comprehensive hospital-based policies and procedures must be in place to comply with JCAHO accreditation standards. The American Academy of Pediatrics and NASPE (Heart Rhythm Society) have also published guidelines for sedation delivered by non-anesthesia personnel.

Care of Patient undergoing Electrophysiologic Study and Radiofrequency Ablation

Table 6 outlines the potential feelings and responses of children undergoing invasive cardiac procedures by age group. Additional stresses associated with electrophysiologic study include the intentional induction of the arrhythmia and the relatively long duration of the procedure. Therefore, regular

TABLE 6. Potential Feelings and Responses to Invasive Procedures

Young child	Older child
Aggression, regression	Anxiety
Injury or harm to person	oss of control
Abandonment/separation from parent	Invasion of privacy
	Bodily harm/death

updates with parents are important to alleviate anxiety.

Table 7 outlines information regarding child psychological preparation prior to an EPS. As mentioned, information-giving is greatly enhanced when it includes sensory information. Table 8 lists some of the sensory information given to children and families. This information will be more reassuring if a nurse in the lab is identified whom the child can address his/her questions and concerns.

Specific counseling for either conscious sedation or general anesthesia is part of the preparation. Older children who will receive IV sedation are told that there is very little pain involved but the procedure can be quite long and they will need to lay very still. They will feel a pinch or a "poke" when the IV is started. They will be given two medicines through the IV: one for pain and another that will make them feel relaxed and sleepy. They are reassured that a nurse will remain with them and will increase the medicine based on the child's self-reported pain. The routine use of EMLA cream has been effective in decreasing discomfort from the local anesthetic given at the site of catheter insertion. Some children will feel the delivery of the radiofrequency energy, and can be told: "It feels like heartburn; like when you eat too much pepperoni pizza." Reassurance that most children report that the procedure is not as "bad as they thought it would be" is often helpful in alleviating anxiety. If general anesthesia is used, information about the induction procedure (what the child will feel) may be beneficial.

Discharge instructions with phone numbers for both week-day and evening hours are an important part of care. Table 9 outlines routines from pre-to post discharge.

Care of the Patient who Requires Device Implantation

Although many parents have some familiarity with device therapy in older relatives or friends, they may express surprise that this would even be a consideration in a child and may interpret it as meaning the child has a poor prognosis. Common stressors encountered during device implantation are listed in Table 10.

Children and families need information about what to expect during the perioperative period. Parents, in particular, need information regarding indications for implantation, device function, pre-operative testing and routines, anesthesia, and the operation itself. Nurses, using both hands-on instruction

TABLE 7. Child Psychological Preparation Prior to an EPS

Young child	Older child (>8 yrs)	Adolescent
Provide minimal information within 24hrs of procedure	Provide basic information about 1 week prior to procedure	Usually involved in decision. Often process begins at time of dx or clinic visit
Allow medical play opportunities. Use child's questions as cue for providing information	If seen day before procedure, offer tour of cath lab, sensory descriptions of experience	If seen day before procedure, offer tour of cath lab, sensory descriptions of experience, lay flat 4–6hrs post procedure
Minimize separation from parents	If very anxious, consider referral to child life specialist	May benefit from learning cognitive-behavioral techniques
Consider general anesthesia	Consider general anesthesia	IV sedation usually adequate

TABLE 8. Sensory Information Given to Children and Families

Visual	Feel (if IV sedation)	Hear	Taste
The EP lab-can describe or make reference to OR's on TV. Tour of lab helpful for some. People wearing masks Monitors/cameras	Cardiac electrodes Arm restraints Sterile drapes "Cold" feel of betadine Brief pain with local anesthetic Need to lay flat during cath and 4–6 hrs afterwards	Beeping monitors Voices of doctors and other personnel May offer music or movies as means of distraction	Won't be able to eat of drink before or during EPS Will be started on clear liquids and then solid foods as tolerated

and printed materials, provide guidance regarding device follow-up including signs and symptoms of device malfunction, use of remote (via transtelephonic transmission) monitoring, and the need for periodic replacement due to battery depletion and appropriate activities for their children. It is our practice to recommend CPR training for all parents; parents with pacemaker-dependent children receive CPR training in the hospital prior to discharge. Education of the children varies, based on developmental stage, among other factors. Children who undergo initial pacemaker implant in infancy or early childhood need ongoing education regarding their heart disease and treatment. It is important to remember that children's social networks may include nontraditional family members, school personnel, and extended family.

Nurses are in a key position to facilitate the child's acceptance. For example, at our center, a school-aged child who needs a first time pacemaker or defibrillator is given a stuffed bear with a "demo" of the device in a pocket. A simple explanation is given such as "this bear is for you. He (or she) has a pacemaker (or defibrillator), too. As you can see, it is very small but it has a big job in helping your heart to beat on time."

Discharge instructions are an important part of the care delivered. Table 11 describes our discharge routines.

Parents are often anxious about complications and long-term outcomes and should have an opportunity to ask sensitive questions without the child present. Parents should also be made aware of social work and other available support systems. Parents have a critical role in providing emotional support to their child and should be prepared and encouraged to participate in normal parenting routines (such as feeding, bathing, playing, reading stories, cuddling) while the child is hospitalized. Reassurance should be given to the parents that they will remain near the child, that they will be encouraged to stay with the

TABLE 9. Processing Routines

Pre-procedure Routine	Post-procedure hospital care	Discharge instructions
NPO solids 6 hours Clear liquids until 3 hours Then NPO If older patient, shave upper leg	6 hour post procedure observation, cont CR monitoring Lay flat 2–4 hours, if venous access only then raise HOB 30 deg. If venous access, BRP after 4 hrs Testing: ECG, Holter, ± echo Tylenol for pain	Remove pressure dressing after 12 hours or in morning Cover site with band-aid X 2 days. May shower POD 1 May return to full activities including competitive sports, swimming in 72hrs

TABLE 10. Stressors Related to Device Implantation

Psychological	Physical
Separation from parents/peers	Hospital routines; monitoring, meds, taking V/S
Fear of death	Blood draws/IV starts
Loss of control	Incision/pain/bandage changes
Anticipation of surgery 2–4 hrs duration	Need for repeated surgeries
Anesthesia induction	If ICD, potential for painful shocks
Induction of arrhythmia	Usually 24 hr hospital stay
Body image changes	
Fear of social rejection	
Device recalls	

child whenever possible, that staff will be supportive of their family's needs, and that the child's pain will be well controlled. Guidance regarding the child's appearance after surgery as well as the impact of hospital experiences on children's behavior both during and after discharge is also valuable.

Psychosocial Adjustments with Arrhythmias and Implanted Pacemakers

A major treatment goal for children with arrhythmias and implanted devices is to facilitate positive child and family adjustments long-term with prevention of avoidable negative psychosocial outcomes. Important illness-related tasks for parents of children with arrhythmias or devices include participation in all aspects of the treatment and preservation of both their child's and their emotional well-being, including preparing for an uncertain future. Adaptive parental strategies include maintaining family integrity through cooperation and maintenance of an optimistic outlook; maintaining social support, self-esteem, and psychological stability; and understanding the medical situation through communication with other parents and healthcare providers.

It is well recognized that children with heart disease are at increased risk for psychological adjustment issues. Although the majority of children make positive adjustments, up to 30% of children with chronic illness may demonstrate secondary psychosocial impairment by age 15. Negative adjustments are likely to occur when stress exceeds the child's adaptive capacities. Yet there is wide variation in children's adaptations and some children are able to make positive adaptations despite extremely stressful circumstances. In this regard, many children with arrhythmias are successfully treated and "cured" of their tachycardia, thus making long-term issues a

TABLE 11. Discharge Routines

Wound care	Activity instructions	Follow-up
Initial dressing removed after 24 hours. Cover with gauze dressing to protect from clothing. Leave open to air as much as possible. Post implant oral antibiotics often advised for 2–3 days.	If new lead(s), avoid overhead arm movement for 6 weeks If transvenous device, rotate affected shoulder gently to avoid muscle stiffness. Arm sling optional.	Wound check in 2 weeks Office visit with device interrogation in 6 weeks.
May shower after 72 hours.	May return to school/work in a week or as able.	Send transtelephonic ECG if symptoms of palpitations, dizziness, or syncope.
Call immediately if s/s of infection.		Discuss triage/access of medical system if problems.
May use sling to support arm first 24–48 hours.		

lesser concern. Children at high risk for emotional or behavioral adjustment problems may benefit from early identification and intervention.

Children with concomitant heart disease and chronic recurrent arrhythmias or devices must cope with multiple chronic stressors. While most children with pacemakers exhibit normal psychological adjustments after pacemaker implantation, problems with fear of social rejection, negative perceptions of having a pacemaker, external locus of control, and altered body image have been described and the emotional aspects of incorporating complex, life-saving devices into body image and sense of self has received little attention. At initial implant and at pacemaker replacement, many questions and concerns often arise. Many parents are unfamiliar with the use of pacemakers in children and may assume that needing a pacemaker means the child's "heart is getting worse" or "won't live long since only old people get pacemaker." Exploration of such misperceptions affords the opportunity to correct overly pessimistic interpretations and to reframe perceptions to include the positive aspects of the implant (e.g., greater energy levels). Conversations with providers about what the pacemaker means for the child's future are important as emotional well-being is predicated on a hopeful outlook; one that permits belief in a personal and meaningful future. Families are generally more relaxed during routine follow-up visits making this an optimal time to further the provider-family relationship, explore lifestyle issues, and affirm successes in the child's life. Nurses can also link families with state and national family support networks and reputable online support groups. Children may benefit from meeting other children with pacemakers in the clinic or by having a "pacemaker pen pal." One child who attends our clinic was feeling isolated because he "never met other kids who have pacemakers." A social gathering was arranged with the child, his parents, and three other families who agreed to participate. A pediatric psychologist facilitated the discussion and the distressed child experienced a more positive mood after the meeting. Attending camps for children with heart problems has also been shown to improve children's adjustments.

An important functional outcome for children with pacemakers is successful integration into the school community. While there is realistic concern about negative stereotyping by school personnel, parents often feel that school personnel should be informed about their child's illness and that health professionals are the most appropriate persons to do this education. To facilitate school integration, we have used a variety of interventions including participation in parent-teacher meetings at the child's school.

Psychosocial Adjustments Post Implantation

Current experience of the medical outcomes for children (4–18 years) after ICD implant indicates that children do return to school post-implant. In "younger" ICD patients (13–49 years) there is a return to overall good health and return to previous activities. However, diminished social interactions, worry and avoidance of exercise, concern about sexual activity, body image concerns, and fear about shock and the possibility of dying remain. It has been our experience that the potential for a life-threatening arrhythmia and receiving a shock are sources of considerable anxiety. While the incidence of severe psychologic problems have been low in young patients post implant who have not received shocks, problems with anxiety, separation anxiety, and school phobia have been common following delivery of a shock. Conscious shocks, particularly repeated conscious shocks, have proven to be very traumatic and we have several patients develop symptoms of traumatic stress disorder (PTSD) including intrusive thoughts re-experiencing the traumatic event, suicidal thoughts, feelings of detachment, and high levels of anxiety. Most patients respond well to counseling and behavioral-cognitive therapies such as relaxation techniques; however, some have required pharmacotherapy, most often antidepressant medication. It is important to note that none of these adolescents had a history of psychiatric problems prior to their ICD implant. Similar to conditioned learning, repeated unpredictable aversive stimuli such as shocks can produce fear and avoidance behavior and can have long-term debilitating effects. This behavior mode, referred to as learned helplessness, has been linked with human depression.

Whenever possible, contacting the family in advance of surgery permits anticipatory guidance and perhaps more importantly demonstrates that the healthcare team aims to provide reliable attention and care to the child and family. Again, overly pessimistic perceptions can be reframed; for example, by introducing the idea that the ICD will provide "a guardian angel" that will permit the child to participate in life more fully.

Support groups and specialty summer camps have shown to be helpful in promoting positive adjustments to chronic illness in a family member. Professionals working with these patients report that many younger patients do not participate in ongoing ICD patient support groups because the groups do not address issues of personal concern such as work, intimate relations, childbirth, school, friends, and dating. Also, the relatively small numbers of young defibrillator patients in any one geographical area have limited the ability to initiate support groups specifically for this population. In an attempt to address these needs, we have collaborated with the medical electrophysiology service at our center to sponsor a yearly "Youth and Young Adult Support Seminar" for ICD patients "thirty-something" and younger. In addition to educational workshops on topics of interest to young patients, there are professionally facilitated support groups divided by age, gender, and role (patient, spouse, parent). Post conference evaluations reveal that the children and adolescents interact with one another and with young adult ICD patients and see first hand that a meaningful future is possible for them.

Because school re-entry after ICD implantation has been a source of considerable concern for parents and school professionals, education of school personnel with parental permission by clinic staff regarding the child's arrhythmia, how the device works, what to do if therapy is given, and development of an emergency plan for school personnel is routine after implantation. During the meetings, it has been helpful to emphasize the protection afforded by the device and that the purpose of undergoing device implantation is to permit the child to live, as much as possible, a normal life.

SUMMARY

Nurse–patient–family interactions afford key opportunities for education, support, and interventions aimed at facilitating positive child and family adjustments for children with arrhythmias. While some situations are relatively minor requiring minimal nursing intervention, others are very demanding requiring an intensive, long-term level of care. As the complexity of care advances, the need for collaboration amongst all care providers involved increases.

SUGGESTED READING

1. LeRoy S. Clinical dysrhythmias after surgical repair of congenital heart disease. *AACN Clinical Issues* 2001;12:87–99.
2. Schurig L, Gura M, Taibi B. Educational guidelines: Pacing and electrophysiology, Armonk, NY: Futura, 1997.
3. Gura M, Bubien RS, Belco KM. North American Society of pacing and electrophysiology. Standards of professional practice for the allied professional in pacing and electrophysiology. *PACE* 2003;26:127–131.
4. Lazurus RS, Folkman S. *Stress, Appraisal, and Coping*. New York, NY: Springer;1984.
5. Bankhead C, Emery J, Qureshi N, et al. New developments in genetics-knowledge, attitudes and

information needs of practice nurses. *Fam Pract* 2001;18:475–486.
6. LeRoy S & Russell M. Long QT syndrome and other repolarization related syndromes. *AACN* 2004;15: 419–431.
7. LeRoy S, Elixon M, O'Brien P, et al. AHA Scientific Statement: Recommendations for preparing children and adolescents for invasive cardiac procedures. *Circulation* 2003:1008;2550.
8. Johnson JE. Effects of accurate expectations about sensations on the sensory and distress components of pain. *J Pers Soc Psychol* 1973; 27:261–275.
9. O'Byrne KK, Peterson L, Saldana L. Survey of pediatric hospitals' preparation programs: evidence of the impact of health psychology research. *Health Psychol* 1997;147–154.
10. Campbell LA, Kirkpatrick SE, Berry CC, et al. Psychological preparation of mothers of preschool children undergoing cardiac catheterization. *Psychol Health* 1992;7:175–185.
11. Campbell LA, Kirkpatrick SE, Berry CC, Lamberti, JJ. Preparing children with congenital heart disease for cardiac surgery. *J Pediatr Psychol* 1995; 20:313–328.
12. Campbell L, Clark M, Kirkpatrick SE. Stress management training for parents and their children undergoing cardiac catheterization. *Am J Orthopsychiatry* 1986;56:234–243.
13. Cassell S. Effect of brief puppet therapy upon the emotional responses of children undergoing cardiac catheterization. *J Consult Psychol* 1965;29:1–8.
14. Melamed BG, Myer R, Gee C, et al. The influence of time and type of preparation on children's adjustments to hospitalization. *J Pediatr Psychol* 1976;1:31–37.
15. Committee on Drugs. Guidelines for monitoring and management of pediatric patients during and after sedation for diagnostic and therapeutic procedures. *Pediatrics* 1992;89:1110–1114.
16. Committee on Drugs; Guidelines for monitoring and management of pediatric patients during and after sedation for diagnostic and therapeutic procedures: addendum. *Pediatrics* 2002;110:836–838.
17. Alpern D, Uzark K, Dick M. Psychosocial responses of children to cardiac pacemakers. *J Pediatrics* 1989;114:494–501.
18. Andersen C, Horder K, Kristensen L, Mickley H. Psychosocial aspects and mental health in children after permanent pacemaker implantation. *Acta Cardiologica* 1994;49:405–418.
19. Utens EM, Versluis-Den Bieman HJ, Verhulst FC, et al. Psychological distress and styles of coping in parents of children awaiting elective cardiac surgery. *Cardiol Young* 2000;10:239–244.
20. Casey FA, Sykes DH, Craig BG, et al. Behavioral adjustment of children with surgically palliated complex congenital heart disease. *Pediatr Psychol* 1996;21:335–352.
21. Lavigne JV, Faier-Routman J. Psychologic adjustments to pediatric physical disorder. *J Pediatr Psychol* 1992;17:133–157.
22. Andrews SG. Informing schools about children's chronic illness; parents' opinions. *Pediatr* 1991;88: 306 1991.
23. Dubin AM, Batsford WP, Lewis RJ, et al. Quality-of-life in patients receiving cardioverter defibrillators at or before the age of 40. *PACE* 1996;19:1555–1559.
24. DeMaso DR, Lauretti A, Spieth L, et al. Psychosocial factors and quality of life in children and adolescents with implantable cardioverter-defibrillators. *Am J Cardiol* 2004;93:582–587.

Index

A
α-cardiac actin, 235
accessory pathways
 evaluation of, 56
Adenocard, *see* adenosine
adenosine, 61, 76, 285–287
alpha agonist agents, 192
amiodarone, 75, 87, 106, 130, 249, 277
antiarrythmic medications, 30, 65–66, 148, 267
 defibrillation thresholds, effects on, 287
 pharmacology of
  β-blockers, 275–277
  class IA, 268–271
  class IB, 271–273
  class IC, 273–275
  class III, 277–283
  class IV, 283–285
  other agents, 285
anticholinergic agents, 193
arrhythmogenic right ventricular dysplasia (ARVD), 229, 236–237
arterial cannulation, for monitoring systemic blood pressure, 46
atenolol, 276
atrial fibrillation
 associated disease, 115
 incidence, 115
 mechanism related to, 116–117
 therapy of, 117–118, 262
atrial flutter, 95–113, 157–158
 atypical
  ECG characteristics, 95–96
  mechanism of, 95
  treatment of, 96–97
 IART
  ECG characteristics, 98–99
  electrophysiologic evaluation, 100–102
  etiology of, 99–100
  incidence of, 97–98

 mapping of, 102–106
 prevention of, 111
 prognosis of, 111–112
 treatment of, 106–111
atrial pacing, 57
atrial switch procedure, 263
atriofascicular fiber, 55
atrioventricular (AV) node, 3, 7, 9, 34, 65, 163
 ablation, 79
 in patients with abnormal cardiac anatomy, 80
atrioventricular block
 class II, indications of, 179
 ECG characteristics, 164–168
 electrophysiologic features, 168–169
 first degree, 165, 253
 prognosis and treatment, 169–170
 second degree, 165–198, 253
 third degree, 173
  clinical manifestation of, 174–177
  disorders causing, 177
  etiology, 173–174
  Fetal, 253
  prognosis of, 178–180
  treatment, 177–178
atrioventricular delay, 5
atrioventricular discordance, 10
atrioventricular nodal reentrant tachycardia (AVNRT)
 cryotherapy, 79–80
 drugs used for treatment of, 75
 electrocardiogram, 70–75
 electrophysiology study, 75–79
 incidence of, 69
 radiofrequency ablation, 75–79
 symptoms of, 69
 treatment of
  antiarrythmic medications, 65–66
  cryotherapy, 79–80
  drugs used for treatment of, 75
  electrocardiogram, 70–75
  electrophysiology study, 75–79

atrioventricular nodal reentrant tachycardia (*Cont.*)
    incidence of, 69
    observation, 65
    prognosis of, 66–67
    radiofrequency ablation, 75–79
    symptoms of, 69
    transcatheter ablation, 66
atrioventricular reentrant tachycardia (AVRT), 51, 248
    diagnostic testing of
        electrophysiology study, 56–65
        exercise treadmill testing, 56
    mechanism of, 53–54
atrioventricular septal defects, 10
atropine, 35, 160
α-*tropomyosin*, 235
automatic atrial tachycardia (AAT)
    clinical presentation, 119–120
    criteria for, 120
    diagnosis of, 120–121
    treatment of, 121–122
automaticity, of cells, 27

B
basket contact mapping, 102
Bazett's formula, 222
beta blockers, 192
beta-blockade treatment, 219
β-*mMHC*, 235
Betapace, *see* sotalol
breath-holding spells, 187
Brevibloc, *see* esmolol
Brugada syndrome, 141, 222
bundle branch block pattern, 62
bystander, 54

C
calcium channels, 23–24
    L-type, 25
cardiac conduction system
    abnormalities
        complete heart block, 237–238
        defects management, 238–239
        genetic testing to detect, 239
        Kearns-Sayre syndrome, 238
        progressive cardiac conduction defect, 238
        short QT syndrome, 238
    development of
        anatomy of, mature, 8
        anatomy of, in congenital heart disease, 10–13
        anatomic development, of conduction tissue, 7–8
        impulse generation, development of, 5
        impulse propagation, development of, 5
        nodal cell development, 7
    electrophysiologic study of, 33
        anesthesia, role of, 47
        conduction through excitable tissue, 36
        considerations in, 41–42
        in developing children, 37–38
        properties of, 34
        SA node evaluation, 34
        techniques in, 42
    physiology of
        action potential, 25–29
        arrythmia formation mechanism, 29–30
        ion channels, 18–25
        resting membrane potential, 17–18
cardiac myocytes, 17
cardiac rhythm, in left atrial isomerism, 12
cardiac ryanodine receptor type 2 (*RYR2*), 229
cardiomyopathies, 232
    dilated (DCM), 234–236
    hypertrophic (HCM), 231–234
CARTOMERGE™ Image Integration Module, 296
cell markers, 7
C-fibers, 192
channelopathies, 220
chaotic atrial tachycardia, *see* multifocal atrial tachycardia
chick embryo, 3
    Hamburger Hamilton (HH) stage 3 of, 3
chloride channels, 24
concealed pathways, 51, 58
congenital heart block, 174
    in childhood, 176–177
    in neonate, 175–176
    in utero, 175, 253
connexins, 3, 7
contact three-dimensional mapping, 103–104
contractile proteins, 3
Cordarone, *see* amiodarone
Corgard, *see* nadolol
Corvert, *see* ibutilide
Covera, *see* verapamil
Cox maze procedure, 118
cryotherapy, 66, 79–80, 118
cystic fibrosis transmembrane conductance regulator (CFTR), 24
cytoskeletal proteins, 3, 7

D
DAD, *see* delayed afterdepolarization, 29
δ-*sarcoglycan*, 235
desmin, 235
device implants
    choices in
        bipolar versus unipolar electrode usage, 199–201
        cardiac resynchronization devices, 203
        dual versus single chamber pacemakers, 201
        epicardial versus endocardial placement, 198
        ICDs, 203

follow up of
   clinic evaluation, 210–212
   CRT follow up, 212–213
   ICD, 213–214
   transtelephonic monitoring, 213
   implantation techniques
      acute electrode testing, 208–209
      coronary sinus lead placement, 206–208
      endocardial electrode placement, 205–206
      epicardial implant techniques, 204–205
      ICD placement, 209–210
   indications for
      bradyarrhythmias, 195–196
      heart failure, 197
      tachyarrhythmias, 196–197
   pacemaker implant
      in children, 178
digitalis glycosides, 285
digoxin, 65, 75, 249, 285
diltiazem, 284–285
disopyramide, 192
D-loop ventricles, 10
dofetilide, 281–282
drug–device interactions, 287
dystrophin, 235

E
EAD, *see* early afterdepolarization, 29
Ebstein's anomaly, 12, 53
echocardiogram, 85
ectopic atrial tachycardia
   electrocardiogram, 33
electrode threshold testing, 211
embolism, 263
entrainment mapping, 102
esmolol, 276
ET-1 (endothelin-1) signaling, 8
exercise treadmill testing, 56

F
familial atrial fibrillation, 230
familial ventricular fibrillation, *see* ion channel disorders
fetal arrhythmia
   cardiace lesions, associated with fetal artial flutter, 249
   diagnosis techniques, 242–244
   drugs for treatment of, 250
   specific, 244
      atrioventricular conduction block, 253–255
      bradyarrhythmias, 252–253
      extrasystoles, 244–245
      supraventricular ectopy, 245–246
      tachyarrhythmias, 246–252
      ventricular ectopy, 246
flecainide, 66, 75, 249, 273–274
Florinef, 192
fludrocortisone, *see* Florinef

Fontan operation, 263

G
gap junctions, 7, 24

H
head-up tilt (HUT) table test, 189
   protocols of, 189–191
hemi-Fontan procedure, 28
heterotaxy syndromes, 12
His bundle electrogram, 56
His-Purkinje system, 3, 6, 8, 34, 64
hypertrophic cardiomyopathy, 258–260
hypoplastic left heart syndrome, 264

I
Intra-atrial reentrant tachycardia, 95–113 *see* atrial flutter
ibutilide, 282–283
ICD therapy, 150, 203
impulse generation, development of, 5
impulse propagation, development of, 5
incisional IART, electrophysiologic evaluation goals for, 100
inderal LA, *see* Propranolol
inderal, *see* Propranolol
intracatheter cooling system, 109
intrinsic deflection, 34
inwardly rectifying channels, 23
ion channel disorders, that enhance automatically
   ARVD2, 229, 230
   CPVT, 229, 230
   FVT, 229, 230
ion channel disorders, that prolong action potential
   Anderson syndrome, 228, 229
   Brugada syndrome, 222
      diagnosis of, 227–228
      patient management of, 228
   long QT syndrome, 217–218
      diagnosis of, 222, 224–225
      drugs avoided in treatment of, 223–224
      drug induced LQTS, 222
      IKr defects, 221
      IKs defects, 218
      patient management, 225–226
      SCN5A defects, 221–222
isoproterenol, 62, 76, 101, 160
Isoptin, *see* verapamil

J
Jervell and Lange-Nielsen syndrome, 219
J-shaped stylets, 206
J-tipped guide wire, 45
junctional tachycardias
   accelerated
      diagnosis, 129

junctional tachycardias (Cont.)
    natural history, 129
    treatment, 130
  congenital JET, 125–127
  junctional ectopic tachycardia (JET), 122
  post-operative JET, 123–125

K
KCNA5 gene, 23
Kearns-Sayre Syndrome, 238

L
left anterior oblique (LAO), 295
left heart obstructive lesions, 263–264
lidocaine, 271–272
line of block, criteria for, 108

M
Magnesium sulfate, 287
Mahaim fibers, 54
  Feature of, 55
  role of, 54
manifest pathways, 51, 58
mexiletine, 272–273
Mexitil, *see* mexiletine
Mobitz I, 165, 254
Mobitz II, 166–168
Mönckeberg's sling, 12
multifocal atrial tachycardia
  antiarrhythmic therapy, 137
  clinical context, 135–136
  electrocardiographic features, 136–137
  incidence of, 136
  mechanism of, 137
  patient management, 137–138
Mustard operation, 28, 36, 98, 107, 110, 263
myocardial bridge, 262–263
myocardial contraction, 3

N
nadolol, 276
neural marker (G1N2), 5
nicorandil, 221
non-contact endocardial mapping, 104–106
normal sinus fetal heart rate, 241
nursing management, of arrhythmic young patients
  definitions and processes, 316
  nursing interventions, for pre-procedure preparation, 319
  nursing plan, 318
  nurse patient interaction, 320–322
  patient care planning, 320–322
  psychosocial rehabilitation, 323–325
  role of, 316, 318
  scope of, 315

P
pacemaker activity, 4–5
pacemaker management, 110
pacemaker therapy, 193
Pacerone, *see* amiodarone
pathway location, role of, 53
persistent junctional reciprocating tachycardia (PJRT)
  ablation of, 87–88
  anatomy of, 83
  cycle length, 63
  diagnosis and treatment of, 247–249
  ECG findings, 83–86
  electrophysiology, 86–87
  mapping of, 87–88
  treatment of, 87
post orthostatic tachycardia syndrome (POTS), 91, 187
potassium channels, 20–23, 65
  voltage-gated, 22
precardiac cells, positions of, 4
procainamide, 249, 269–271
Procan SR, *see* procainamide
progressive cardiac conduction defect, 238
Pronestyl, *see* procainamide
propafenone, 274–275
propranolol, 35, 275–276
P-waves, 12, 96, 108, 137

Q
QRS transition, 53
Quinidex, *see* quinidine
quinidine, 249, 268–269

R
radiofrequency ablation, 75–79, 107, 290–292, 293
refractory period, 27, 40
  effective, accessory pathway, 60
response cells
  rapid, 25–26
  slow, 26
retrograde P-waves, 53
right anterior oblique (RAO), 295
Romano-Ward syndrome, 218
Rythmol, *see* propafenone

S
salt tablets, 191
sarcoplasmic reticulum (SR) stores, 229
SCN5A gene, 20, 221, 222
seizures, 187, 261
Senning procedures, 263
serotonin reuptake inhibitors, 193
short QT syndrome, 238
sick sinus syndrome, 153
  ECG characteristics, 153–157
    atrial flutter/fibrillation, 157–158
    sinus arrhythmia, 155
    sinus bradycardia, 154

sinoatrial node reentrant tachycardia, 157
    sinus pauses and sinus arrest, 156–157
  electrophysiologic features, 158–159
    sinoatrial conduction time, 158–159
    sinus node recovery time, 159
  prognosis of, 161–162
  treatment of, 159–161
sinoatrial (SA) node, 3, 7–8
  recovery time (SNRT), 34
  reentrant tachycardia, 157
sinoatrial conduction time, 158–159
sinoatrial exit block, 156
sinus arrhythmia, 155
sinus bradycardia, 154, 252
sinus node recovery time, 159
sinus pauses and sinus arrest, 156–157
sinus tachycardia, 247
sodium channels, 19–20, 65
sotalol, 66, 106, 249, 281
spironolactone, 221
splanchnic layer, of mesoderm, 4
sudden infant death syndrome (SIDS), 257
sudden unexpected cardiac death (SCD)
  in acquired coronary artery anomalies, 263
  in Brugada syndrome patients, 261
  causes of, in young, 259
  in commotion cordis, 264
  in congenital long QT syndrome patients, 260–261
  in congenital coronary artery anomalies, 262–263
  definition of, 257
  in hypertrophic cardiomyopathic patients, 258–260
  in myocarditis patients, 260
  in pulmonary hypertension patients, 264
  statistics of, 257
  in structural congenital heart disease, 263–264
  in Wolff-Parkinson-White syndrome, 261–262
supraventricular tachycardia (SVT), 51, 247
see atrial fibrillation
see atrioventricular nodal reentrant tachycardia (AVNRT)
see atrioventricular reentrant tachycardia (AVRT)
see Intra-atrial reentrant tachycardia (IART)
see persistent junctional reciprocating tachycardia (PJRT)
syncope, 230, 261
  diagnosis of, 183–185
  neurocardiogenic, 185
    clinical presentation of, 187–189
    pathophysiology of, 185–186
    prognosis of, 193–194
    treatment of, 191–193

T

tachycardia mapping, 64, 66
tachycardia–bradycardia, 156
tachycardias, development of
  abnormal automaticity, 29
  atrioventricular reentrant, see atrioventricular reentrant tachycardia
  automatic atrial tachycardia (AAT), see automatic atrial tachycardia (AAT)
  junctional tachycardias, see junctional tachycardias
  intra-atrial reentrant tachycardia (IART), see atrial flutter
  multifocal atrial tachycardia (MAT), see multifocal atrial tachycardia
  persistent junctional reciprocating tachycardia (PJRT) see persistent junctional reciprocating tachycardia (PJRT)
  reentry, 29
  sinoatrial reentrant, 91–93
    criteria for diagnosis of, 92
  supraventricular tachycardia (SVT), see supraventricular tachycardia (SVT)
  triggered activity, 29
  ventricular tachycardia (VT), see ventricular tachycardia (VT)
Tafazzin gene, 235
Tambocor, *see* flecainide
Tendon of Todaro, 8
Tenormin, *see* atenolol
tetralogy of Fallot, 263
Thorel's pathway, 9
Tikosyn, *see* dofetilide
transcatheter ablation, 66, 149–150
  of arrhythmic patients with CHD, 311–313
  of atrial flutter, 305–310
  of cardiac arrhythmias
    approach to atrial ectopic tachycardias, 301–303
    approach to AVNRT, 304
    approach to leftsided pathways, 297–298
    approach to rightsided pathways, 298–301
    cryoablation, 292–293
    indications for, 289–290
    pericardial approach, 303
    radiofrequency ablation, 290–292, 310, 313
    techniques of, in children, 293–296
  of non-ischemic ventricular tachycardia, 310
transcutaneous pacing, 160
transesophageal atrial pacing, 97
transposition great arteries, 11
transseptal catheter delivery, into superior vena cava, 46–47
triangle of Koch, 9–11, 80, 100
tricuspid atresia, 11
troponin T, 235
twin atrioventricular nodes, 11

U

utero detection, 175

V
vagomimetic drugs, 285
ventricular tachycardia (VT), 251
  clinical investigation of, 145–146
ventricular tachycardia (VT) (*Cont.*)
  clinical presentation of, 141
  definition of, 139–141
  diagnosis of, 143–145
  electrocardiographic findings, 141–143
  surgical management of, 148–151
  treatment of, 146–148

ventriculoatrial discordance, 11
verapamil, 65, 249, 283

W
Wenchebach's bundle, 9
Wenckebach cycle length, 59, 61
Wenckebach, *see* Mobitz I
Wolff-Parkinson White syndrome, 51–53, 60–61, 65, 67, 115, 234

X
Xylocaine, *see* lidocaine

Made in United States
North Haven, CT
08 February 2024